Numerically Solving Polynomial Systems with Bertini

SOFTWARE • ENVIRONMENTS • TOOLS

The SIAM series on Software, Environments, and Tools focuses on the practical implementation of computational methods and the high performance aspects of scientific computation by emphasizing in-demand software, computing environments, and tools for computing. Software technology development issues such as current status, applications and algorithms, mathematical software, software tools, languages and compilers, computing environments, and visualization are presented.

Software, Environments, and Tools

Daniel J. Bates, Jonathan D. Hauenstein, Andrew J. Sommese, Charles W. Wampler, *Numerically Solving Polynomial Systems with Bertini*

Uwe Naumann, *The Art of Differentiating Computer Programs: An Introduction to Algorithmic Differentiation*

C. T. Kelley, *Implicit Filtering*

Jeremy Kepner and John Gilbert, editors, *Graph Algorithms in the Language of Linear Algebra*

Jeremy Kepner, *Parallel MATLAB for Multicore and Multinode Computers*

Michael A. Heroux, Padma Raghavan, and Horst D. Simon, editors, *Parallel Processing for Scientific Computing*

Gérard Meurant, *The Lanczos and Conjugate Gradient Algorithms: From Theory to Finite Precision Computations*

Bo Einarsson, editor, *Accuracy and Reliability in Scientific Computing*

Michael W. Berry and Murray Browne, *Understanding Search Engines: Mathematical Modeling and Text Retrieval, Second Edition*

Craig C. Douglas, Gundolf Haase, and Ulrich Langer, *A Tutorial on Elliptic PDE Solvers and Their Parallelization*

Louis Komzsik, *The Lanczos Method: Evolution and Application*

Bard Ermentrout, *Simulating, Analyzing, and Animating Dynamical Systems: A Guide to XPPAUT for Researchers and Students*

V. A. Barker, L. S. Blackford, J. Dongarra, J. Du Croz, S. Hammarling, M. Marinova, J. Waśniewski, and P. Yalamov, *LAPACK95 Users' Guide*

Stefan Goedecker and Adolfy Hoisie, *Performance Optimization of Numerically Intensive Codes*

Zhaojun Bai, James Demmel, Jack Dongarra, Axel Ruhe, and Henk van der Vorst, *Templates for the Solution of Algebraic Eigenvalue Problems: A Practical Guide*

Lloyd N. Trefethen, *Spectral Methods in MATLAB*

E. Anderson, Z. Bai, C. Bischof, S. Blackford, J. Demmel, J. Dongarra, J. Du Croz, A. Greenbaum, S. Hammarling, A. McKenney, and D. Sorensen, *LAPACK Users' Guide, Third Edition*

Michael W. Berry and Murray Browne, *Understanding Search Engines: Mathematical Modeling and Text Retrieval*

Jack J. Dongarra, Iain S. Duff, Danny C. Sorensen, and Henk A. van der Vorst, *Numerical Linear Algebra for High-Performance Computers*

R. B. Lehoucq, D. C. Sorensen, and C. Yang, *ARPACK Users' Guide: Solution of Large-Scale Eigenvalue Problems with Implicitly Restarted Arnoldi Methods*

Randolph E. Bank, *PLTMG: A Software Package for Solving Elliptic Partial Differential Equations, Users' Guide 8.0*

L. S. Blackford, J. Choi, A. Cleary, E. D'Azevedo, J. Demmel, I. Dhillon, J. Dongarra, S. Hammarling, G. Henry, A. Petitet, K. Stanley, D. Walker, and R. C. Whaley, *ScaLAPACK Users' Guide*

Greg Astfalk, editor, *Applications on Advanced Architecture Computers*

Roger W. Hockney, *The Science of Computer Benchmarking*

Françoise Chaitin-Chatelin and Valérie Frayssé, *Lectures on Finite Precision Computations*

Numerically Solving Polynomial Systems with Bertini

Daniel J. Bates
Colorado State University
Fort Collins, Colorado

Andrew J. Sommese
University of Notre Dame du Lac
Notre Dame, Indiana

Jonathan D. Hauenstein
North Carolina State University
Raleigh, North Carolina

Charles W. Wampler
General Motors R&D
Warren, Michigan

siam.
Society for Industrial and Applied Mathematics
Philadelphia

10 9 8 7 6 5 4 3 2 1

Partial royalties from the sale of this book are placed in a fund to help students attend SIAM meetings and other SIAM-related activities. This fund is administered by SIAM, and qualified individuals are encouraged to write directly to SIAM for guidelines.

Library of Congress Cataloging-in-Publication Data

Bates, Daniel J. (Daniel James), 1979- author.
Numerically solving polynomial systems with Bertini / Daniel J. Bates, Colorado State University, Fort Collins, Colorado, Jonathan D. Hauenstein, North Carolina State University, Raleigh, North Carolina, Andrew J. Sommese, University of Notre Dame, Notre Dame, Indiana, Charles W. Wampler, General Motors R&D, Warren, Michigan.
pages cm. – (Software, environments, and tools)
Includes bibliographical references and index.
ISBN 978-1-611972-69-6
1. Polynomials–Data processing. 2. Geometry, Algebraic–Data processing. I. Title.
QA161.P59B38 2013
512.9'422028553–dc23 2013025584

We thank our families
for their sacrifices and support as we worked on this book
and the developments described within it, especially Bertini.

We dedicate this book to them.
Sarah, Rachel, and Luke
Julie, Morgan, Leah, and Kristyn
Rebecca, Rachel, Ruth, Nick, and Caleb
Vani, Megan, and Anne

Contents

List of Figures xi

Conventions xiii

Preface xv

I Isolated Solutions 1

1 Polynomial Systems 3
- 1.1 An Introduction to Polynomial Systems and Their Solution Sets . . . 4
- 1.2 Mathematical Background* . . . 16

2 Basic Polynomial Continuation 23
- 2.1 Homotopy Functions . . . 24
- 2.2 Path Tracking . . . 33
- 2.3 A User-Defined Homotopy* . . . 35
- 2.4 Bertini Guide: Basic Continuation . . . 39

3 Adaptive Precision and Endgames 41
- 3.1 Some Basic Facts . . . 41
- 3.2 Adaptive Precision . . . 42
- 3.3 Endgames for Singular Solutions . . . 47

4 Projective Space 55
- 4.1 Naturally Arising Projective Spaces . . . 56
- 4.2 Paths to Infinity . . . 59
- 4.3 Taming Infinity: Homogenization-Dehomogenization . . . 61
- 4.4 Calculating Infinity: Projective Patches . . . 62
- 4.5 Multihomogeneous Structures . . . 64
- 4.6 Numerical Accuracy* . . . 65
- 4.7 Normalization Effect of Homogenization* . . . 68

5 Types of Homotopies 71
- 5.1 Multihomogeneous Homotopy . . . 72
- 5.2 Linear Product Homotopy* . . . 80
- 5.3 Polyhedral Homotopy* . . . 82

*Sections marked with * are technical in nature or discuss methods not available in Bertini, so they may not be of interest to all readers.

5.4 Regeneration . . . 83
5.5 Examples . . . 90
5.6 Bertini Guide: Multihomogeneous and Regeneration Homotopy . . 100

6 **Parameter Homotopies** **101**
6.1 Parameter Homotopy Basics . . . 102
6.2 Parameter Homotopy in Bertini . . . 104
6.3 Examples . . . 107
6.4 Nonsquare Systems . . . 109
6.5 Nested Parameter Spaces . . . 111
6.6 Symmetry . . . 112
6.7 Advanced Refinements . . . 114
6.8 Bertini Guide: Parameter Homotopy . . . 117

7 **Advanced Topics about Isolated Solutions** **119**
7.1 Sharpening Using Newton's Method During a Homotopy Run . . . 120
7.2 Sharpening Module . . . 121
7.3 Evaluation and Newton Iterations . . . 125
7.4 Parameter Homotopies for Root Sharpening . . . 129
7.5 User-Defined Homotopies . . . 131
7.6 Determining the Rank of a Matrix . . . 135
7.7 Scaling Polynomial Systems . . . 137
7.8 Rounding Errors in Polynomial Construction . . . 138
7.9 Bertini Guide: Newton Iterations and Sharpening . . . 139

II Positive-Dimensional Solution Sets **141**

8 **Positive-Dimensional Components** **143**
8.1 Irreducible Components of an Algebraic Set . . . 144
8.2 Witness Sets and the Numerical Irreducible Decomposition . . . 150
8.3 Sampling a Component . . . 157
8.4 Membership Testing . . . 157
8.5 A Brief Note on the Expected Accuracy of Coefficients . . . 158
8.6 Bertini Guide: Positive-Dimensional Solving . . . 159

9 **Computing Witness Supersets** **161**
9.1 Rank of a System . . . 162
9.2 Randomization and Bertini's Theorem* . . . 163
9.3 Working Dimension by Dimension . . . 165
9.4 Cascade Methods . . . 167

10 **The Numerical Irreducible Decomposition** **171**
10.1 Junk Removal . . . 171
10.2 Pure-Dimensional Decomposition . . . 173

11 **Advanced Topics about Positive-Dimensional Solution Sets** **179**
11.1 Printing a Witness Set . . . 179
11.2 Computing Specific Codimensions . . . 181
11.3 Multiplicity* . . . 184
11.4 Macaulay Dual Spaces* . . . 186

11.5 Local Dimension and Mechanism Mobility 187
11.6 Deflation of Irreducible Components 189

III Further Algorithms and Applications 193

12 Intersection 195
12.1 Intersecting Irreducible Components 196
12.2 Regeneration Extension 202
12.3 Generalized Intersection 203
12.4 Fiber Products 203

13 Singular Sets 209
13.1 Rank-Dropping Conditions 210
13.2 Isosingular Sets* 214
13.3 Bertini Guide: Isosingular Stabilization Test 221

14 Real Solutions 223
14.1 Real versus Complex Dimension 224
14.2 Isolated Solutions 225
14.3 Certification of Nonsingular Isolated Real Roots 226
14.4 Cell Decomposition of the Real Solution Set 227
14.5 A Point on Each Connected Component 237
14.6 Real Numerical Irreducible Decomposition* 240
14.7 Software Summary 241

15 Applications to Algebraic Geometry 243
15.1 Invariants 243
15.2 Secant Varieties 245
15.3 Developing and Testing Conjectures 246
15.4 Symbolic versus Numerical Computation 248

16 Projections of Algebraic Sets 251
16.1 Technical Details on Constructible Sets and Slicing* 252
16.2 Pseudowitness Sets 253
16.3 Membership Tests Using Pseudowitness Points 256
16.4 Decomposing Images of Algebraic Sets 258
16.5 Further Applications of Pseudowitness Sets* 259
16.6 Bertini Guide: Computing Pseudowitness Sets 260

17 Big Polynomial Systems Arising from Differential Equations 261
17.1 Polynomial Systems from Discretizing Differential Equations 262
17.2 Computing Solutions to Big Polynomial Systems 266
17.3 Software Summary 269

IV Bertini Users Manual 271

A Bertini Quick Start Guide 273
A.1 Installing Bertini 273
A.2 The First Run 273

A.3 Basic Format for Input Files . . . 274
A.4 Declaring Variables and Subfunctions . . . 275
A.5 Positive-Dimensional Examples . . . 276
A.6 User-Defined and Parameter Homotopies . . . 277

B Input Format **281**
B.1 Syntax for Input Files . . . 281
B.2 The `start` File . . . 286
B.3 Other Input Files . . . 287
B.4 Command-Line Input for Sampling and Sharpening . . . 288

C Calling Options **289**
C.1 Basic Bertini Calls . . . 289
C.2 Interactive Modes . . . 289
C.3 Automation of Bertini Calls . . . 291

D Output Files **293**
D.1 Zero-Dimensional Runs . . . 293
D.2 Positive-Dimensional Runs . . . 296

E Configuration Settings **301**
E.1 Track Type Settings . . . 301
E.2 Multiple Precision Settings . . . 304
E.3 Adaptive Precision Settings . . . 305
E.4 Tracker Settings . . . 306
E.5 Endgame Settings . . . 309
E.6 Security Settings . . . 311
E.7 Postprocessing Settings . . . 312
E.8 User-Defined and Parameter Homotopy Settings . . . 314
E.9 Numerical Irreducible Decomposition Settings . . . 315
E.10 Regeneration Settings . . . 317
E.11 Miscellaneous Settings . . . 319

F Tips and Tricks **323**
F.1 Formulating Polynomial Systems . . . 323
F.2 Solving Tips . . . 325
F.3 Additional Postprocessing . . . 327

G Parallel Computing **329**

H Related Software **331**
H.1 Interfaces between Bertini and Other Software . . . 331
H.2 Modules Using Bertini as a Library . . . 332
H.3 Software Packages Similar to Bertini . . . 332

Bibliography **335**

Software Index **347**

Subject Index **349**

List of Figures

1.1 The solution set $\mathbf{V}(x^2-1, y^2-4)$ consists of four points: $(\pm1, \pm2)$. . . . 5

1.2 Intersection of circles of the form $(x-c)^2+y^2=1$ with the y-axis, $x=0$, for several values of $c\in[0,1]$. . . . 12

1.3 Two algebraic curves, each with two real components. The one on the left has one complex component, whereas the one on the right factors into two complex components. . . . 18

2.1 Solution paths for Example 2.1 sliced at three values of t. . . . 26

2.2 In Example 2.1, the solution of $h(z,s)=0$ for z is a three-sheeted cover over s with four ramification points. . . . 27

2.3 Distance between nearest pair of roots in Example 2.1. . . . 28

2.4 The gamma trick: (a) the family of arcs and (b) the arc for $\gamma=0.4+0.77i$, which misses bad points for Example 2.1. . . . 29

2.5 Cutaway view of Example 2.1 showing the paths followed. . . . 29

2.6 Euler–Newton predictor-corrector path tracking. . . . 35

2.7 Path around the unit circle in $\mathbb{C}$. . . . 36

2.8 Real and imaginary parts of z for $z^2-s=0$ with the path around the unit circle in s highlighted. . . . 38

2.9 The path $(z(t),t)$ for the `AroundTheCircle` example. . . . 38

3.1 Endgame operating zone. . . . 49

3.2 A homotopy that has two paths converging to a singular point at $t=0$. 50

3.3 One step of the power-series endgame. . . . 51

3.4 Cauchy endgame: the path $(p(t),t)$ closes up after tracking t around a triangle twice. . . . 53

4.1 The real projective line, $\mathbb{P}^1_{\mathbb{R}}$, with two patches. . . . 62

5.1 Newton polygons for (5.9), (5.10), (5.11). . . . 81

5.2 Mixed volume for (5.12). . . . 83

5.3 Lagrange point coordinates. Solid arrows indicate gravitational and centrifugal forces. When m_3 is small, its effect on m_1 and m_2 can be ignored. . . . 91

5.4 Lagrange points. There are five equilibrium locations for the small third body. . . . 94

5.5 Vector diagram of a four-bar linkage. . . . 96

6.1 Stewart–Gough platform and notation. . . . 108

14.1 The curve $y^2 = x^2(x-1)$, whose real points consist of a curve and an isolated point at the origin. 224
14.2 A polyhedron with seven faces. 228
14.3 Cell decomposition of a sphere. 229
14.4 Cell decomposition of a limaçon curve. 232
14.5 Cell decomposition of a sphere with projection onto a plane. 234
14.6 Cell decomposition of a torus. Projection from high angle. 236
14.7 Cell decomposition of a torus. Projection from low angle. 237
14.8 Real part of a sextic curve with critical points. 240

17.1 Stencil of points. 265
17.2 Grid with 1901 gridpoints: $(N,M) = (100,20)$. 265

Conventions

The following notational conventions are used in this book.

1. $\mathbb{C}$ is the set of complex numbers, $\mathbb{R}$ is the set of real numbers, $\mathbb{Z}$ is the set of integers, and $\mathbb{Z}_{\geq 0}$ (resp., $\mathbb{Z}_{>0}$) is the set of nonnegative (resp., positive) integers.

2. We often use $:=$ when we are stating a definition, e.g., the disk of radius r in the complex plane $\mathbb{C}$ around a point x is defined

$$\Delta_r(x) := \left\{ z \in \mathbb{C} \;\middle|\; |z - x| < r \right\}.$$

   In pseudocode statements of algorithms, we use the same symbol for copying the right-hand result to the left, e.g., $k := k + 1$ increments k by one.

3. We use $\mathbb{C}^* := \mathbb{C} \setminus 0$, the set of all complex numbers except zero.

4. The N-dimensional complex Euclidean space is

$$\mathbb{C}^N := \underbrace{\mathbb{C} \times \cdots \times \mathbb{C}}_{N \text{ times}} = \{(z_1, \ldots, z_N) \mid z_i \in \mathbb{C}\}.$$

5. When it is clear from the context, we let 0 denote the origin of a vector space.

6. Often when using indices, we refer to the objects being discussed as a sequence with first and last elements. For example, we might write a Jacobian matrix as

$$\begin{bmatrix} \frac{\partial f_1}{\partial z_1} & \cdots & \frac{\partial f_1}{\partial z_N} \\ \vdots & \ddots & \vdots \\ \frac{\partial f_n}{\partial z_1} & \cdots & \frac{\partial f_n}{\partial z_N} \end{bmatrix}.$$

   This is an abuse of notation in the case $N = 1$ or $n = 1$. Rather than avoiding the abuse and obscuring things we usually leave the reader to fill in the special cases, e.g., in the example just given with $N = n = 1$ we would mean

$$\left[\frac{\partial f_1}{\partial z_1} \right], \text{ not } \begin{bmatrix} \frac{\partial f_1}{\partial z_1} & \frac{\partial f_1}{\partial z_1} \\ \frac{\partial f_1}{\partial z_1} & \frac{\partial f_1}{\partial z_1} \end{bmatrix}.$$

7. The notation $f : \mathbb{C}^N \to \mathbb{C}^n$ means that f is a system of n functions in N complex variables, or more precisely, f is a map from $\mathbb{C}^N$ to $\mathbb{C}^n$. When f is differentiable, we use Jf to mean the Jacobian matrix of f, that is, the $n \times N$ matrix appearing in item 6.

8. We use multidegree notation. For example, if $z_1,\dots,z_N$ are indeterminates, and $\alpha=(\alpha_1,\dots,\alpha_N)$ is an N-tuple of nonnegative integers, then z^α denotes $z_1^{\alpha_1}\cdots z_N^{\alpha_N}$. We let $|\alpha|:=\sum_{j=1}^N \alpha_j$ and $\alpha!=\alpha_1!\cdots\alpha_N!$.

9. $\mathbb{C}[z]$ is the set of polynomials in z with coefficients in $\mathbb{C}$. Similarly, $\mathbb{C}[z_1,\dots,z_N]$ is the set of multivariate polynomials

$$p(z_1,\dots,z_N)=\sum_{\alpha\in\mathscr{A}} c_\alpha z^\alpha$$

with $\mathscr{A}$ a finite set of N-tuples of nonnegative integers and c_α complex numbers. We typically denote $p(z_1,\dots,z_N)$ by $p(z)$. The degree of polynomial $p(z)$ is $\{\max|\alpha| \mid c_\alpha \neq 0\}$.

10. We use $p^{(j)}(z):=\frac{d^j p(z)}{dz^j}$, the jth derivative of p with respect to z.

11. The symbol \ is the "setminus" operator, that is,

$$A\setminus B:=\{z\in A \,\big|\, z\notin B\}.$$

12. For a set A, $\#A$ is the cardinality of A.

13. For a subset A of a topological space, $\overline{A}$ denotes the closure of A.

14. For real numbers a,b, we denote open and closed intervals of the real line as

$$[a,b]=\{x\in\mathbb{R} \mid a\le x\le b\}, \qquad (a,b]=\{x\in\mathbb{R} \mid a< x\le b\},$$
$$[a,b)=\{x\in\mathbb{R} \mid a\le x< b\}, \qquad (a,b)=\{x\in\mathbb{R} \mid a< x< b\}.$$

15. A point x in N-dimensional complex projective space, $\mathbb{P}^N$, is often written via homogeneous coordinates enclosed in square brackets, i.e., $[z_0,z_1,\dots,z_N]\in\mathbb{P}^N$. (Projective space is explained in Chapter 4.) Unfortunately, this means that a real interval $[u,v]\subset\mathbb{R}$ and a point on the projective line $[u,v]\in\mathbb{P}^1$ have the same notation. The distinction between the two will be clear from the context.

16. We use Hardy's big O notation. The expression $f(x)=g(x)+O(h(x))$ as $x\to a$ means that there exists a constant $C>0$ such that $|f(x)-g(x)|\le Ch(x)$ for all x sufficiently near a but not necessarily equal to a. The most typical choices for a are 0 and ∞, and the most typical choice for $h(x)$ is x^s for some nonnegative integer s.

17. We use $\square$ to designate a random or unspecified complex number.

Preface

Systems of polynomial equations are a common occurrence in problem formulations in engineering, science, and mathematics. Solution sets of such systems, i.e., algebraic sets, are well-behaved, and effective algorithms have been developed in recent years to numerically compute and manipulate them. This field is called *numerical algebraic geometry*. Moreover, there is free software implementing these algorithms both for single computers and for computer clusters. As the developers of Bertini [22], we focus on this open-source software package in this book.

The goal of this book is threefold:

1. Explain enough background on solution sets of polynomial systems so that the reader has a basic understanding of the geometry of such sets and can understand both the data types and the breadth of the various computations of numerical algebraic geometry.

2. Show the reader how to use Bertini to solve polynomial systems efficiently and interpret the output correctly.

3. Provide a detailed manual of commands and settings for Bertini.

In addition to simple introductory problems, we present examples of polynomial systems arising in applications and treat each as a case study to show how the software can be used to deduce different sorts of information needed by a user. Files for all of the examples in this book can be found at the website for this book, `www.siam.org/books/se25`, where one may also find a link to the Bertini software website, `bertini.nd.edu`.

Applications

Polynomial systems have been solved numerically in many fields, including mechanism and robot kinematics, chemistry, computer-aided design (CAD), systems of nonlinear differential equations (ordinary and partial), differential algebra, economics, automatic control systems, magnetization, string theory, algebraic geometry, and others. Let us begin by discussing the polynomial systems that occur in three of these areas. These and other applications are considered later in the book.

In the theory of mechanisms, a large set of problems leads to polynomial systems. For example, for a serial-link robotic arm, the equations relating the robot's hand position to the rotations of its joints and its physical parameters, such as the length of the links making up the arm, are naturally polynomials. Typically the robot can reach a given hand location in multiple poses, each given by a set of joint rotations. Finding all such poses is a straightforward application of polynomial solving software [171, 181].

Of course, mechanisms have uses far beyond robotics, e.g., medical instruments, windshield wipers, airplane landing gear, metal-stamping presses, and all manner of devices with jointed moving parts fall under the same heading. Any time rigid parts are connected by the most common types of joints—from simple hinges to ball-and-socket joints—their motion is well-modeled by a system of polynomial equations [37, 146, 126, 105, 184]. These equations might be used to analyze the motion of a proposed design (mechanism analysis) or to find a design that produces a desired motion (mechanism synthesis). Synthesis problems can be especially challenging, as they often involve several copies of the basic polynomial system, each copy expressing the requirement to reach a different position along the desired motion and all to be solved simultaneously.

Another source of polynomial systems arises in chemistry (see, e.g., [131]) and in combustion engineering [3]. Each equation of the form

$$2H_2 + O_2 \rightleftharpoons 2H_2O$$

leads to a polynomial relation between the molar quantities arising in the equation, e.g., $[H_2]$, $[O_2]$, and $[H_2O]$ in the above equation. The condition for equilibrium between these species takes the form

$$k[H_2O]^2 = [H_2]^2[O_2],$$

where k is an appropriate rate constant. After writing one such equation for each possible reaction between species in a chemical system and adding mass conservation equations according to the initial concentrations of the species, one has a system of polynomials whose solutions are all possible equilibrium states for the reaction.

In many cases, discretization of nonlinear differential equations leads to systems of polynomials. Methods based on a polynomial solver such as Bertini give a general approach to boundary value problems for single ordinary differential equations (ODEs) [4, 5, 6]. For systems of ODEs, such methods yield solutions not easily found by other methods [74]. Systems of nonlinear partial differential equations (PDEs) arising in such areas as tumor growth can lead to large polynomial systems with thousands of variables. The methods in this book yield new solutions and much useful numerical information for such problems; see, e.g., [75, 76, 77, 78, 79, 82, 81]. These methods, and in particular Bertini's endgame routines to compute singular solutions, have led to a homotopy method based on weighted essentially nonoscillatory (WENO) schemes [122] for solving steady-state problems of hyperbolic conservation laws. This advance is an order-of-magnitude improvement over the usual approaches [80] for moderate-sized problems and is an even greater improvement as the size of the problems increases.

Software

Advances in methodology and in computer speed now make problems that once seemed intractable easy and straightforward to solve. Consider, for example, the nine-point path synthesis problem for four-bar linkages, which we discuss in detail in §5.5.2. This was first posed in 1923 [9], but it took until 1962 for a few solutions to be found by a precursor to modern polynomial continuation [148, 149]. The first computation of complete solution sets in 1992 [182, 183], using continuation, took weeks of expert formulation of the equations and 300 hours of mainframe CPU time. As of 2010, a computer running Bertini on one processor* could solve an initial example of the problem in 8 hours, and a parallel computer with 65 cores* could do it in under 8 minutes [93]. After an initial solve, just eight cores of the same parallel computer can find all 1442 triplets of solutions of any

*2.4-GHz Opteron processor, 64-bit Linux.

*Eight nodes of 2.33-GHz quad-core Xeon 5410 plus 1 manager, 64-bit Linux.

other instance of the nine-point problem using a parameter homotopy (as described in §§6.6 and 6.11) in just 8 seconds! This kind of computing power is now available in many consumer-grade personal computers. Due to improved algorithms such as regeneration (§5.4), the user no longer needs to be an expert in continuation to get results like these.

Our software package, Bertini, offers easy access to the latest advances in homotopy methods. A key feature is the use of adaptive multiprecision, which makes the computations robust and accurate. Also, Bertini supports both serial processing on a single CPU and parallel processing,* allowing the user to take advantage of the availability of multicore technology in personal computers and larger installations with hundreds of CPUs.

Starting from the first release of Bertini version 1.0 in April 2008 up to the current release (version 1.4) used in this book, there has been a steady growth in the Bertini user community. It has become apparent that an up-to-date and thorough manual would sustain this growth by helping new users get started more quickly and by helping experienced users get the most out of all that Bertini has to offer. It is in that spirit that we write this book.

Related work

Two of the authors of this book, Sommese and Wampler, already published a full-length monograph on the subject of polynomial continuation [161]. However, that book concentrates on presenting the theory behind the algorithms, while this new book concentrates on practical usage of the algorithms. Although we follow roughly the same order of development, this book is very different from its predecessor. Here, we develop material in terms of applications and case studies, which we solve completely using Bertini. We try to convey what can be done, how to do it with Bertini, and how this can answer questions in engineering and science. Compared to [161], we give relatively little information on why the algorithms work and only enough information on how they work to enable the user to understand the optional settings that can be tweaked to optimize performance. We also take the opportunity to give the reader a glimpse of newer developments in the field, some of which are fully implemented in Bertini and some that can be explored by interfacing to Bertini through its facility for "user-defined homotopy."

For academics interested in the field, the combination of [161] with this book provides a full complement of material, in both theory and practice, suitable for a course in numerical algebraic geometry. Mechanical engineers might prefer to use the survey article [184] to overview the relationship between algebraic kinematics and numerical algebraic geometry and use this book to equip students with the latest computational tools in the field, a combination that could be called "numerical algebraic kinematics."

Overview of this book

Part I begins by introducing in Chapter 1 the main objects of interest, namely, polynomial systems and their solution sets. We focus on isolated solutions of polynomial systems (positive-dimensional solution sets become the main focus of attention starting in Chapter 8), using a number of simple examples to illustrate the key concepts, including a first discussion of singularities and multiplicity. We note that our methods depend on some random selection of constants and discuss how this makes the algorithms work correctly with "probability one." We give a brief introduction to straight-line programs for evaluation of polynomials and Bertini's facilities to make it easy to input polynomials in this form.

*Bertini's parallel facilities are primarily based on the message passing interface (MPI).

We also discuss in the Chapter 1 situations where the number of equations is not equal to the number of variables. This is known as a *nonsquare system*. We treat the case of both too few and too many equations. The treatment of systems with excess equations again requires the use of random numbers, so we discuss the related topic of genericity and what it means in practice.

We also give a brief overview of some algebraic geometry background involving algebraic sets, randomization, and Bertini theorems. These and similar sections throughout the book are marked with a star (*) to indicate that they may be skipped by those who only need access to the core computational routines for polynomial solving. These sections are included to help those who desire a deeper understanding of the theoretical underpinnings of numerical algebraic geometry.

In Chapter 2, we discuss the fundamental continuation method and issues that arise in using it. We start this initial discussion with isolated solutions defined by N polynomials in N variables. Many polynomial systems arising in practice are in this format or can easily be put into this format. We show how to run examples with Bertini using the simplest algorithm in the toolset, the total-degree homotopy. We discuss some of the basic components of continuation, e.g., ODE predictors and a Newton corrector, and how settings in Bertini control these components.

In Chapter 3, we consider two of the most important advances implemented in Bertini: adaptive precision and endgames. Both of these algorithms are useful in treating "bad" (singular) and nearly bad points that we encounter during continuation. Adaptive precision allows us to move beyond near singularities without losing accuracy. Endgames allow us to accurately compute singular solutions without relying on computations at or very near the singularity.

In Chapter 4, we discuss projective space, which provides a completion of Euclidean space. This makes the notion of solutions at infinity meaningful and straightforward for computation.

Chapter 5 covers the main alternatives to the total-degree homotopy. We discuss the Bézout number of solutions and why acknowledging the different structures of systems significantly decreases the computation needed to solve a polynomial system. Up to this point, if the reader has been running the example problems in Bertini, he or she will have already experienced multihomogeneous continuation by simply placing variables into logical groups. With one change in the settings, the same input file will invoke the most recent homotopy type, the regeneration homotopy. For a casual user, it may be enough to know what switch to flip, but a more serious user will find it worthwhile to understand all the homotopy types that are available and how they may sometimes offer advantages or suffer from disadvantages. By understanding the homotopies in more detail, the reader will be better able to monitor the runtime progress of each homotopy and can understand the statistical information provided with each computational run.

In Chapter 6, we introduce parameter spaces. Many problems in engineering and science need to be solved many times for different values of parameters. Recalling the serial-link robot example mentioned earlier, it is easy to envision scenarios where one may wish to find the arm poses that achieve many successive hand positions. One of the important properties of continuation methods is that if a problem is solved once generically (at a random point in the complex parameter space), the resulting solutions can be used to initiate successive solves for different parameter values. Many times the successive solves take a small fraction of the time that was needed for the initial solve. Indeed, speedups by a factor of 100 are common; see, e.g., [133, 134, 182]. We also discuss complex and real solutions. Even though real solutions are often the object of interest, continuing real solutions alone from the solution at a generic value of the parameters will not generally lead to all the real solutions for other values of the parameters, i.e., complex solutions may continue to real solutions and vice versa.

Chapter 7 covers a number of advanced topics related to the computation of isolated solutions. One benefit of numerical algebraic geometry is that approximations to solutions may be computed to extremely high accuracy with little work. This is called endpoint sharpening and is covered in detail in this chapter. There are also brief discussions on how Bertini determines matrix ranks and how polynomial systems may be scaled for more efficient computation.

Part II treats positive-dimensional solution sets of polynomial systems, beginning with Chapter 8, in which we discuss different ways of representing such sets, including the classical irreducible decomposition of an algebraic set and its numerical analogue, the numerical irreducible decomposition. We give a careful discussion of witness sets and supersets. Witness sets are the data structure we use to represent positive-dimensional sets, and witness supersets are the first step in their computation. We discuss how Bertini can be used to sample a component once a witness set has been found and to determine whether a given solution point lies on a particular component.

With the concepts of witness sets and witness supersets in place, Chapter 9 discusses how witness supersets are computed. We discuss how one can obtain a witness superset for components at each dimension independently. When one wants to find solution components at every dimension, cascade algorithms are often a good choice, as described in this chapter. All of this can be accomplished with no knowledge of the inner workings of Bertini and via a single Bertini call, so this chapter may not be essential for some readers.

In Chapter 10, we describe algorithms used to convert a witness superset for the union of all solution components in a given dimension into a union of witness sets, one for each irreducible component. Even though many users will be content using the default settings for this conversion, it is useful for some users to have a more thorough understanding of the techniques happening behind the scenes, especially if they push the limits of what Bertini can handle.

Chapter 11 wraps up the treatment of positive-dimensional solutions with brief discussions of some advanced topics. We explain how to ask Bertini to print out detailed information concerning a witness set and how to compute solution sets at specific dimensions instead of performing a full decomposition of all dimensions. We also discuss a special software module that, given a polynomial system and a solution point, analyzes the derivatives of the system to determine the local dimension of the solution set. For example, this is of interest in kinematics, where one may have a mechanism in one assembly configuration and would like to know the number of degrees of freedom with which it can move from that pose. The method used is highly related to the concept of multiplicity and to the deflation method for reducing multiplicity, so these are also discussed.

In Part III, the book moves beyond the irreducible decomposition and looks at some of the things that can be done once witness sets are in hand. Many of these algorithms have not yet been fully implemented in Bertini, but users can explore them via user-defined homotopies. We mainly include these topics to give a view of the current state of research in the area.

First, in Chapter 12, we look at intersecting components. A more general operation is to intersect the projection of components, which gives rise to the idea of a fiber product. Fiber products are of particular interest as a means of finding exceptional sets: places where the solution set of a parameterized system increases in dimension. One application is in finding exceptional mechanisms, known as overconstrained mechanisms in the kinematics literature.

Chapter 13 looks at the issue of singularities on positive-dimensional sets. We see how to specify rank-dropping conditions and how this can be used to find isosingular sets: sets of points that share a common singularity structure. This prepares us for Chapter 14.

In Chapter 14, real solutions are considered. The standard methods of numerical algebraic geometry work over the complex numbers, and the problem of finding real solutions is surprisingly difficult. This is an area that is under active development. We discuss briefly the numerical approach to solving these systems and what "solving" means.

Next, Chapter 15 looks at some applications to algebraic geometry and, in particular, ways to compute invariants of algebraic sets, such as the geometric genus of a curve and characteristic classes. This chapter will be of particular interest to algebraic geometers.

In Chapter 16, we discuss how numerical algebraic geometry deals with images of algebraic sets under algebraic maps. The numerical approach to this situation, which may be called "numerical elimination theory," is both easier to use and applicable to larger problems than classical elimination theory.

One of the relatively new application areas for polynomial systems is in the solution of systems of nonlinear PDEs. These typically lead to very large systems with hundreds to thousands of variables. To follow even a single path of such a system can be challenging, requiring techniques such as sparse linear algebra. Chapter 17 touches on what has been done in this area.

In the appendices of Part IV, we give a thorough users manual for Bertini, beginning with a "Quick Start Guide" in Appendix A. We highly recommend that the reader install and experiment with Bertini while reading the book.

Supplementary materials

A link to a webpage containing supplementary materials for this book can be found at `www.siam.org/books/se25`. In particular, the various short Bertini input files appearing in the text are available there along with input files for longer systems that are not printed verbatim here, such as ones for the six-revolute (6R) robot inverse problem, the Stewart–Gough forward kinematics problem (and its several Griffis–Duffy variants), and several more.

Acknowledgments

This book could not have come to fruition without the various forms of support we have received from a number of generous individuals and organizations.

First, we wish to express our gratitude to the funding agencies that have financially supported so much of our work: the National Science Foundation, the Air Force Office of Sponsored Research, and the Department of Defense. We have also benefitted from the support of various research institutes, including the Fields Institute, the Institute for Mathematics and Its Applications (IMA), and Institut Mittag-Leffler. We also are grateful to our employers during the course of writing this book—Colorado State University, General Motors, North Carolina State University, Texas A&M University, and the University of Notre Dame—for providing various opportunities to work on this book and the developments described within it.

Several people read and commented on various drafts of this manuscript. Their contributions improved the final quality of this book, and their efforts are greatly appreciated. Those individuals are Daniel Baker, Noah Daleo, Brent Davis, Eric Hanson, Wenrui Hao, Matthew Niemerg, and Francesco Pancaldi.

Part I

Isolated Solutions

Chapter 1
Polynomial Systems

Polynomial systems show up throughout mathematics, engineering, and the sciences. They can be found in kinematics, chemistry, PDEs, string theory, economics, optimal control, algebraic geometry, and many other fields. As often as this sort of problem arises, it may come as a surprise that scalable and parallelizable methods for solving these problems are only now being developed.

Linear systems of equations have been solved effectively in practice for decades. On the other end of the spectrum, general nonlinear problems are so varied that there are few effective general approaches. Polynomial systems are one step up from linear systems. While theory about polynomial systems has been developed over the past two centuries (under the name *algebraic geometry*), the time for efficient, numerical computation of the solutions of a polynomial system is now.

This book aims to convey the basic notions of the field of *numerical algebraic geometry*, which includes numerical and symbolic-numeric methods for computing and manipulating the solution sets of systems of polynomial equations. The authors of this book have developed the free, open-source software package Bertini [22], in which many of the algorithms described in this book have been implemented. Accordingly, this book serves not only as an introduction to the field but also as a tutorial and a users manual (particularly the appendices) for Bertini. Our hope is that after reading this book, you will be comfortable with the language and main concepts of numerical algebraic geometry and will be able to use Bertini effectively for tackling your own problems.

In this first chapter, we introduce the main objects of interest, namely, polynomial systems and their solution sets. In particular, this chapter focuses on polynomial systems with zero-dimensional solutions sets (i.e., points), whereas Chapter 8, the introductory chapter for Part II of this book, focuses on those with positive-dimensional solution sets (curves, surfaces, etc.). Since different users of numerical algebraic geometry may want different levels of explanation, both this chapter and Chapter 8 consist of two main sections, one for readers interested in gaining a passing familiarity with the field and an optional one for those interested in more formal statements. Readers interested in full generality may want to have a copy of [161] handy while reading this book.

1.1 ▪ An Introduction to Polynomial Systems and Their Solution Sets

The fundamental problem in this book is the computation of the solution set of a system of polynomial equations

$$f(z) := \begin{bmatrix} f_1(z_1,\dots,z_N) \\ \vdots \\ f_n(z_1,\dots,z_N) \end{bmatrix} = 0, \tag{1.1}$$

where N does not necessarily equal n. If all these functions were linear, the theory of linear algebra and accompanying methods of numerical linear algebra would apply. However, in the setting of polynomials, the analogous pairing is the theoretical field of algebraic geometry with the computational tools of numerical algebraic geometry.

The first subsection below consists of several examples chosen to illustrate the differences between the linear and polynomial settings and to indicate a variety of geometric settings that arise in the study of polynomial systems. We also see how to use Bertini for these basic examples, though a more thorough "Quick Start Guide" can be found in Appendix A. Terms related to some of these situations are introduced in §1.1.2, while §1.1.4 indicates the standard method of handling overdetermined systems, i.e., systems for which $n > N$. More formal statements and explanations of these topics may be found in §1.2. The final subsection describes the use of straight-line programs, which can often be used to increase efficiency.

1.1.1 ▪ Polynomial Systems versus Linear Systems

A linear function in the N variables $x_1, x_2, \dots, x_N$ has the form

$$\square x_1 + \square x_2 + \cdots + \square x_N + \square,$$

where each $\square$ designates an unspecified complex number.[4]

Nonlinear polynomials are more complicated. For example, a degree 2 polynomial in the two variables x, y has the form

$$\square x^2 + \square xy + \square y^2 + \square x + \square y + \square.$$

More generally, a degree d polynomial in N variables has a total of $\binom{N+d}{N}$ monomials of degree no more than d.[5] As illustrated by the next several examples, the solution sets of polynomial systems display a much richer set of phenomena than are encountered in solving linear systems.

Example 1.1. As a first example, consider

$$f(x,y) = \begin{bmatrix} x^2 - 1 \\ y^2 - 4 \end{bmatrix}.$$

---

[4] A complex number is of the form $a + bi$, where $i := \sqrt{-1}$. If $b = 0$, the complex number is *real*; otherwise, it is *nonreal*.

[5] To see this, consider that a monomial of degree at most 8 in three variables can be represented by a string of length 11, using ◦◦|◦|◦◦◦|◦◦ to mean $1^2 x^1 y^3 z^2 = xy^3 z^2$, for example. In this manner, there is a one-to-one correspondence between the monomials of degree at most d in N variables and strings of length $d + N$ having d entries of ◦ and N entries of |, for $\binom{N+d}{N}$ possibilities in all.

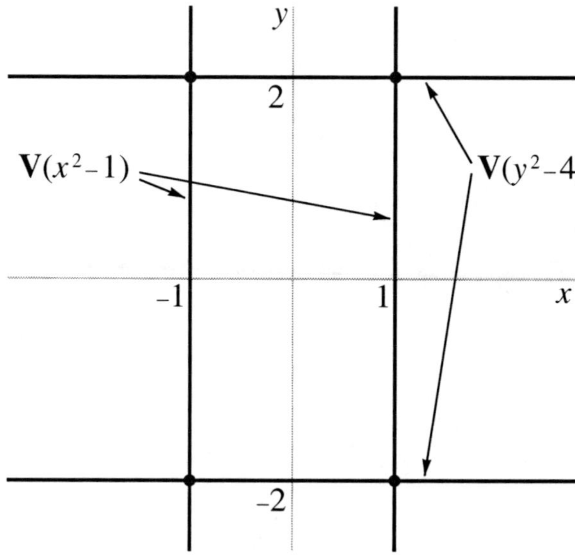

Figure 1.1. *The solution set* $\mathbf{V}(x^2-1, y^2-4)$ *consists of four points:* $(\pm 1, \pm 2)$.

If the two polynomials of f were linear rather than degree 2, then we would expect there to be a single point as the solution set. However, as shown in Figure 1.1, the solution set of this pair of polynomials, i.e., the set of points (x,y) that simultaneously cause both polynomials to evaluate to 0, consists of the four points $(x,y)=(\pm 1, \pm 2)$. Notice that in this figure and its caption, we use for the first time the notation

$$\mathbf{V}(f_1,\ldots,f_n) := \{x \in \mathbb{C}^N \mid f_1(x) = \cdots = f_n(x) = 0\}, \tag{1.2}$$

where each f_i, $i=1,\ldots,n$, is a polynomial in N unknowns.

Let's try solving $x^2-1=y^2-4=0$ with Bertini. If you have not yet set up Bertini, please consult the "Quick Start Guide" in Appendix A for assistance.

Bertini primarily uses files for input and output, with the polynomial system described by an input file. In this case, let's create a file named `input` as follows:

```
%input: our first input file
variable_group x,y;
function f,g;

 f = x^2 - 1;
 g = y^2 - 4;
END;
```

The character `%` is the comment character in Bertini. It causes Bertini to ignore any text after `%` until the end of the line. Here and throughout this book, the first comment will give the name of the input file under which the example is provided on the website for this book. (A link to the example files can be found at `www.siam.org/books/se25/`.) Notice that in the input we must declare our variable names and function names, then define the functions, with each line ending with a semicolon. The file must always end with `END;`. There are many variations on, and extensions of, this basic format, as described throughout the book and especially in Appendix B.

When you call Bertini, in this case by typing

```
>> bertini input
```

at the command line prompt >>, you will see screen output similar to the following:

```
NOTE: You have requested to use adaptive path tracking.  Please make sure that you have
set up the following tolerances appropriately:
CoeffBound: 7.863656000000e+00, DegreeBound: 2.000000000000e+00
AMPSafetyDigits1: 1, AMPSafetyDigits2: 1, AMPMaxPrec: 1024

Tracking path 0 of 4
Finite Solution Summary
NOTE: nonsingular vs singular is based on condition number and identical endpoints

              | Number of real solns |  Number of nonreal solns |  Total
------------------------------------------------------------------------------------
Nonsingular   |  4                   |  0                       |  4
Singular      |  0                   |  0                       |  0
------------------------------------------------------------------------------------
Total         |  4                   |  0                       |  4


Finite Multiplicity Summary

  Multiplicity  |  Number of real solns |  Number of nonreal solns
------------------------------------------------------------------------------------
  1             |  4                    |  0


------------------------------------------------------------------------------------
The following files may be of interest to you:

main_data:              A human-readable version of the solutions - main output file.
raw_solutions:          A list of the solutions with the corresponding path numbers.
raw_data:               Similar to the previous, but with the points in Bertini's
                          homogeneous coords along with more info about the solutions.
real_finite_solutions:  A list of all real finite solutions.
finite_solutions:       A list of all finite solutions.
nonsingular_solutions:  A list of all nonsingular solutions.
singular_solutions:     A list of all singular solutions.
------------------------------------------------------------------------------------

Paths Tracked: 4
```

This output shows that there are four nonsingular real solutions, as expected. Notice the list of files at the end of the screen output. Each contains information about the solutions or about the run, in varying levels of detail and focusing on various subsets of the set of all solutions. To see the actual solutions, look in `main_data`, which has a great deal of information about each solution. Before looking at actual files, we note that each run of Bertini will produce slightly different values due to the random numbers used in the solving process. One can find the seed for the random number generator in `main_data`, which is controlled by the configuration `RandomSeed`. The simpler files `real_finite_solutions`, `finite_solutions`, or `nonsingular_solutions` just list the solutions, all of which are identical for this example:

```
4

9.999999999999999e-01 1.387778780781446e-17
2.000000000000001e+00 2.220446049250313e-16

 1.000000000000000e+00 -6.938893903907228e-18
-1.999999999999997e+00  1.776356839400250e-15

-9.999999999999999e-01  0.000000000000000e+00
```

```
2.000000000000004e+00 -2.331468351712829e-15

-1.000000000000000e+00 0.000000000000000e+00
-2.000000000000000e+00 8.326672684688674e-17
```

Here, each of the four blocks corresponds to one of the solutions, and each of the two lines in each block corresponds to a variable (in the same order as they are declared in the `input` file). Each line has two floating-point numbers, which are the real and imaginary parts of the corresponding coordinate. Thus, the first of the four blocks corresponds to the solution $(x,y)=(1,2)$.

Notice that the first imaginary part, `1.387778780781446e-17`, is very near zero, though it does not look small unless you check the exponent, `-17`. Many more details about Bertini's various output formats and files may be found in Appendix D. ∎

A full-rank linear system of N equations in N unknowns has a unique solution, and if the coefficients of the system are all real, so is the solution. In contrast, a polynomial system with all real coefficients may have all real, all complex, or a combination of real and complex solutions.

Example 1.2. The solution set of

$$f(x,y)=\begin{bmatrix} x^3-1 \\ y^2-4 \end{bmatrix}$$

consists of two real points $(x,y)=(1,\pm 2)$ and four complex ones,

$$(x,y)=\left(-\frac{1}{2}\pm\frac{\sqrt{3}}{2}i,\pm 2\right),$$

where $i=\sqrt{-1}$ is the imaginary unit. ∎

In this book, and also in Bertini, we focus primarily on complex solution sets since the underlying theory applies primarily to $\mathbb{C}$. Finding real solution sets is generally more difficult, as discussed in Chapter 14, which is dedicated to this topic.

Example 1.3. Slightly more generally, considering the system

$$g(x,y)=\begin{bmatrix} x^m-1 \\ y^n-4 \end{bmatrix},$$

for $m,n\in\mathbb{Z}_{>0}$, we see that there are $m\cdot n$ points in the solution set, consisting of the mth cyclic roots of unity in x and the nth cyclic roots of 4 in y. Accordingly, the total number of points in the solution set of $g(x,y)$ is

$$m\cdot n=\deg(g_1)\cdot\deg(g_2).$$

This is no coincidence as the number of isolated points in the solution set of a polynomial system is always bounded above by the product of the degrees of the polynomials. This is one version of Bézout's theorem, which is described and used later in this book. Notice that the linear case ($m=n=1$) results in $m\cdot n=1\cdot 1=1$ solution, as is well known from linear algebra. ∎

Given a linearly dependent pair of linear functions in two variables, the corresponding solution set is a line (rather than a point, as above). Similarly, with more linear functions and more variables, it is possible to build a linear system that has a single higher-dimensional solution set. A similar phenomenon occurs with polynomial functions, though linear dependence does not play a role.

Example 1.4. Let

$$f(x,y)=\begin{bmatrix} xy \\ x(y-1) \end{bmatrix}.$$

Both polynomials of f include the factor x, so $x=0$ (the y-axis) is part of the solution set of this polynomial system. In fact, by inspection, the points on the y-axis are the only solutions of $f(x,y)=0$.

However, the polynomials are linearly independent as vectors in the vector space of polynomials of degree less than or equal to 2, i.e., there are no nonzero constants $a,b\in\mathbb{C}$ such that

$$a(xy)+b(xy-x)=0.$$

We describe in Part II of this book how to use Bertini to find such positive-dimensional solution sets. In particular, Chapter 8 includes a gentle introduction to positive-dimensional solution sets. For now, we simply notice their existence. ■

Even more troubling than these departures from our linear intuition is the fact that polynomial systems may have solution components in different dimensions!

Example 1.5. Let

$$f(x,y)=\begin{bmatrix} x(x-1) \\ x(y-1) \end{bmatrix}.$$

The goal is to determine the points (x,y) for which these polynomials evaluate to zero simultaneously. Again, $x=0$ (a line) will do this, as will $x=y=1$ (a point), and it is clear that those are the only solutions. This goes to show that while any linear problem always has a solution set that is a single component of a given dimension, solutions sets of polynomial systems may have multiple solution components in multiple dimensions. ■

Another noticeable difference between the linear and polynomial settings is that nonlinear polynomial systems may have nonlinear solution sets.

Example 1.6. The real solution set of the polynomial $f(x,y,z)=x^2+y^2+z^2-1$ is the unit (two-dimensional) sphere, while the complex solution set is a surface that contains the real unit sphere. (More on the relationship between real and complex solutions can be found in Chapter 14.) ■

Of course, nonlinear polynomial systems can also have linear solution sets, as we already saw in Example 1.4.

Finally, the solution sets of polynomial systems can be singular, that is, the sets can have *multiplicity* greater than 1. For polynomials in one variable, this corresponds to double roots, triple roots, and the like.

Example 1.7. The univariate polynomial $p(x)=x^2$ has one solution, the point $x=0$. As a set, this is the same as the solution set of $q(x)=x$, but they differ in that the solution of $p(x)=0$ is singular; it is a double root.

Similarly, consider the polynomial $f(x,y)=(x-y)^2$. The solution set is clearly just a line, $y=x$. Although this is the same as the solution of $g(x,y)=x-y$, there is a fundamental difference between the two. Each of the points in the solution set of $f(x,y)=0$ is singular, whereas for $g(x,y)=0$ each point is nonsingular. ■

Singularities are the topic of the next section, and they will be discussed with more detail in the optional section §1.2.2.

1.1.2 ▪ Singular Solutions and Multiplicity

Polynomial systems can have solution sets that are zero-dimensional (points) or positive-dimensional (curves, surfaces, etc.). For the following discussion, let's restrict our attention to those having only zero-dimensional, or isolated, solutions. Isolated solutions can be either *nonsingular* (also called *regular*) or *singular*.

We are surely all familiar with this dichotomy in the case of a single univariate polynomial. For example, $p(z)=(z-1)(z-2)(z-3)$ has three nonsingular solutions, while $q(z)=z^2$ has one singular solution and $r(z)=z^2(z-1)$ has one of each. Said differently, the solutions of $p(z)=0$ each have multiplicity one, while the solution of $q(z)=0$ has multiplicity 2. The *multiplicity* of the solution z^* of a polynomial $p(z)$ is k if the kth derivative of $p(z)$ is the first derivative that is nonzero at z^*, i.e., $p^{(j)}(z^*)=0$ for $0\leq j<k$ but $p^{(k)}(z^*)\neq 0$. Singular solutions are simply those that have multiplicity greater than one. Multiplicity is discussed in greater detail in Chapter 11.

Things get a bit more complicated in the multivariate, several polynomial setting. There is still a dichotomy (singular versus nonsingular) for isolated solutions, but the definitions are more technical, in part because the definition of multiplicity is more technical. If we want to think in terms of derivatives, we now have several partial derivatives for each of several polynomials. For various reasons it no longer makes sense to define multiplicity in terms of the raw number of partial derivatives that vanish. Instead, we think of the rank of the matrix of partial derivatives evaluated at the point in question, as described in §1.2.2. There, we say that a solution is *nonsingular* if this matrix has full rank; otherwise, it is singular. For now, it is enough to know that a polynomial system is somehow degenerate and troublesome at a singular solution but nice at a nonsingular solution.

Particularly troublesome for numerical methods is the fact that Newton's method can have undesirable behavior near singular solutions. Recall that Newton's method, also known as the Newton–Raphson method, is an iterative algorithm for computing numerical approximations to solutions of nonlinear systems of equations. Briefly, given a guess z_i for the solution of $f(z)=0$, Newton's method produces the next guess as

$$z_{i+1}:=z_i-[Jf(z_i)]^{-1}f(z_i) \tag{1.3}$$

with Newton residual $||z_{i+1}-z_i||$. For a polynomial system $f(z)$, a nonsingular solution z^*, and an initial point z_0 adequately near z^*, the iterates of Newton's method will converge *quadratically* to z^*. In other words, Newton's method will rapidly hone in on z^* once we are adequately near it.

This is often not the case for singular solutions. It is often the case that no matter how close you start Newton's method to a singular solution, it will not converge quadratically, and in special cases it may in fact diverge.

Example 1.8. Let us consider three polynomial systems, each with two equations and two variables, which exhibit three very different behaviors with Newton's method.

First, consider

$$f(x,y) = \begin{bmatrix} (x-1)(y-3) \\ (x-2)(y-4) \end{bmatrix}.$$

This system clearly has four solutions, including the nonsingular solution $(2,3)$. Starting at $(2.2, 2.7)$, the first several Newton residuals are

```
3.605551e-01
5.656854e-02
2.093131e-03
3.094729e-06
6.772176e-12
3.242959e-23
7.436488e-46
3.910396e-91
```

The number of accurate digits approximately doubles at each iteration—this is quadratic convergence!

Now consider

$$g(x,y) = \begin{bmatrix} (x-2)^2 \\ (y-3)^2 \end{bmatrix}.$$

The point $(2,3)$ is clearly a solution, but it is now singular. Running Newton's method from the same starting point, $(2.2, 2.7)$, we find the first several Newton residuals are

```
3.605551e-01
1.802776e-01
9.013878e-02
4.506939e-02
2.253470e-02
1.126735e-02
5.633674e-03
2.816837e-03
1.408418e-03
7.042092e-04
```

Newton's method seems to be converging, but the convergence is linear rather than quadratic. In fact, the number of accurate digits increases at a rate of about one digit per three to four Newton iterations.

Finally, consider a famous example of Griewank and Osborne [68]:

$$h(x,y) = \begin{bmatrix} \frac{29}{16}x^3 - 2xy \\ y - x^2 \end{bmatrix}.$$

This system has exactly one solution, $(0,0)$, which is singular. Running Newton's method starting from $(0.1, 0.1)$, the first several Newton residuals are

```
1.414214e-01
9.947362e-02
6.606173e-02
1.355358e-01
1.415362e-01
9.331831e-02
```

```
2.546796e-01
2.720871e-01
1.759505e-01
5.808247e-01
```

These residuals are not looking so good. Not only is there no quadratic convergence, convergence seems to be doubtful! In fact, the Newton residual for the 100th iteration is `3.621184e+29`, suggesting that Newton's method diverges.

Even starting closer to the solution, at $(0.001, 0.001)$, Newton's method also diverges (with the Newton residual for the 100th iteration being `3.433950e+25`)! In fact, this famous example of Griewank and Osborne has been shown to diverge from any starting point other than exactly the origin [68]. ■

The take-away message of all this is that numerical methods, including those in this book, behave nicely near nonsingular solutions but often poorly near singular solutions. Some of the most recent innovations in the field of numerical algebraic geometry are methods for handling singular solutions, including adaptive precision path tracking (see §3.2), singular endgames (see §3.3), and deflation (see §11.6). These techniques tame singular solutions at some extra computational cost, with worse singularities (i.e., higher multiplicity) requiring more computational resources for accurate computation.

Example 1.9. Consider the polynomial system

$$f(x,y) = \begin{bmatrix} (x-c)^2 + y^2 - 1 \\ x \end{bmatrix}.$$

The first polynomial corresponds to a circle of radius 1 centered at $(x, y) = (c, 0)$ in the (real) plane, while the second is just the y-axis. For $c = 0$, the Bertini input file (including c as a constant) is as follows:

```
% CircleAndYaxisC0.input
function f,g;
variable_group x,y;
constant c;

c = 0;
f = (x-c)^2+y^2-1;
g = x;

END;
```

Running Bertini produces the following screen output:

```
Finite Solution Summary

NOTE: nonsingular vs singular is based on condition number and identical endpoints

              | Number of real solns | Number of nonreal solns | Total
------------------------------------------------------------------------
Nonsingular   | 2                    | 0                       | 2
Singular      | 0                    | 0                       | 0
------------------------------------------------------------------------
Total         | 2                    | 0                       | 2
```

The file `main_data` reports the two nonsingular solutions as $(0, \pm 1)$, as expected.

By changing $c = 0$ to $c = 1$ in the `input` file, the pertinent part of the screen output becomes

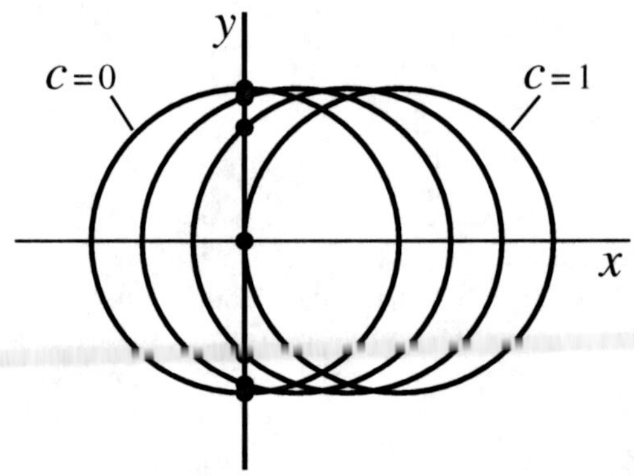

Figure 1.2. *Intersection of circles of the form $(x-c)^2+y^2=1$ with the y-axis, $x=0$, for several values of $c \in [0,1]$.*

```
Finite Solution Summary

NOTE: nonsingular vs singular is based on condition number and identical endpoints

                | Number of real solns | Number of nonreal solns | Total
------------------------------------------------------------------------------
Nonsingular     | 0                    | 0                       | 0
Singular        | 1                    | 0                       | 1
------------------------------------------------------------------------------
Total           | 1                    | 0                       | 1
```

Notice that there is now one singular solution, as the two solutions in the intersection of the circle and the line have merged. In this case, `main_data` correctly reports the origin as the solution and contains the following lines:

```
Estimated condition number: 2.236129410015832e+13
Multiplicity: 2
```

While these lines may not mean much to you yet, they are both clear signals that this is an approximation of a singular solution. Singular solutions have infinite condition numbers, so approximations of singular solutions have very large condition numbers. Also, isolated singular solutions have multiplicity greater than 1. The fact that the solution has multiplicity 2 is in concordance with the behavior illustrated in Figure 1.2, where the two nonsingular roots, $(0,\pm 1)$, for $c=0$ appear to merge into a singular double root as $c \to 1$. (A more detailed discussion of how solutions can behave as a parameter is varied, and how this can be used to construct efficient homotopies, appears in Chapter 6.) ∎

Since Part I of the book is focused on isolated solutions, we postpone thinking about what it means for a point on a positive-dimensional component to be singular until Part II.

1.1.3 ▪ Numerical Accuracy and Function Residuals

Before moving on, we should point out that even nonsingular solutions can cause us some numerical difficulties. Given a polynomial system $f(z)=0$, we aim to compute a complex vector $\widehat{z}$ which is "close" to a solution z^* of $f(z)=0$. By "close" we mean

$$||z^* - \widehat{z}|| < \epsilon \tag{1.4}$$

for a suitable small positive number ϵ, using, for example, the Euclidean norm.

Even though it is natural to think of a numerical approximation of a solution *as* a solution, it is important to keep in mind that it may not have all the qualities that a solution has. One item in particular is that if z^* is an approximate solution of $f(z)=0$, then $f(z^*)$ may not be "small" in any useable way unless the user takes efforts to ensure this, regardless of whether it is singular or nonsingular.

Example 1.10. Consider the polynomial

$$p(z) = z^{10} - 30z^9 + 2. \tag{1.5}$$

Using Bertini, one of the approximate solutions in the file `finite_solutions` is

```
3.000000000000142e+01 -4.742872761198669e-13
```

After converting this to a complex number, namely,

$$z^* = 30.00000000000142 - 4.742872761198669 \cdot 10^{-13} i,$$

where $i = \sqrt{-1}$, we find, to four digits, that $p(z^*) = 29.95 - 9.34i$, a value that is clearly not zero!

One potential reaction is to simply round this approximation off to 30 since they are equal up to 13 digits. However, $p(30) = 2$.

Instead of rounding off, let's try computing the solution to greater accuracy. The setting `FinalTol`, which defaults to `1e-11`, controls the accuracy to which Bertini attempts to find answers. (The relationship of `FinalTol` to accuracy is discussed further in §4.6, but it is an approximation to the absolute or relative error, whichever is smaller.) To change `FinalTol` from its default, we introduce a configuration block to the input file:

```
% degree10.input
CONFIG
 FinalTol:1e-15;
END;
INPUT
 variable_group z;
 function p;
 p =  z^10 - 30*z^9 + 2;
END;
```

This structure of combining a configuration block, `CONFIG ... END;`, followed by an input block, `INPUT ... END;`, is necessary whenever any setting is changed from its default value. To evaluate the polynomial at a point (or points), use the configuration settings:

```
CONFIG
 TrackType:-4;
 MPType:1;
 Precision:128;
END;
```

and issue the command

```
>> bertini <input_file_name> <points_file_name>
```

Table 1.1. *Solution estimates and function values for* (1.5).

Final Tol	z^* Real		Imaginary	Accuracy estimate	$\lvert p(z^*)\rvert$
1e-11	30.00000000000142	+	-4.7e-13i	9.4e-12	31.371
1e-11	30.00000000000021	+	-1.2e-12i	6.7e-12	25.706
1e-15	29.99999999999989	+	1.7e-15i	3.9e-13	0.1688
1e-15	29.99999999999990	+	-3.5e-15i	1.1e-16	0.0768
1e-19	29.9999999999998983894731343124	+	6.2e-24i	1.9e-19	0.32e-9
1e-19	29.9999999999998983894731645752	+	-1.5e-23i	9.4e-20	0.42e-9

With these settings, Bertini evaluates the polynomial (`TrackType:-4`) in fixed multiprecision (`MPType:1`) using 128-bit arithmetic (`Precision:128`), which is about 38 decimal places. Further discussion of the use of configurations may be found in Appendix A, and Appendix E provides a list of all configurations and their allowable and default settings.

In this fashion, we may run the example at several different tolerances to get the results shown in Table 1.1. Repeating the run twice at each tolerance setting shows that the estimate in the accuracy of z^*, which can be found in the output file `main_data`, is in reasonable agreement with the difference between runs, although the agreement for the last two runs is a few digits better than the accuracy estimate predicts. The main point to notice, however, is that the magnitude of $p(z^*)$ is consistently much larger than the reported accuracy in the solution.

What is going on with this example? Even though the real solution near 30 is nonsingular, the moderately high degree of 10 causes the polynomial to have a large derivative in the vicinity of $z = 30$: $p'(30) \approx 2 \cdot 10^{13}$. Therefore, if $\hat{z}$ is the actual root near 30, then for z^* near 30, we have

$$|p(z^*)| \approx 2 \cdot 10^{13} \cdot |\hat{z} - z^*|.$$

So even if the approximate root is close to the actual solution, this large factor significantly magnifies the error. For more on this issue see [161, §5.3]. ∎

1.1.4 ▪ Randomization: Handling Overdetermined Systems

The methods described later in this book require us to have polynomial systems for which the number of polynomials, n, is equal to the number of equations, N, i.e., $N = n$. We refer to these as *square systems*. The point of this brief section is to convey the standard way of handling an overdetermined system, i.e., $n > N$. Underdetermined systems ($n < N$) necessarily have positive-dimensional solution components and are handled in Part II of the book.

Given polynomials $f_1, f_2, \ldots, f_n$, a random linear combination of these polynomials has the form

$$\square f_1 + \square f_2 + \cdots + \square f_n,$$

where each $\square$ is a randomly chosen complex number. The standard way to reduce a polynomial system with n polynomials down to a system with $N < n$ polynomials is to replace $f_1, \ldots, f_n$ with N random linear combinations of the polynomials.

Said differently, we choose a random matrix $A \in \mathbb{C}^{N \times n}$ and replace the system

$$f = \begin{bmatrix} f_1 \\ f_2 \\ \vdots \\ f_n \end{bmatrix}$$

with the system

$$A \cdot f.$$

We often refer to this as "squaring up" the system.

The key property is that all the isolated solutions of f are also isolated solutions of $A \cdot f$. This is guaranteed by Bertini's theorem (Theorem 1.15) with probability one (see §1.2.3). In fact, $A \cdot f$ could have more isolated solutions than f. These "extra" solutions of $A \cdot f$ can be detected and discarded as they will not satisfy f. (As the discussion in §1.1.3 showed, some care in setting tolerances needs to be taken to ensure a correct decision on whether a solution satisfies the equations.)

The "probability-one" qualifier in the key property above means that everything is fine for almost every choice of random constants, though there are a few (a measure zero set of them among all possible choices of random constants) for which the isolated solutions of f are not isolated solutions of $A \cdot f$.

Example 1.11. Let $p(z) = (z+1)(z-1)$ and $q(z) = z(z-1)$. Clearly, the solution set of $p(z) = q(z) = 0$ consists of the point $z = 1$.

Let's randomize the system down to one equation,

$$2p(z) - 3q(z) = 2(z+1)(z-1) - 3z(z-1) = (2-z)(z-1),$$

which has two solutions, $z = 1, z = 2$, i.e., we picked up an extra solution. For $z = 1$, both $p(z)$ and $q(z)$ evaluate to 0. For $z = 2$, we have $p(2) = 3$ and $q(2) = 2$ for which the linear combination evaluates to zero. This point would move as we changed the numbers used to create the linear combination. In fact, since for almost any other choices of random complex constants, the linear combination would be a degree 2 polynomial in z, it will necessarily have two roots (by the fundamental theorem of algebra), one of which must be $z = 1$.

Note that the random complex constants chosen here are neither random nor complex and were chosen for simplicity of exposition. ∎

Squaring up an overdetermined system may also have an impact on any positive-dimensional components of $\mathbf{V}(f)$. Those effects are described in Chapter 8.

1.1.5 ▪ Straight-Line Programs

Very often polynomials are given as nested expressions. For example, consider the 12th-degree polynomial

$$p(z) = (z^3 + z + 1)^4 - 5(z^2 + 2)^3 + 6.$$

The natural way to evaluate this is by evaluating the subexpressions $a = z^3 + z + 1$ and $b = z^2 + 2$ and then evaluating $a^4 - 5b^3 + 6$. This prescription is called a straight-line program[6] for the evaluation of $p(z)$. The expressions $a = z^3 + z + 1$ and $b = z^2 + 2$ are called subfunctions.

[6]Technically, a straight-line program is a representation of $p(z)$ as a sequence of single unary or binary operations. Although Bertini does in fact make use of them in this sense for efficient evaluation and differentiation, such details are unnecessary for most users, so we use the term straight-line program more loosely, as above.

For such a simple polynomial, the difference between evaluating by means of the straight-line program and evaluating the expanded version of the polynomial is not great, but for large systems the differences in computation are often considerable, e.g., consider the straight-line program for evaluating the system for the nine-point problem §5.5.2.

Very often polynomial systems arising from physical models are naturally constructed in a straight-line form. Since evaluation of polynomials takes up a large part of the computational cost in any polynomial solver, it is important to take advantage of straight-line representations of polynomials. A second important reason to use straight-line programs is that expansion can introduce numerical inaccuracy.

Older solvers, e.g., the one described in [131], expected polynomials to be entered as coefficients of the monomials. Bertini has facilities to make it easy to enter polynomials for straight-line program evaluation. For example, the input file for solving the polynomial $p(z) = 0$ above may be written

```
% subfunctions.input
INPUT
 variable_group z;
 function p;

 a = z^3 + z + 1;
 b = z^2 + 2;
 p = a^4 - 5*b^3 + 6;
END;
```

Note that the subfunctions a and b do not have to be declared, though they do need to precede any expression, e.g., `p = a^4 - 5*b^3 + 6;`, which contains them.

Running the program we get the summary

```
Finite Solution Summary

NOTE: nonsingular vs singular is based on condition number and identical endpoints

                | Number of real solns  |  Number of nonreal solns  |  Total
--------------------------------------------------------------------------------
Nonsingular     |         2             |             10            |   12
Singular        |         0             |             0             |   0
--------------------------------------------------------------------------------
Total           |         2             |             10            |   12


Finite Multiplicity Summary

  Multiplicity  |  Number of real solns  |  Number of nonreal solns
--------------------------------------------------------------------------------
        1       |          2             |         10
```

1.2 ▪ Mathematical Background*

This section and the other technical sections later in the book are, like this one, optional. You can safely skip this and the other sections marked with a star (*) if you neither need nor want the careful definitions and statements that support the algorithms of numerical algebraic geometry.

We have chosen to include these technical sections to provide a bit more rigor for those readers uncomfortable with the looseness of the gentle statements made throughout the text. For example, we introduced the concept of a singular point earlier in this chapter without providing a formal definition. This section contains that and other such definitions and statements. It is by no means intended as an introduction to algebraic geometry or a replacement for the much more thorough numerical algebraic geometry text [161].

1.2.1 ▪ Algebraic Sets and Irreducible Components*

In this section, we define what we mean by an *algebraic set*, describe the decomposition of such a set into *irreducible components*, and close with a brief discussion of the two underlying topologies that we use throughout the book.

Algebraic sets

Solution sets of polynomial systems are more properly called *algebraic sets*. More precisely, an *affine complex algebraic set* (respectively, *affine real algebraic set*) $\mathbf{V}(f_1,\ldots,f_n)$ is the locus of solutions on $\mathbb{C}^N$ (respectively, $\mathbf{V}_{\mathbb{R}}(f_1,\ldots,f_n)$, the locus of solutions on $\mathbb{R}^N$) of a system of polynomials as in (1.1) with complex (respectively, real) coefficients. Note that $\mathbf{V}(f_1,\ldots,f_n)$ and $\mathbf{V}_{\mathbb{R}}(f_1,\ldots,f_n)$ are the underlying sets: all multiplicity information is thrown away, e.g., $\mathbf{V}(z^3)=\mathbf{V}(z)=\{0\}$.

In high school analytic geometry, we are exposed to many real algebraic sets, e.g., circles, ellipses, parabolas, lemniscates, spheres, and so on. The link between the *real* solution set of a *real* polynomial system and the system of polynomials is weak. *Complex* algebraic sets are much nicer than real algebraic sets as the links between *complex geometry*, *complex numerical analysis*, and the underlying algebra is very tight.

As a first example of the quality of the link between complex geometry and algebra, consider $z^2+1=0$. There are no solutions $z\in\mathbb{R}$ but two in $\mathbb{C}$, $\pm i$. By the fundamental theorem of algebra, any degree d one-variable polynomial has exactly d roots, counting multiplicities. Hilbert's Nullstellensatz, the generalization of this to several variables, says that if $\mathbf{V}(f_1,\ldots,f_n)\subset\mathbb{C}^N$ is empty, then there exist polynomials $g_1,\ldots,g_n$ such that

$$f_1 g_1+\cdots+f_n g_n=1.$$

For example, if $n=1$, then $f_1=1/g_1$, from which it may be seen that f_1 is a constant.

To give further illustrations of the quality of the complex geometry/algebra linkage, we first need the concept of dimension for algebraic sets. To do this we first define a *manifold point* $p^*=(p_1^*,\ldots,p_N^*)$ of $X:=\mathbf{V}(f_1,\ldots,f_n)$ (respectively, $X_{\mathbb{R}}:=\mathbf{V}_{\mathbb{R}}(f_1,\ldots,f_n)$ for a real polynomial system) to be a point $p^*\in X$ with a neighborhood $U\subset X$ (respectively, $p^*\in X_{\mathbb{R}}$ with a neighborhood $U\subset X_{\mathbb{R}}$) such that for some mapping $\Phi(z_1,\ldots,z_N)$, Φ_U maps U one-to-one and onto a neighborhood of the origin in $\mathbb{C}^k$ (respectively, $\mathbb{R}^k$). We refer to the set of manifold points of algebraic set X as X_{reg}. The *complex dimension* $\dim_{p^*}X$ of X at p^* is defined to be k, and the *real dimension* $\dim_{\mathbb{R},p^*}X_{\mathbb{R}}$ is also defined to be k. Note that the usual dimension from topology of $U\subset X$ is $2k$, whereas that of $U\subset X_{\mathbb{R}}$ is k, the difference stemming from the fact that each complex coordinate has both a real and an imaginary part. We often refer to the complex dimension as the *dimension*, and we usually refer to the dimension from topology as the real dimension.

Irreducibility

Now let us turn to the notion of irreducibility for algebraic sets. A polynomial $f(z_1,\ldots,z_N)$ is said to be *irreducible* if f cannot be factored as $f=gh$, where g and h are positive degree polynomials. For example, one can show that

$$y^2+(x+2)x(x-1)$$

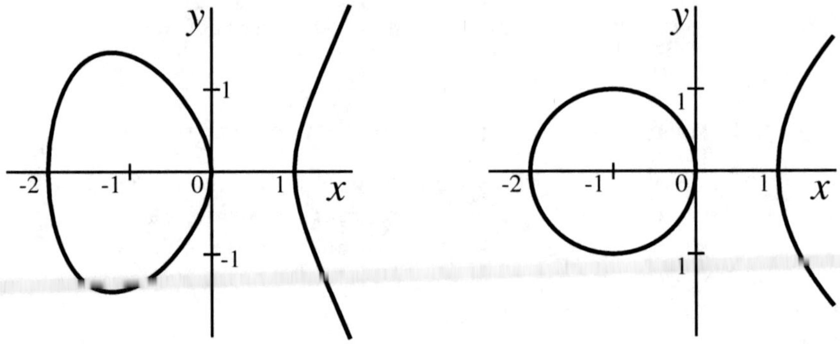

Figure 1.3. *Two algebraic curves, each with two real components. The one on the left has one complex component, whereas the one on the right factors into two complex components.*

cannot be factored, so it is irreducible. It may be shown that a polynomial f is irreducible if and only if $\mathbf{V}(f)_{\text{reg}}$ is connected. Therefore, over the complex numbers $\mathbf{V}(y^2+(x+2)x(x-1))$ must be connected. This is false over the reals, and indeed, the left-hand panel of Figure 1.3 shows that the real points of this same set, i.e., $\mathbf{V}_{\mathbb{R}}(y^2+(x+2)x(x-1))$, have two connected components. This is in contrast to the zero set of the factored polynomial

$$\left((x+1)^2+y^2-1\right)\left(y^2-4x+4\right),$$

which has two complex components and, as shown in the right-hand panel, also has two real components. Complex components that contain no real points at all also exist, such as x^2+1. In short, one cannot judge reducibility by looking at real components, and real points on the same complex component may not be connected in real space. But all the points in an irreducible complex component are path connected, which is an important property we use in homotopy methods.

For X any algebraic set, any connected component of X_{reg} has an affine complex algebraic set as its closure in the usual topology. This is not necessarily the case for the real points $X_{\text{reg}} \cap \mathbb{R}^N$, as the closure in the usual topology of, say, the left-hand loop of $\mathbf{V}_{\mathbb{R}}(y^2+(x+2)x(x-1))$ in Figure 1.3 is just that loop, and this loop is not an algebraic set, as there is no polynomial whose solution set is just that loop.

It may be shown (see [176, Chapter 1]) that every polynomial f may be factored as $f = p_1^{m_1} \cdots p_s^{m_s}$, where m_j are positive integers and p_j are irreducible polynomials. This gives a decomposition

$$\mathbf{V}(f) = \bigcup_{j=1}^{s} \mathbf{V}(p_j), \tag{1.6}$$

which may be shown to be the decomposition of $\mathbf{V}(f)$ as the union of the closures of the connected components of $\mathbf{V}(f)$.

Geometrically, we say an affine complex algebraic set $X = \mathbf{V}(f_1,\dots,f_n) \subset \mathbb{C}^N$ is *irreducible* if X_{reg} is connected. Algebraically this corresponds exactly to the radical of the ideal in $\mathbb{C}[z_1,\dots,z_N]$ generated by $f_1,\dots,f_n$ being prime (see [44, 140]). We have little need for such algebraic language in this book and therefore do not give definitions of ideal, prime ideal, radical of an ideal, etc., which the reader can find covered nicely in [44, 140].

It can be shown that X_{reg}, the set of manifold points of X, is the complement in X of an affine complex algebraic set, and X_{reg} is dense in X in the usual topology from $\mathbb{C}^N$.

For an arbitrary $p^* \in X$, we define the dimension X at p^*, denoted $\dim_{p^*} X$, to be the maximum dimension of the irreducible components that go through p^*. The dimension (i.e., complex dimension) for the full set X is then defined as

$$\dim X := \max\{\dim_{p^*} X \mid p^* \in X\}. \tag{1.7}$$

The same density statement holds for manifold points of $X_{\mathbb{R}}$ and we define $\dim_{p^*} X_{\mathbb{R}}$ similarly. We refer to an affine complex algebraic set X as *pure-dimensional* if $\dim X = \dim_{p^*} X$ for all $p^* \in X$.

For any nonconstant polynomial f, the complex solution set $\mathbf{V}(f)$ is of dimension $N-1$ at every point of $\mathbf{V}(f)$. In contrast, the real solution set of a single real polynomial may be any dimension, e.g., consider $\mathbf{V}_{\mathbb{R}}(z_1^2 + \cdots + z_i^2)$ for each $i = 1, \ldots, N$, where the real solution set is just the origin. For more on real versus complex dimension, see §14.1.

The generalization to a system of polynomials of the factorization of a single polynomial into irreducible factors is the decomposition of a complex algebraic set into irreducible complex algebraic sets. This decomposition is called the *irreducible decomposition*. Precisely, given a polynomial system $f_1, \ldots, f_n$ as in (1.1) with $X = \mathbf{V}(f_1, \ldots, f_n)$, there is a decomposition

$$X := \bigcup_{i=0}^{\dim X} X_i,$$

where for each i, X_i is either empty or a pure i-dimensional affine complex algebraic set. Let $\mathscr{I}$ denote the set of the i with $X_i \neq \emptyset$. For each $i \in \mathscr{I}$, there is a finite nonempty set $\mathscr{I}_i$ such that

$$X = \bigcup_{i \in \mathscr{I}} X_i = \bigcup_{i \in \mathscr{I}} \bigcup_{j \in \mathscr{I}_i} X_{i,j}, \tag{1.8}$$

where each $X_{i,j}$ is an *irreducible* i-dimensional affine complex algebraic set and where no $X_{i,j}$ is contained in the closure of $X \setminus X_{i,j}$. There are more refined decompositions that take into account the multiplicity structure (see [53]).

Among the first major successes of numerical algebraic geometry were making sense of this decomposition numerically and the development of the algorithms to do the numerical version of this decomposition. This *numerical irreducible decomposition* [160, 154, 155, 156, 157, 158] is discussed in detail in Part II, where we will see how Bertini efficiently and reliably computes it.

The Zariski topology and the Complex topology

As subsets of Euclidean space, algebraic sets $X \subset \mathbb{C}^N$ inherit the *classical topology* (also called the *complex* or *usual topology*), which is the topology on X with open sets that are unions of the intersections of X with the open balls with positive radii centered at points of $\mathbb{C}^N$. There is a far weaker but extremely useful topology on X called the *Zariski topology*, whose open sets are taken to be intersections of X with sets of the form $\mathbb{C}^N \setminus Y$, where Y is an affine complex algebraic subset of $\mathbb{C}^N$. Open sets in the Zariski topology are open sets in the complex topology, but the reverse is far from true. For example, the Zariski open sets of $\mathbb{C}$ coincide with the sets which are complements of finite sets, so these are larger than any finitely bounded open set in the complex topology. The usefulness of Zariski open sets stems (as we will see in §1.2.3) from the fact that many conditions automatically hold true for Zariski open sets, giving much stronger conclusions than those that hold generally for open sets in the complex topology.

There are two key facts about the complex and the Zariski topologies. First, if U is a Zariski open subset of a complex algebraic set, then the closure in the Zariski topology and the closure in the complex topology coincide. For this reason we may safely refer to the closure $\overline{U}$ of a Zariski open subset U of a complex algebraic set without specifying the topology. Also, if X is irreducible, then any nonempty Zariski open subset $U \subset X$ is connected (and even path connected). In fact, such a subset U is the complement in X of a real-two-codimensional (or even lower-dimensional) subset Z of X, where Z thus has Lebesgue measure zero.

1.2.2 ▪ Singularities and Multiplicity*

The system $f(z) = 0$ from (1.1) is called *square* if $n = N$, i.e.,

$$f(z) := \begin{bmatrix} f_1(z_1,\ldots,z_N) \\ \vdots \\ f_N(z_1,\ldots,z_N) \end{bmatrix} = 0. \tag{1.9}$$

A solution $z^* \in \mathbb{C}^N$ is said to be *isolated* if there is an open ball around z^* such that the only solution in this ball is z^*, i.e., there exists $r > 0$ such that the only solution contained in

$$B_r(z^*) := \{z \in \mathbb{C}^N \mid |z - z^*| < r\}$$

is z^*. A solution z^* is said to be *nonsingular* if the *Jacobian* of f at z^*, namely,

$$J(f, z^*) := \begin{bmatrix} \frac{\partial f_1(z^*)}{\partial z_1} & \cdots & \frac{\partial f_1(z^*)}{\partial z_N} \\ \vdots & \ddots & \vdots \\ \frac{\partial f_n(z^*)}{\partial z_1} & \cdots & \frac{\partial f_n(z^*)}{\partial z_N} \end{bmatrix}, \tag{1.10}$$

has rank N. Otherwise, z^* is *singular*. Every nonsingular solution is isolated, though not every isolated solution is nonsingular, e.g., $f(z) = z^2$ has a singular solution, the double root $z = 0$. Multiplicity is discussed in some detail in Chapter 11.

There are two distinct notions of singularities, i.e., nonsmooth points, for the solution set of a system of polynomial equations such as (1.9) on $\mathbb{C}^N$:

1. the singularities of the underlying set $\mathbf{V}(f)$;

2. the singularities of the underlying set $\mathbf{V}(f)$ equipped with the multiplicity structure coming from the system. We let $f^{-1}(0)$ denote the set $\mathbf{V}(f)$ with multiplicity information, made precise below.

Singularities in the first sense are always singularities in the second sense, but not vice versa.

Before we precisely define this second notion of singularity, let us consider a few examples.

Example 1.12. Let $f(z)$ be the polynomial z^k on $\mathbb{C}$ for some positive integer k. Clearly $\mathbf{V}(z^k) = \{0\}$, and this is smooth as a set. The multiplicity structure in this case is the point 0 counted k times. $\mathbf{V}(f)$ is smooth, but $f^{-1}(0)$ is smooth only when $k = 1$, i.e., when the multiplicity of the point $z = 0$ with respect to $f(z) = 0$ is one. Note that this is the same as the condition that the derivative of z^k is nonzero at the origin. ∎

More generally, if $p^* \in \mathbb{C}^N$ is an isolated solution of the system $f(z)$ in (1.9), then p is a smooth point of $\mathbf{V}(f)$ but is a smooth point of $f^{-1}(0)$ precisely when the multiplicity of p^* is one.

Now let's consider some positive-dimensional sets.

Example 1.13. Let $f(z,w) = z^k w^j$, where j,k are some positive integers. The set $\mathbf{V}(f)$ consists of the union of two one-dimensional components: the z and w axes. As a set, $\mathbf{V}(f)$ is smooth everywhere except at the origin, where the two one-dimensional components intersect. The multiplicity of the z-axis is j and the multiplicity of the w-axis is k. Taking into account multiplicity, the points on the z-axis minus the origin are smooth only if $j = 1$, and similarly, the points on the w-axis minus the origin are smooth only if $k = 1$. ∎

The key to describing the singular points of $f^{-1}(0)$ for a system as in (1.9) is to take a point p^* of $Z = \mathbf{V}(f)$ with $\dim_{p^*} Z = \ell$ and define p^* to be a singular point of $\mathbf{V}(f)$ either if p^* is contained in two distinct components of Z or if p^* as a solution of f restricted to a general $(N-\ell)$-dimensional linear space through p^* has multiplicity at least two.

Let us give one last example.

Example 1.14. Let

$$f(z,w) = \begin{bmatrix} zw \\ w^2 \end{bmatrix}.$$

Note that $\mathbf{V}(f)$ is the z-axis and is hence smooth. On the other hand $f^{-1}(0)$ is smooth away from the origin. Given a general line through the origin $t \to (at, bt)$ for general constants a, b, the system pulls back to abt^2, b^2t^2, and thus the origin is a singular point of $f^{-1}(0)$. ∎

1.2.3 ▪ Probability-One Results for Nonsquare Systems*

Many times algebraic geometry guarantees that certain statements are *generically* true. Before we give an exact definition of what we mean, let's give a typical example and indicate how we might use genericity to deduce useful information.

Consider a not identically zero univariate polynomial $p(z)$ of degree d. We know that p has exactly d roots counting multiplicities. Thus, for all $z \in \mathbb{C}$ outside of a finite set of points, $p(z) \neq 0$. How might we use this? Consider the question of whether a degree d polynomial $p(z)$ is identically zero. This is not a difficult question but is similar to the nontrivial problem of deciding whether a polynomial in several variables is identically zero on an irreducible algebraic set. We might choose a "random" point z^* (say, from the constant probability density on the unit disk) and evaluate $p(z^*)$. If this number is nonzero, then $p(z)$ is clearly not identically zero, while if the number *is* zero, then with *probability one* we conclude $p(z)$ is identically zero.

There are two perspectives from which to consider this procedure.

One is from the level of the model we have of algebraic geometry with exact arithmetic and infinite precision and accuracy. We give definitions and basic theory from this point of view in this subsection and present an algorithm (for solving a polynomial system using a total-degree homotopy) based on this viewpoint in §2.1.4.

The other perspective is from the level of actual computation where numbers have a limited number of digits, polynomial systems are represented only approximately on the computer, and evaluation is not exact. From this perspective, there is a probability, which can be significantly different from zero, that a "probability-zero" event occurs. This perspective and how Bertini deals with inexactness will be discussed after adaptive precision is discussed in §3.2.

We say that a statement $\mathscr{P}(x)$, parameterized by x belonging to a nonempty irreducible complex algebraic set, is *generically true* or *true with probability one* if $\mathscr{P}(x)$ is true for all x in a nonempty Zariski open subset of X. This definition is motivated by the fact that if X is irreducible, then a nonempty Zariski open subset $U \subset X$, which is the complement in X of a real-two-codimensional subset, is connected. In fact, $\overline{U} = X$ and the Lebesgue measure of $X \setminus U$ is zero.

Consider the statement that given a polynomial $p(z)$ of degree d and a random point $x \in \mathbb{C}$, then with probability one, p is the zero polynomial if $p(x) = 0$. Here the statement $\mathscr{P}(x)$ is

if $p(x) = 0$, then p is the zero polynomial.

Another example comes from Bertini's theorem and its many generalizations. The Bertini software package, which depends critically on the use of these theorems in almost every computation, is named, accordingly, after Eugenio Bertini.

Theorem 1.15 (Bertini's theorem). *Suppose that $f_1(z),\ldots,f_n(z)$ are polynomials defined on a nonempty Zariski open subset U of an irreducible affine complex algebraic set $X \subset \mathbb{C}^N$ with the property that given any point $x \in U$, there is at least one j such that $f_j(x) \neq 0$. Then, there is a nonempty Zariski open subset $\mathscr{U} \subset \mathbb{C}^n$ such that for $\lambda = (\lambda_1,\ldots,\lambda_n) \in \mathscr{U} \subset \mathbb{C}^n$, $f_\lambda(z) = \lambda_1 f_1(z) + \cdots + \lambda_n f_n(z)$ has the property that $f_\lambda^{-1}(0)$ is either empty or smooth of pure-dimension $N-1$, i.e.,*

1. $Z_\lambda := \mathbf{V}(f_\lambda(z))$ *is empty or pure-dimension* $N-1$; *and*
2. $\mathbf{V}\left(f_\lambda(z), \frac{\partial f_\lambda(z)}{\partial z_1},\ldots,\frac{\partial f_\lambda(z)}{\partial z_N}\right) = \emptyset$.

In other words, if a set of polynomials has no common solution, a general linear combination of them will either be empty or define a smooth (and, in particular, multiplicity one) hypersurface.

There are many variants of this statement. For example, under the assumptions on the f_i of the theorem, induction yields that there is a nonempty Zariski open subset $\mathscr{U}$ of the $k \times n$ matrices $\mathbb{C}^{k\times n}$ such that for $A \in \mathscr{U}$ and with

$$f(z) = \begin{bmatrix} f_1(z_1,\ldots,z_N) \\ \vdots \\ f_n(z_1,\ldots,z_N) \end{bmatrix}$$

it follows that $Z_A = \mathbf{V}(A \cdot f(z))$ is either empty or of pure-dimension $N-k$.

Another variant says exactly what was indicated earlier in this chapter: Randomizing a polynomial system so that it is square can only add (never remove) isolated solutions. Another variant, connected to positive-dimensional solution sets, is described in Chapter 8.

Chapter 2
Basic Polynomial Continuation

This chapter describes the fundamental computation of numerical algebraic geometry: homotopy continuation. We follow the historical development of the subject by starting with the problem of finding a finite set $\mathscr{S}$ of solutions of a square system,

$$f(z) := \begin{bmatrix} f_1(z_1,\dots,z_N) \\ \vdots \\ f_N(z_1,\dots,z_N) \end{bmatrix} = 0, \tag{2.1}$$

defined for $z \in \mathbb{C}^N$, with the property that $\mathscr{S}$ contains every isolated solution of $f(z) = 0$. This problem is fundamental in several ways:

1. Using the solution of this problem, it is easy to solve the analogous problem for a possibly nonsquare system (1.1).
2. Computing the full solution set of $f(z) = 0$, i.e., computing the possibly positive-dimensional solution set $\mathbf{V}(f)$, is to a large extent built from the computation of isolated solutions of associated polynomial systems arising from the polynomial system (1.1).
3. Many of the issues that arise in the application of continuation to the computation of isolated solutions also arise in the algorithms to compute the possibly positive-dimensional solution set of $f(z) = 0$.
4. Many applications only require computing isolated solutions (though this is changing as people realize that more general information is often within reach).

In §2.1, we codify the properties we need in a homotopy to compute a finite set of solutions of $f(z) = 0$ containing all the isolated solutions of $f(z) = 0$. To make everything concrete we use the explicit total-degree homotopy of §2.1.4.

In §2.2, we discuss in detail the fundamental computation of numerical algebraic geometry, tracking a single path. Path tracking uses numerical methods for solving differential equations and Newton's method. In addition to its use in finding solution points, accurate tracking, at which Bertini excels, is the key to computation of monodromy, which underlies the manipulation of positive-dimensional algebraic sets, and to other calculations underlying applications of algebraic geometry.

We introduce some syntax in this chapter for commanding Bertini to track paths. As we progress through the book, more of the syntax and the accompanying concepts will be explained. We strongly suggest browsing Appendix B while reading this chapter.

2.1 ▪ Homotopy Functions

A general approach to finding solutions of a system of general equations, say, A, which could be a system of polynomials or a system of partial differential equations, etc., is to take a system B of “similar” equations with known solutions and deform B and its solutions to A and its solutions. This approach is called *homotopy continuation* or *continuation* and the deformation is accomplished by means of a parameterized family of equations, which specializes to A and B for different values of the parameters. The parameterized family of equations is called a *homotopy*. In this generality, what “similar” means is unclear and much can go wrong, but in the case of polynomials, theory and experience have given rise to broad classes of excellent homotopies.

To find the isolated solutions of (2.1), the strategy is to find all the isolated solutions of a simple system $g(z)$ and then, using homotopy continuation, deform the system $g(z)$ and its solutions to $f(z)$ and its solutions. If the start system $g(z)$ and the deformation rule—the homotopy—are chosen appropriately, this process will lead to all the solutions of $f(z)$. It is well understood how to construct such homotopies, and several of the most effective constructions are built into Bertini. In addition, Bertini has a facility allowing the user to specify a user-defined homotopy.

In this section, we will codify what is needed so that a homotopy may be used to find the solutions of a polynomial system. We also explain the *total-degree homotopy* in §2.1.4. First, though, we begin with an intuitive description of the three-step process of solving a polynomial system via homotopy continuation.

2.1.1 ▪ Intuitive Discussion of Homotopy Continuation

There are three basic steps for solving a polynomial system $f(z)$ (which we call the *target system*) via homotopy continuation:

1. **Build and solve a *start system* g(z).** The idea is to construct a polynomial system related to $f(z)$ (for example, with the same degrees) that we can solve easily. We refer to the solutions of $g(z)=0$ as *startpoints* for a reason that will become clear in a moment. The total-degree homotopy is the simplest choice and is described in §2.1.4.

2. **Construct a homotopy between f(z) and g(z).** The simplest form for this is $H(z,t)=t\cdot g(z)+(1-t)\cdot f(z)$ with t a new parameter, so that we know the solutions of $H(z,1)=g(z)=0$ and want to compute the solutions of $H(z,0)=f(z)=0$. The key observation (described in more detail below) is that the solutions vary continuously and will not, with probability one, collide in the interval $t\in(0,1]$. Thus, we get *paths* connecting the startpoints (solutions of the start system, $g(z)$) to the *endpoints* (a superset of the solutions of the target system, $f(z)$). Each path is the image of $[0,1]$ under a continuous map, say, $p(t)$; that is, $H(p(t),t)=0$ for $t\in[0,1]$.

3. **Follow the paths from $t=1$ to $t=0$.** We use predictor-corrector methods to march along the paths from $t=1$ to $t=0$, as discussed in §2.2. In fact, we track most of the way to $t=0$ with predictor-corrector methods, then switch to more powerful *endgames*, which are described in more detail in §3.3. Some paths could approach a point “at infinity” as $t\to 0$, in which case the target system has fewer isolated solutions than the start system. We refer to such paths as *divergent*. Other paths could come together at $t=0$, a special situation that forces the use of endgames.

Before explaining how these steps are carried out on general systems, let us explore a simple example in depth. This simple example illustrates the main phenomenon that can happen in general and shows how certain basic tricks lead to a procedure that succeeds with probability one.

Example 2.1. Suppose we wish to solve $f(z) = 0$ for the following polynomial in one variable:

$$f(z) = -2z^3 - 5z^2 + 4z + 1.$$

Of course, there exist specialized methods for solving univariate polynomials, and in the case of a cubic, such as this example, one could apply the cubic formula, which gives the solutions in terms of square-root and cube-root operations. (Caution: The cubic formula is fine for symbolic work but can be problematic numerically.) We choose a low-degree univariate polynomial for illustration so that we can show all details of the continuation procedure.

The first step is to choose a start system. It suffices here to choose any cubic polynomial with three distinct roots that we can solve easily. Let's choose

$$g(z) = z^3 + 1.$$

The roots of $g(z)$ are $z = -e^{2k\pi i/3}$, $k = 0,1,2$, that is, $\mathbf{V}(g) = \{-1, (1 \pm \sqrt{3}i)/2\}$, where $i = \sqrt{-1}$.

The next step is to construct a homotopy to deform $g(z)$ into $f(z)$. Let's use a *linear homotopy* between them:

$$h(z,s) = s\,g(z) + (1-s)f(z). \tag{2.2}$$

Notice that $h(z,1) \equiv g(z)$ and $h(z,0) \equiv f(z)$.

To solve for the roots of $f(z)$, we will move continuously from $s = 1$ to $s = 0$ and track the solutions of $h(z,s) = 0$, starting from the startpoints, $\mathbf{V}(g)$.

It is important to note that s is a complex variable, so there are an infinite number of continuous one-real-dimensional paths going from $s = 1$ to $s = 0$. A natural choice would seem to be the real line segment, $[0,1]$, but as we will see in a moment, that choice is problematic. Instead, let's consider the family of all circular arcs that connect $s = 1$ to $s = 0$. For each $\gamma \in \mathbb{C} \setminus \mathbb{R}$, the one-real-dimensional curve given by

$$q(t) = \frac{\gamma t}{\gamma t + (1-t)}, \qquad t \in [0,1], \tag{2.3}$$

is a circular arc. Notice that $q(1) = 1$ and $q(0) = 0$, so by setting $s = q(t)$, we have an arc that connects $s = 1$ to $s = 0$, as desired.

Following one of these circular arcs, our homotopy becomes $h(z, q(t)) = 0$. Substituting (2.3) into (2.2) and clearing the denominators (which is valid, since $\gamma t + (1-t) \neq 0$ for $t \in [0,1]$), we obtain the homotopy

$$H(z,t) = \gamma t\, g(z) + (1-t) f(z).$$

Choosing $\gamma = 0.40 + 0.77i$, we have that as t goes from 1 to 0 on the real line segment $[0,1]$, the solutions of $H(z,t) = 0$ are nicely behaved; as shown in Figure 2.1, we have three smooth paths that never intersect. These are easy to track numerically from $V(g)$ to the solutions we seek:

$$\mathbf{V}(f) = \{-3.0942, -0.2028, 0.7969\} \quad \text{(approximate)}.$$

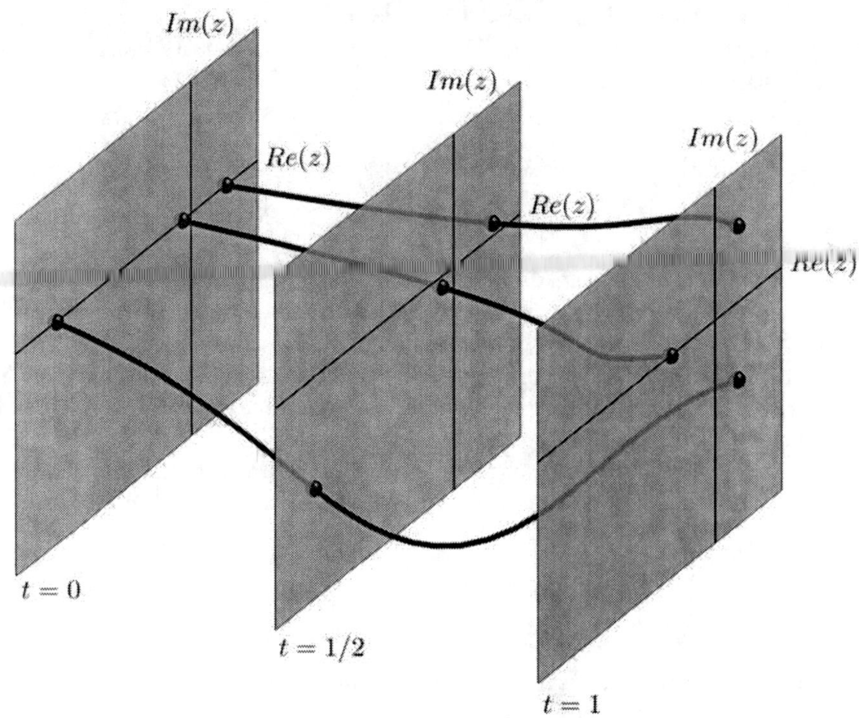

Figure 2.1. *Solution paths for Example* 2.1 *sliced at three values of t.*

Notice that the continuation paths are nonreal for most of the homotopy, but they may be real at the start and end. ∎

The reader may wonder how to choose γ to get this nice behavior. The short answer is that *almost any* choice of $\gamma \in \mathbb{C}$ works! More precisely, all that is required is for γ to avoid falling on any of a finite number of rays from the origin in $\mathbb{C}$. Accordingly, if we choose γ *at random* in $\mathbb{C}$, the homotopy defines three smooth paths with *probability one.* For numerical reasons, we generally choose γ to be near magnitude one. In fact, even if we restrict γ to the unit circle by choosing $\gamma = e^{i\theta}$ for random real θ, the procedure still succeeds with probability one, because the finite number of bad rays in $\mathbb{C}$ strike the unit circle in a finite number of points that we must avoid. The use of (2.3) to randomly choose a good path between $s = 1$ and $s = 0$ is known as the "gamma trick." See Chapter 6 for more on this.

It is important to note that this good behavior is not special to this particular example. The same basic principle works for any polynomial and indeed, with the right choice of start system, for any polynomial system.

2.1.2 ▪ Detailed Analysis of the Example*

In Example 2.1 above, we made a claim, without substantiation, that the procedure would succeed with probability one. To understand why this is true, we may examine the example in detail and in particular determine exactly where the homotopy fails to define

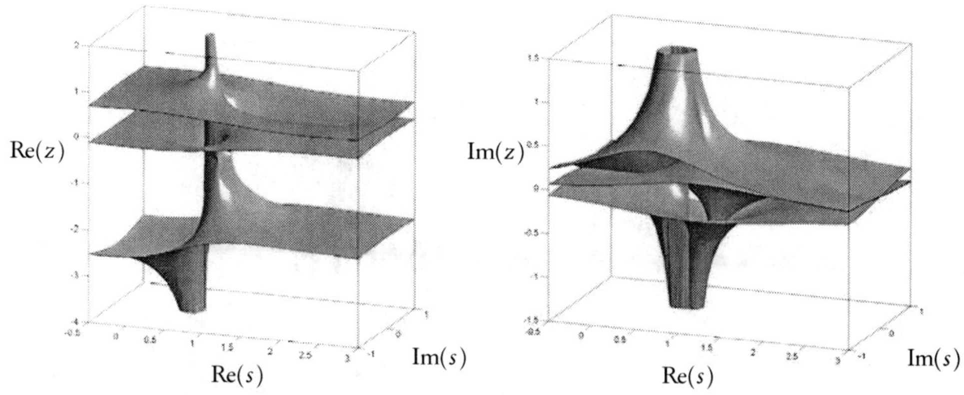

(a) (b)

Figure 2.2. *In Example* 2.1, *the solution of* $h(z,s) = 0$ *for* z *is a three-sheeted cover over* s *with four ramification points.*

smooth paths.[7] In the process, we will gain an intuitive understanding of why all the homotopies we use in this book succeed with probability one.

The key is to look at how the solutions of $h(z,s) = 0$ behave as s varies over the complex numbers. For any given value of s, say, $s = s^* \in \mathbb{C}$ (i.e., $s^* = a + bi$, $a, b \in \mathbb{R}$), we can solve the corresponding univariate equation, $h(z,s^*) = 0$, for z. For almost all values of s^*, $h(z,s^*)$ is a cubic equation with three distinct roots, but there are a few spots where there are just two distinct roots. Varying $s = a + bi$ over the range $a \in [-0.5, 3]$, $b \in [-1, 1]$, we get the surfaces shown in Figure 2.2, where the real and imaginary parts of z are shown as functions of s. We see, as expected, that $\mathbf{V}(h)$ consists of three sheets that touch each other occasionally. While the sheets of the real and imaginary parts appear to intersect along curves, they rarely intersect at a common value of s. To illustrate this fact, Figure 2.3 shows the distance between the nearest pair of solution points, both as a surface and a contour plot. That is, it plots

$$\Delta(z) := \min_{\substack{i,j \in \{1,2,3\} \\ i \neq j}} |z_i - z_j|$$

as a function of s, where z_1, z_2, z_3 are the three solutions to $h(z,s) = 0$. Notice that there are four isolated values of s where a pair of roots merges into a double root. The valleys around these points are extremely narrow, with walls that approach the vertical at the bottom. This means that stepping away from the singular values of s ever so slightly causes the double root in z to separate nicely.

In addition to the existence of double points, Figure 2.2 indicates that there is a region where one of the three roots grows large. In fact, there is a value of s over which one of the roots diverges to infinity.

Let's carefully examine the anomalous points where there are fewer than three distinct roots. Any place where two or more roots meet, the system is singular, meaning that $\partial h / \partial z = 0$. In other words, this occurs at the solutions to the system of two polynomials

[7] It is not necessary to perform this detailed analysis to construct a working homotopy; we do it here to show why the simple construction that we ultimately use is a good one.

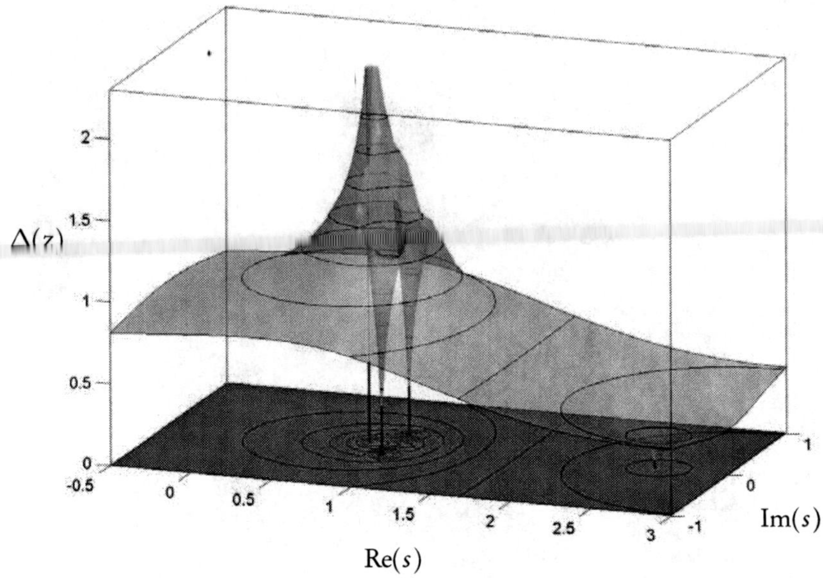

Figure 2.3. *Distance between nearest pair of roots in Example* 2.1.

in two variables

$$\begin{bmatrix} h(z,s) \\ \frac{\partial h}{\partial z}(z,s) \end{bmatrix} = \begin{bmatrix} sg(z)+(1-s)f(z) \\ sg'(z)+(1-s)f'(z) \end{bmatrix} = 0,$$

where g' and f' are the derivatives of g and f, the start and target systems, respectively. A little algebra shows that this system has solutions only at the roots of

$$f(g'-f')+f'(g-f)=0,$$

which is a fourth-degree equation (the degree 5 terms cancel). Specifically, $h(z,s)$ has double roots where

$$s \in \{0.756,\ 2.535,\ 0.921 \pm 0.235i\} \quad \text{(approximate)}.$$

The values of (z,s) where multiple roots occur are called the *ramification points* of the three-sheeted cover over s, and the values of s for these are called *branch points*.

In addition to the ramification points where solutions merge, a root may also disappear by diverging to infinity. This happens where, considering s as a parameter, the leading cubic term of $h(z)$ disappears. The coefficient of z^3 in $h(z,s)$ is $1s-2(1-s)$, which vanishes at $s = 2/3$.

According to the above, there are three distinct, finite roots of $h(z,s)$ for every value of s except for five points in the s-plane: the four branch points and the one point of divergence. Consequently, any continuous one-real-dimensional path in the s-plane that goes from $s = 1$ to $s = 0$ while avoiding these five bad points will define three smooth paths.

Now, let's consider the family of circular arcs defined by (2.3). As γ varies, we get different arcs or, more precisely, there is a unique arc for each value of $\angle\gamma \in (-\pi,\pi]$. (To first order near $t = 0$, $s \approx \gamma t$, so $\angle\gamma$ is the angle of the tangent to the arc at $s = t = 0$.) Figure 2.4(a) shows the family of arcs from which we choose, and Figure 2.4(b) shows the

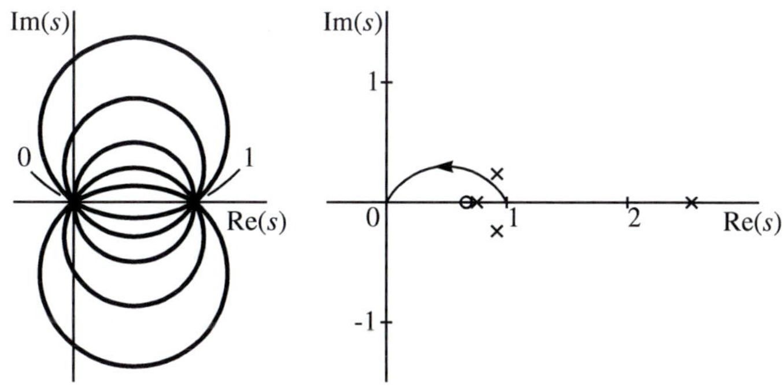

(a) (b)

Figure 2.4. *The gamma trick:* (a) *the family of arcs and* (b) *the arc for* $\gamma = 0.4+0.77i$*, which misses bad points for Example* 2.1.

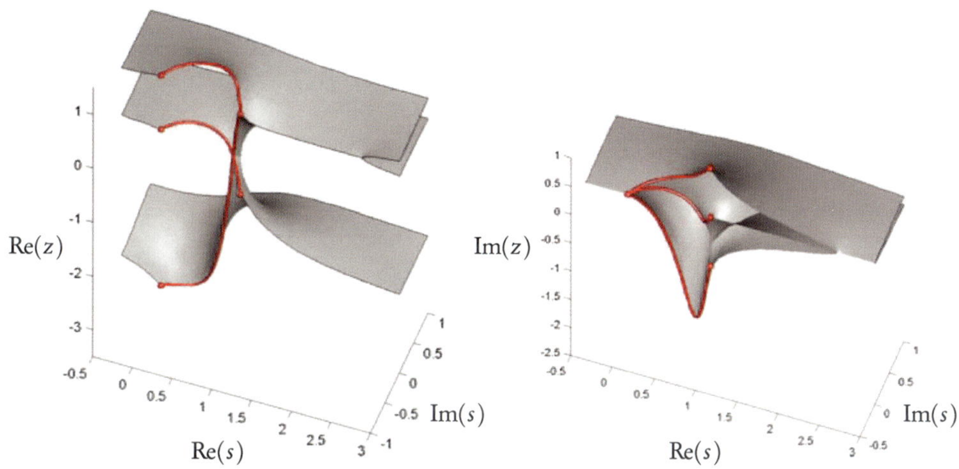

(a) (b)

Figure 2.5. *Cutaway view of Example* 2.1 *showing the paths followed.*

particular arc for $\gamma = 0.4 + 0.77i$ that we have used for the example at hand. On this plot, the ramification points are marked as $\times$ and the divergence point is marked as $\circ$. The key property that makes the chosen homotopy path suitable is that it does not pass through any of the ramification points or the point of divergence. Clearly, almost any of the arcs in the family have this property, which is why a random choice of γ gives a good homotopy with probability one.

To see the result of this choice of arc between $s = 1$ and $s = 0$, Figure 2.5 shows the same three-sheeted function as Figure 2.2 except the sheets are trimmed back to reveal, in red, the paths followed as t goes from 1 to 0. Although a crossing can be seen in the imaginary parts at $s = 0$, the real and imaginary parts of two distinct paths are never simultaneously equal. Thus, when the real and imaginary parts of the paths are plotted versus t, we get our original view of them in Figure 2.1, where the paths are clearly smooth and distinct.

While the solution paths are smooth for all but a finite number of angles of γ, in numerical work we track them only approximately. If the path in s passes too close to a branch point, the corresponding solutions in z can move closer than the approximation, thereby having the potential for the path tracker to confuse them and fail to successfully track both paths past the near singularity. We will discuss later how to make a path tracker that resists this kind of mistake by using a variable step size (§2.2) and extra precision (Chapter 3). Even so, in the presence of numerical approximation, the finite number of bad angles for γ in exact arithmetic broaden into small failure zones near the bad values. This notwithstanding, the extreme narrowness of the valleys in Figure 2.3 is typical, making the failure zones tiny. Accordingly, the probability of choosing an acceptable γ is very high.

A final note: One might be tempted to try to avoid the use of randomness by solving for the branch points and points of divergence and then picking a path that avoids them. This is a losing game, because solving for these points is harder than solving the original problem. But without doing any detailed analysis whatsoever, we know that the number of ramification and divergence points is finite, so a random arc misses them with probability one. That and the narrowness of the bad zones are the properties that make continuation using a random γ a robust solution approach.

2.1.3 ▪ Homotopies

By a homotopy, we mean an appropriately smooth function

$$H(z,t):\mathbb{C}^N \times [0,1] \to \mathbb{C}^N. \tag{2.4}$$

For the purposes of this book, H is obtained from the composition of a family of systems $\mathcal{H}(z;s)$ (parameterized by s in an open subset $U \subset \mathbb{C}^M$) with a path $s = q(t)$. To be precise, we have

1. a mapping $\mathcal{H}(z;s):\mathbb{C}^N \times U \to \mathbb{C}^N$, where U is an open subset of some complex Euclidean space $\mathbb{C}^M$ and where each of the coordinate functions is polynomial in the z variables with coefficients that are complex analytic functions in the s variables; and
2. a differentiable map $q:[0,1] \to U$, which is the path parameterized by t going from $s=1$ to $s=0$ in the parameter space.

A typical example of a homotopy is

$$H(z,t) = (1-t)(z^2-5z+5) + t(z^2-4). \tag{2.5}$$

This is an example of a total-degree homotopy, which is a class of homotopies discussed in §2.1.4.

It may easily be shown that for each $t \in (0,1]$, there are two distinct solutions of $H(z,t)=0$.[8] Given the two distinct roots of $H(z,t)=0$ for $t \in (0,1]$, it may be shown that there are two differentiable functions $p_1(t), p_2(t)$ (corresponding to the paths leading from the startpoints to the endpoints), which both map $(0,1]$ to $\mathbb{C}$ and such that

- $H(p_j(t),t)=0$ for $j=1,2$;
- $p_1(1)=2$ and $p_2(1)=-2$, i.e., the points $\{p_1(1),p_2(1)\}$ are the solutions of $H(z,1)=0$, i.e., the startpoints; and

[8] This is also true for this special example for the closed interval $[0,1]$, but in general due to singular solutions of $H(z,0)=0$, the solutions are sure to be distinct only for $t \in (0,1]$.

- the limits of the $p_j(t)$ as $t \to 0$ exist and equal the roots of $H(z,0)=0$. These are the endpoints.

Tracking or *path tracking* is the numerical process of approximating the paths, thereby obtaining approximations of the solutions of $H(z,0)=0$ from the known approximations to the solutions of $H(z,1)=0$.

A *good homotopy* for the system (2.1) and a set of D distinct solutions $\mathscr{S}_1$ of $g(z) := H(z,1)=0$, i.e., startpoints, is a system of infinitely differentiable functions

$$H(z,t)=\begin{bmatrix} H_1(z_1,\ldots,z_N,t) \\ \vdots \\ H_N(z_1,\ldots,z_N,t) \end{bmatrix} \tag{2.6}$$

such that

1. for any choice of $t \in [0,1]$, $H(z,t)$ is a system of polynomials;
2. given any of the distinct D startpoints $w \in \mathscr{S}_1$, there is a smooth map $p_j(t) : (0,1] \to \mathbb{C}^N$ satisfying $p_j(1) = w$ (without loss of generality, we hereafter denote by w_j the w associated to p_j); and
3. p_j is smooth on $(0,1]$, and for each $t^* \in (0,1]$:
   (a) the associated paths don't cross, i.e., there do not exist two integers j,k satisfying $1 \le j < k \le D$ and $t^* \in (0,1]$ such that $p_j(t^*) = p_k(t^*)$; and
   (b) the points $p_j(t^*)$ are smooth isolated solutions of $H(z,t^*)=0$.

In addition, we say that (2.6) is a *good homotopy* for the system (2.1) if we can choose a set of D distinct solutions $\mathscr{S}_1$ of $g(z) := H(z,1) = 0$ such that, letting $||v||_2$ denote the two-norm of a vector v, the set of finite limits

$$\mathscr{S}_0 := \left\{ z \in \mathbb{C}^N \;\middle|\; ||z||_2 < \infty \text{ and } z = \lim_{t\to 0} p_j(t) \right\}$$

contains every isolated solution of $f(z)=0$.

Though it is obvious from the above properties, we list for emphasis the critical defining property of a homotopy:

4. $H(p_j(t),t)=0$ for all $t \in (0,1]$ and $1 \le j \le D$.

To make everything concrete, we consider the total-degree homotopy. This was the first homotopy extensively used in the solution of polynomial systems.

2.1.4 ▪ Total-Degree Homotopies

Assume we wish to solve a polynomial system (2.1). We construct a good homotopy

$$H(z,t)=(1-t)\begin{bmatrix} f_1(z_1,\ldots,z_N) \\ \vdots \\ f_N(z_1,\ldots,z_N) \end{bmatrix} + \gamma t \begin{bmatrix} g_1(z_1,\ldots,z_N) \\ \vdots \\ g_N(z_1,\ldots,z_N) \end{bmatrix} = 0 \tag{2.7}$$

for solving (2.1) as follows.

Let $d_i = \deg f_i$. First, choose a system of N polynomials $g_1,\ldots,g_N$ of degrees $d_1,\ldots,d_N$, respectively, such that the system $g(z)=0$ is easy to solve and has exactly $D := d_1 \cdots d_N$ solutions $\mathscr{S}_1$. For example, we may take $g_i(z) = z_i^{d_i} - 1$, in which case $\mathscr{S}_1$ is the set

$$\left\{ \left(e^{(j_1/d_1)2\pi\sqrt{-1}},\ldots,e^{(j_N/d_N)2\pi\sqrt{-1}} \right) \,\middle|\, 0 \le j_1 \le d_1 - 1,\ldots,0 \le j_N \le d_N - 1 \right\}.$$

Now choose a random complex number $\gamma \neq 0$. In practice, we take a number following the uniform distribution in a small band around the unit circle. The use of a random number is called the *gamma trick*, and it functions here just as in the univariate example of Example 2.1. It was first introduced in [132]. (See also §6.1 for the *general gamma trick* from [161, Lemma 7.1.3].) With this trick, theory guarantees with probability one that $H(z,t)$ is a good homotopy. The logic behind the theory is exactly as discussed in the detailed examination of Example 2.1 in §2.1.2.

Given an isolated solution z^* of $f(z)=0$, the total-degree homotopy guarantees that the number of paths ending at z^* is equal to the *multiplicity* of z^* as a solution of $f(z)=0$: see §1.2.2 and §11.4 for a full definition and more details about multiplicity. This is true for all standard homotopies, e.g., the multihomogeneous homotopies of §5.1. Total-degree homotopies are the simplest of all homotopies to construct and are the default method used by Bertini to solve a polynomial system.

2.1.5 ▪ Reduction of Continuation to Path Tracking

The following natural questions immediately arise:

1. How do we use $H(z,t)=0$ to track from $t=1$ to $t=0$?
2. How do we construct homotopies so that they have trackable paths leading to a finite number of solutions, with all isolated solutions of the system among these?

This subsection gives an overview of the answer to the first question with the details encompassing a large portion of the remainder of this chapter. There are many choices to be made when implementing a tracker. In Bertini, reasonable default values for these choices are made, but there are configuration settings that allow the user to make different choices.

There are many ways to construct homotopies to satisfy the second question, increasingly so as the number of variables increases. We presented total-degree homotopies in §2.1.4. Though these homotopies may always be set up so they work, they are often far from optimal. Improvements to these homotopies, which follow from the theory in Chapter 4, are presented in Chapter 5.

Let's assume we have a differentiable (i.e., adequately nice) map $p(t):(0,1]\to\mathbb{C}$ with $w=p(1)$ a solution of $H(z,1)=0$ and with $p(t)$ a multiplicity one solution of $H(z,t)=0$ for each $t\in(0,1]$. The key to computing the possibly infinite value $\lim_{t\to 0} z(t)$, which we denote s_0, is the Davidenko differential equation [45, 46] obtained by differentiating $H(z(t),t)=0$ with respect to t:

$$\frac{\partial H}{\partial z}\frac{\mathrm{d}z}{\mathrm{d}t} + \frac{\partial H}{\partial t} = 0. \tag{2.8}$$

Solving this differential equation to find roots of $H(z,0)=0$ breaks into two regimes:

1. solving for $t\in[\epsilon,1]$ with $\epsilon \ge 0$ small, but not too small (see §2.2); and
2. taking the limit to estimate $\lim_{t\to 0} z(t)$ (§3.3).

If the endpoint to which a path $z(t)$ is heading as $t \to 0$ is nonsingular, then ϵ may safely be taken to be 0. If the root is singular, the choice of ϵ is not straightforward. Unfortunately, since it is difficult to predict the multiplicity at $t = 0$ given some $t > 0$, we will always assume the worst, i.e., that each endpoint is possibly singular.

Remark 2.2. We track from $t = 1$ to $t = 0$ and not from $t = 0$ to $t = 1$ because there are considerably more floating-point numbers near $t = 0$ than $t = 1$. Since singular endpoints (see §3.3) require many digits of accuracy, this is useful.

Remark 2.3. We mentioned the possibility that a path might diverge to infinity as $t \to 0$. This can be made numerically tractable through the use of projective space (see Chapter 4), which has the effect of representing points as the ratio of homogeneous coordinates. Thus, endpoints at infinity become computable as ratios where the denominator approaches zero.

2.2 ▪ Path Tracking

While knowledge of the material in this section is unnecessary for running Bertini, these concepts are fundamental to virtually everything that Bertini does.

Arguably the most important operation in numerical algebraic geometry is the tracking of a path. To make matters concrete, assume that we have the following:

1. a family of functions on $\mathbb{C}^N$

$$H(z;q) = \begin{bmatrix} H_1(z_1,\ldots,z_N;q_1,\ldots,q_M) \\ \vdots \\ H_N(z_1,\ldots,z_N;q_1,\ldots,q_M) \end{bmatrix} = 0 \tag{2.9}$$

   such that each H_i is polynomial in the variables $z \in \mathbb{C}^N$ and analytic in the parameters $q \in \mathbb{C}^M$; and

2. differentiable maps $\phi : t \in [0,1] \to q \in \mathbb{C}^M$ and $\psi : t \in [0,1] \to z \in \mathbb{C}^N$ with

   (a) $H(\psi(t), \phi(t)) = 0$ for $t \in (0,1]$ and

   (b) the Jacobian of H with respect to $z_1,\ldots,z_N$ has rank N for the points $(\psi(t), \phi(t))$ with $t \in (0,1]$.

In practice, we construct H and the map ϕ in a such a way that we are guaranteed ψ exists with $\psi(1) = p_0$, a prescribed point. The objective is to compute $p^* = \psi(0)$, i.e., the solution of the target system corresponding to this particular path. This operation is called path tracking.

For simplicity assume that $M = 1$ and $q_1 = t$. We are usually in this situation. Let us denote $\psi(t)$ by $z(t)$. We have the Davidenko differential equation [45, 46]

$$\frac{\partial H(z(t),t)}{\partial t} + \sum_{i=1}^{N} \frac{\partial H(z(t),t)}{\partial z_i} \frac{dz_i(t)}{dt} = 0 \quad \text{with} \quad z(1) = p_0. \tag{2.10}$$

Thus tracking a path is equivalent to solving this initial value problem.

Let $JH(z,t)$ denote the Jacobian matrix with respect to the variables z

$$JH := \frac{\partial H}{\partial z} := \begin{bmatrix} \frac{\partial H_1}{\partial z_1} & \cdots & \frac{\partial H_1}{\partial z_N} \\ \vdots & \ddots & \vdots \\ \frac{\partial H_N}{\partial z_1} & \cdots & \frac{\partial H_N}{\partial z_N} \end{bmatrix}$$

evaluated at (z,t), and let

$$z(t) := \begin{bmatrix} z_1(t) \\ \vdots \\ z_N(t) \end{bmatrix}$$

denote the solution of (2.10). With this notation, we may rewrite the Davidenko differential equation as

$$\frac{\partial H(z(t),t)}{\partial t} + JH(z(t),t) \cdot \frac{dz(t)}{dt} = 0, \tag{2.11}$$

or since $JH(z(t),t)$ is invertible in the path,

$$\frac{dz(t)}{dt} = -[JH(z(t),t)]^{-1} \frac{\partial H(z(t),t)}{\partial t}. \tag{2.12}$$

There are many marching methods for solving initial value problems, e.g., check almost any numerical analysis textbook [165, 39]. There are two special features of our initial value problem:

1. The path constructed by the marching method would, if perfectly exact, satisfy an explicit function $H(z(t),t) = 0$, so the function $H(z,t) = 0$ may be used to "correct" the marching method result.

2. We know that the path is on an algebraic curve, and therefore we can construct a local model of the curve near p^*: this local model is particularly useful in the case when p^* is a singular solution of $H(z,0) = 0$.

If p^* is singular as a solution of $H(z,0) = 0$, the Jacobian of H with respect to the variables becomes singular (rank deficient) as $t \to 0$. We use different computation methods for path tracking when t is far from zero versus when it is near zero; we call the tracking near zero the *endgame*, a topic discussed in §3.3. Of course, this breakup of "near" and "far" is somewhat arbitrary.

The remainder of the chapter discusses different tracking algorithms. We will present these in the manner in which they are used before the endgame, but these are also fundamental in the endgame. A basic reference on path tracking methods is [7].

2.2.1 ▪ First-Order Tracking (Euler–Newton)

We solve (2.12) by a marching method starting at $t_0 = 1$ with p_0 as the initial value and successively compute the approximations $p_1, p_2, \ldots$ at values $t_0 > t_1 > t_2 > \cdots > 0$. The simplest method is *Euler's method*, which computes the approximation

$$p_{i+1} = p_i - JH(p_i,t_i)^{-1} \frac{\partial H(p_i,t_i)}{\partial t} \Delta t_i,$$

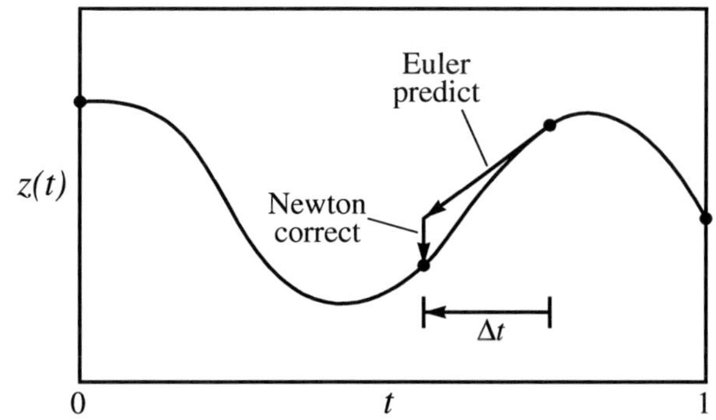

Figure 2.6. *Euler–Newton predictor-corrector path tracking.*

where $\Delta t_i = t_{i+1} - t_i$. Geometrically, this corresponds to predicting along the line tangent to the solution path from the current point on the path.

Since each point on the path should satisfy $H(z,t) = 0$, we may use a *predictor-corrector* strategy. This means we follow the prediction to p_{i+1} by a correction using *Newton's method* for $H(z, t_{i+1})$ starting with $z_0 = p_{i+1}$. Recall from (1.3) that Newton's method uses the iterative formula

$$z_{i+1} = z_i - [JH(z_i, t_{i+1})]^{-1} H(z_i, t_{i_1}). \tag{2.13}$$

One or two iterations of Newton's method will usually make a much more accurate prediction of $z(t_{i+1})$ (see Figure 2.6), so we replace p_{i+1} with this corrected value before initiating the next predict-correct cycle.

In practice, each step size Δt_i is chosen adaptively. If an Euler–Newton step "fails," i.e., if the residual of the final Newton correction is larger than the desired tracking accuracy (i.e., the corrector didn't correct well enough), Δt_i is halved and this shorter step is attempted. In Bertini, the default number of Newton iterations is two, which is explained further in §E.4. Conversely, if several (the default being five) consecutive steps have been successful, Δt_i is doubled, up to a prescribed maximum value (the default being 0.1). Though simple, this adaptive step size scheme works very well.

2.2.2 ▪ Higher-Order Methods

The main advantage of using higher-order methods is that the step sizes may be larger. Several such methods, such as `rkf45`, the default predictor of Bertini, have well-understood and efficient error estimation built into them [20]. Bertini has a number of different ordinary differential equation (ODE) predictors available; see §E.4.3. There are many other settings, e.g., the maximum number of steps allowed (before and during the endgame) plus the minimum and maximum allowed step sizes. These may be looked up in §E.4, which covers tracker settings.

2.3 ▪ A User-Defined Homotopy*

For most polynomial solving, a Bertini user simply chooses a type of homotopy, perhaps even accepting the default total-degree homotopy, and lets Bertini handle the details, such as constructing and solving a start system. But some users may wish to define their own homotopies while taking advantage of Bertini's powerful tracking routines to solve them.

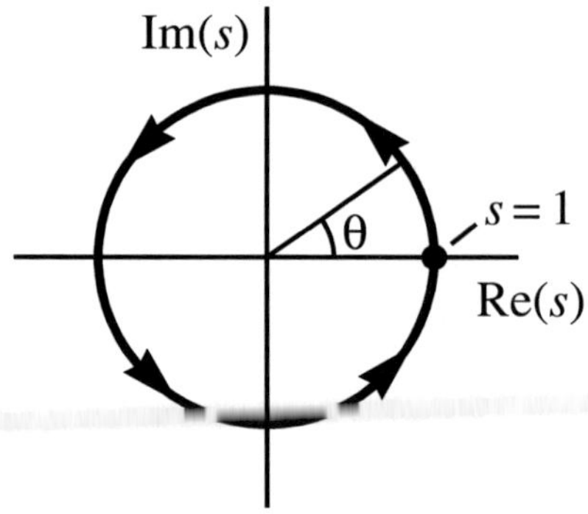

Figure 2.7. *Path around the unit circle in* $\mathbb{C}$.

This means the user must input the formulas defining the homotopy and provide the list of startpoints in numerical form. Then, Bertini can track all the paths and report their endpoints.

To illustrate this process, let's consider tracking a single solution path of $H(z,s) = z^2 - s$ as s goes clockwise around the unit circle starting at $s = 1$ (see Figure 2.7). To proceed, two files are needed:

1. the input program (much like the programs already seen) and
2. a file listing startpoints for the homotopies.

If you have appropriate files called `input` and `start` in the folder you are working in, you can just type at the command prompt >>

```
>> bertini
```

Using explicit file names `AroundTheUnitCircle.input` and `AroundTheUnitCircle.start`, the command would be

```
>> bertini AroundTheUnitCircle.input AroundTheUnitCircle.start
```

Let us list the input file and then explain the syntax:

```
%AroundTheUnitCircle.input
CONFIG
 UserHomotopy:1;
END;

INPUT
 variable z;
 function H;
 parameter q1,q2;
 pathvariable t;

 q1 = cos(2*Pi*(1-t));
 q2 = sin(2*Pi*(1-t));
 s = q1 + I*q2;
 H = z^2 - s;
END;
```

The configuration `UserHomotopy:1` tells Bertini that the file is a user-defined homotopy and there will be a start file containing the startpoints. Without this configuration, Bertini would build the start system, solve it, then track each path, as with the examples in Chapter 1. For a user-defined homotopy, Bertini expects a family of systems

$$f(z;q) := \begin{bmatrix} f_1(z_1,\ldots,z_N;q_1,\ldots,q_m) \\ \vdots \\ f_N(z_1,\ldots,z_N;q_1,\ldots,q_m) \end{bmatrix} = 0 \tag{2.14}$$

defined on $\mathbb{C}^N \times U$, where U is an open subset of $\mathbb{C}^m$ and where each f_j is a polynomial in the z variables with coefficients that are functions of the q parameters. These coefficient functions can be any expression in the q parameters and constants that is a polynomial or an expression in terms of the analytic functions exp(), sin(), cos(), and tan(), where the arguments of the functions are again polynomials or analytic functions, ad infinitum. Notice that the constants `Pi` and `I` are predefined as π and $\sqrt{-1}$, respectively.

The line `variable z;` tells Bertini the names of the variables for the family of systems, while the line `parameter q1,q2;` tells Bertini the names of the family's parameters. The line `pathvariable t;` gives the path variable for the path in the parameter space, here specified by

```
q1 = cos(2*Pi*(1-t));
q2 = sin(2*Pi*(1-t));
```

The difference between the (unique!) path variable and the parameters is that the path variable t goes from 1 to 0 while the parameters are functions of t. The most common choice (and the correct one if this distinction is unclear) is to have just one parameter, say, s, set equal to the path variable:

```
parameter s;
pathvariable t;
s = t;
```

In the case at hand, the net result is that $s = \cos\theta + i\sin\theta$ with θ starting at 0 and going to 2π. Thus, s goes once around the unit circle, as we set out to do.

The start file specifies the startpoints for z when $t = 1$. For our example, there are two solutions $\{1,-1\}$ for $z^2 - 1 = 0$, and the start file specifies which subset of these roots we wish to track paths from. In this case, let's just track from $z = 1$ by providing the following start file:

```
1

1 0;
%AroundTheUnitCircle.start
```

In a start file, the first nonempty line must contain a single integer specifying how many points Bertini should track from. The next lines give the coordinates of the points. Bertini expects a complex point, so we have `1 0`, meaning $1+0*\sqrt{-1}$. Comments are as before, but in start files, they must be given after all the data (and in fact, the % sign is not needed since Bertini stops reading after it gets to the number of points it was told to expect).

The file `main_data` lists assorted information plus the points at the end of the paths. In this case, we have a single endpoint:

```
-1.000000000000000e+00 2.504507018441515e-16.
```

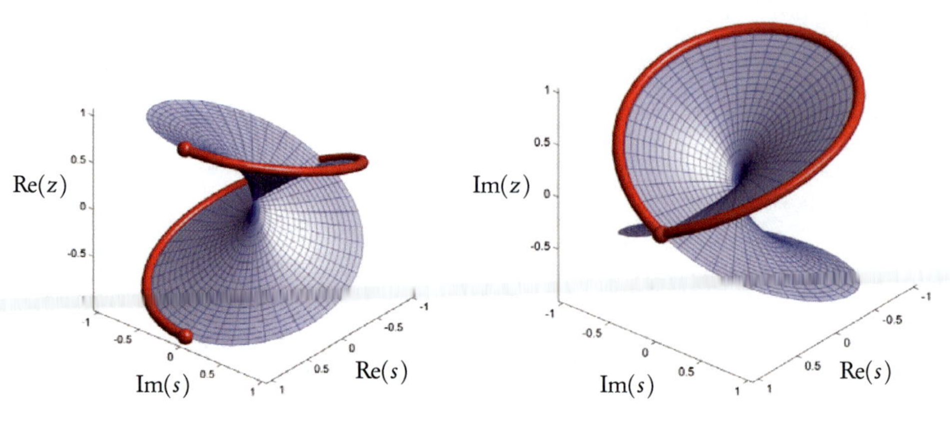

(a) (b)

Figure 2.8. *Real and imaginary parts of z for* $z^2 - s = 0$ *with the path around the unit circle in s highlighted.*

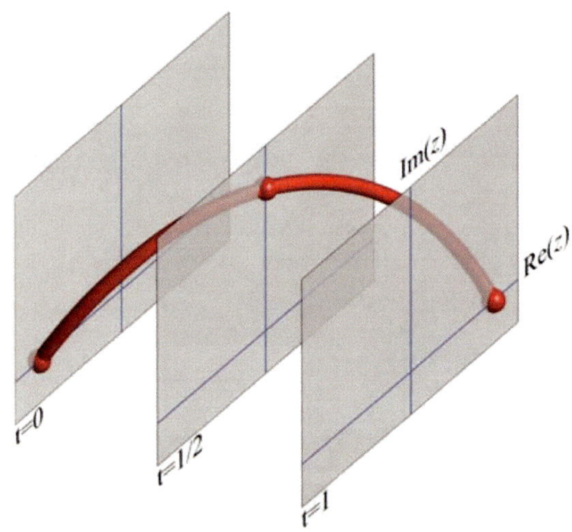

Figure 2.9. *The path* $(z(t), t)$ *for the* `AroundTheCircle` *example.*

So as illustrated in Figures 2.8 and 2.9, we have found that as s goes around the unit circle, starting and ending at $s = 1$, the starting solution $z = 1$ comes back to $z = -1$. This is as expected, since $s = e^{i\theta}$ for θ going from 0 to 2π; hence $z = \sqrt{s} = e^{i\theta/2}$ ends at $e^{i\pi} = -1$.

Since Bertini also accepts the complex exponential function, we could have illustrated the same path using the following input file:

```
%AroundTheUnitCircleAlt.input
CONFIG
 UserHomotopy:1;
END;

INPUT
 variable z;
```

```
  function H;
  parameter s;
  pathvariable t;

  s = exp(2*Pi*I*(1-t));
  H = z^2 - s;
END;
```

2.4 ▪ Bertini Guide: Basic Continuation

The simplest use of Bertini is the total-degree homotopy for finding isolated roots.

Total-Degree Homotopy Summary

Input files
  `input` = run configuration & system definition
Output files
  Endpoints include all **isolated solutions**, sorted in various files.
Key configurations:
  `FinalTol`
Key declarations:
  `function`
  `variable_group`
Notes:
  Using just 1 variable group always gives a total-degree homotopy.
Execution:
  `>> bertini <input>`

We also briefly touched on using Bertini to evaluate a polynomial system at a given set of points.

Function Evaluation Summary

Input files
  `input` = run configuration & system definition
  `start` = list of points to evaluate at
Output files
  Evaluations are listed in the file named `function`
Key configurations:
  `TrackType:-4;`
  `MPType`
  `Precision`
Key declarations:
  `function`
  `variable_group`
Execution:
  `>> bertini <input> <start>`

Finally, we gave an example of a user-defined homotopy.

User-Defined Homotopy Summary

Input files
`input` = run configuration & system definition
`start` = list of points to track paths from
Output files
Endpoints are sorted in various files.
Key configurations:
`UserHomotopy:1;`
Key declarations:
`pathvariable`
`parameter`
Execution:
```
>> bertini <input> <start>
```

Chapter 3

Adaptive Precision and Endgames

Singularities are the major source of ill-conditioning for polynomial systems. In this chapter we discuss the algorithms needed to deal with these issues.

For many users from engineering and science, most of the concepts of this chapter will not be needed directly. Indeed, whether singularities are of interest to the user or not, they may well be present. Fortunately, the algorithms used in Bertini (with their default settings) will often handle them transparently. Accordingly, we have a first section, §3.1, dealing with basic facts that will be sufficient for many users, whereas the rest of the chapter is more advanced.

In §3.2, we discuss the need for more than double precision. Probability-one algorithms can fail because of ill-conditioning. This ill-conditioning usually arises because bad sets, which have measure zero with exact numbers, become nonnegligible when a finite number of digits are employed in computation. The use of adaptive multiprecision in Bertini goes a long way towards solving these problems. In this section we give the basic ideas underlying the algorithms used.

In §3.3, we discuss algorithms called *endgames* that are used to effectively compute highly singular solutions at the end of a homotopy path. These algorithms are a good illustration of the need for multiprecision.

3.1 ▪ Some Basic Facts

Recall from Chapter 1 that isolated solutions can be either nonsingular or singular and that each solution comes with an invariant called multiplicity. In this section, we give some intuition regarding multiplicity.

Let us begin with a polynomial in one variable,

$$p(z) = z^k + a_1 z^{k-1} + \cdots + a_k,$$

of degree $k \geq 1$. It is well known that it can be factored as a product

$$p(z) = \prod_{j=1}^{r} (z - z_j)^{\mu_j},$$

where the z_j are distinct points in $\mathbb{C}$ and the μ_j are positive integers, whose sum is $\mu_1 + \cdots + \mu_r = k$. The integer μ_j is called the *multiplicity* of z_j as a root of $p(z) = 0$, or equivalently the multiplicity of z_j with respect to p.

Given a total-degree homotopy $H(z,t)$ as in §2.1.4, it is reasonable to hope that exactly μ_j paths have the limit z_j as $t \to \infty$. This is indeed true.[9]

Fix a square system

$$f(z) := \begin{bmatrix} f_1(z_1,\ldots,z_N) \\ \vdots \\ f_N(z_1,\ldots,z_N) \end{bmatrix} = 0$$

and an isolated solution z^* of $f(z) = 0$. There is a generalization of the multiplicity of a single polynomial for an isolated solution z^* of the system $f(z) = 0$; see §1.1.2 and §11.3. Given a good homotopy $H(z,t)$ for $f(z)$, which is a polynomial system in the variables $z_1,\ldots,z_N,t$ on $\mathbb{C}^N \times \mathbb{C}$ (as most good homotopies are), the number of paths ending at an isolated solution z^* of $f(z) = 0$ equals the multiplicity of z^*. This fact follows from Theorem 3.1 (see [154]), which is in turn a special case of Grauert's direct image theorem [72].

Technical statement regarding multiplicity and number of paths*

Theorem 3.1. *Given a system*

$$H(z;q) := \begin{bmatrix} H_1(z_1,\ldots,z_N;q_1,\ldots,q_M) \\ \vdots \\ H_N(z_1,\ldots;q_1,\ldots,q_M) \end{bmatrix} = 0$$

of N complex analytic functions defined for $(z;q)$ in an open set $U \subset \mathbb{C}^N \times \mathbb{C}^M$, and a point (z^*,q^*), which is an isolated solution of multiplicity μ of $H(z;q^*) = 0$, it follows that for any sufficiently small $\epsilon > 0$, there is a $\delta > 0$ such that for any point q' with the Euclidean distance $|q' - q^*| < \delta$, the sum of the multiplicities of the solutions of $H(z;q') = 0$ with Euclidean distance less than ϵ from z^* is μ.

Although the multiplicity is a number with a heavy topological flavor, we need a more elaborate technique (see §11.3) to define it in general. Notice that Theorem 3.1 applies only to square systems. One might hope that a similar result would hold for nonsquare systems with more equations than unknowns, but there is no natural way to view an isolated solution of such a system as the limit of μ points, where μ is the multiplicity of the solution. In particular, for $N \geq 3$, nonsquare polynomial systems exist with the zero scheme at an isolated solution of the system not equal to a fiber in a flat family of zero schemes over an irreducible algebraic set with at least one fiber consisting of isolated nonsingular solutions. For more about this result of Iarrobino, which is well beyond the scope of this book, see [65].

3.2 ▪ Adaptive Precision

Recall that *precision* refers to the number of digits (or bits, though we prefer to stick to the decimal convention) used to represent a floating-point number. *Accuracy* refers to

[9] If the system to be solved is overdetermined and thus randomized down to a square system in order to build $H(z,t)$, then the number of paths with endpoint z_j equals the multiplicity of z_j as a solution of the *randomized* system. From Bertini's theorem, this could be more than the multiplicity of z_j as a solution of the original, overdetermined system, so the number of paths going to z_j will always be an upper bound (possibly a sharp one) on the multiplicity of z_j.

the number of accurate digits in some computed quantity. Many numerical methods are implemented with a fixed precision, e.g., MATLAB uses double precision throughout. Bertini is different: it uses a regular, low precision level (double precision, around 16 digits) as much as possible but increases precision on the fly as necessary.

In this section, we answer two questions:

1. Why do we need more than double precision? If we only want answers to a few decimal places, why bother with higher precision? After all, higher precision is computationally expensive and it complicates coding (though this is hidden from a user of Bertini).

2. Given a convincing positive answer to the first question, why not simply work with some large number of digits at all times?

After answering these questions, we give a brief description of the basic ideas underlying the adaptive precision algorithms of Bertini.

3.2.1 ▪ The Need for Extra Precision

There are two phenomena driving the need for extra precision. One is the well-known problem of ill-conditioning, which can arise in any numerical context, even linear systems, but is even more frequently a problem for high-degree polynomial systems. The second concerns numerically approximating the probability-one success rate that theoretically holds when exact computation is assumed. We touch on these one at a time here.

Ill-conditioning and precision

We know from Wilkinson's work (see [42]) that when we try to solve the linear system

$$Av = b,$$

where b is a given N-vector and A is a given $N \times N$ matrix, we may lose up to approximately $\log_{10}(\text{cond}(A))$ digits of accuracy. This is not quite precise, but the point is clear. If the condition number of A is 10^9 and computations are done in double precision, which employs approximately 16 digits, then the answer computed for v can be trusted only to about 7 digits. If we want more digits of accuracy than that, the computation must be done in higher precision. For a homotopy $H(z,t)$, the role of the matrix A is played by the Jacobian JH of H with respect to the z-variables while b is H itself, with both H and JH evaluated as we track a path. This is because path tracking accuracy is determined by Newton's method in the predictor-corrector cycle, as discussed in §2.2.1.

For large polynomial systems having perhaps thousands of variables and equations or even small systems with polynomials of more than moderate degree (e.g., degree 10 under modest conditions), double precision is simply not sufficient. In Chapter 17, we will see that the Jacobians of good solutions of the polynomial systems arising from discretizations of systems of differential equations may have condition numbers of 10^9, which is large enough to prevent secure continuation if only double precision is used. Examples such as (1.5) show that for even modest polynomials of degree 10, double precision evaluation of the polynomial at an approximation of a root that is given accurately to full double precision may yield a result that is far from zero.

Probability one and precision

Numerical algebraic geometry is based on algorithms that depend on certain bad phenomena happening with probability zero; that is, for random choices of certain parameters, bad phenomena don't happen. One example is a random choice of γ in the gamma trick in §2.1.4 guaranteeing that good homotopy paths exist.

Unfortunately, the finite set of approximate complex numbers used by computers is a far cry from the continuum of complex numbers in that the pool of numbers from which we make a random choice is not infinite. Worse is the fact that all floating-point computations are approximate, so in a sense, in finite precision the loci of bad phenomenon expand from thin, measure-zero sets to a fuzzy region of poor or unreliable performance. For example, while in a continuum model our homotopy paths have a zero probability of encountering a singularity, in finite precision they might pass close enough that ill-conditioning effects make the computations untrustworthy.

Given this situation, how can we make computations reliable? The saving grace is that by increasing precision sufficiently, one may shrink the regions of poor performance as close as desired to the theoretical measure-zero set where a continuum model says failure happens.

For example, a typical probability-one statement is that a degree d polynomial $p(z)$ is identically zero if $p(z^*) = 0$ for a random complex number z^*. The logic behind this is that since $p(z)$ vanishes at no more than d points, the complement of this set is dense. On the other hand, if $p(z^*) \neq 0$ for any test point z^*, then clearly $p(z)$ is not identically zero. So with probability one, we can decide whether $p(z)$ is identically zero with a single evaluation at a random point z^*. The probability that a random number generator will choose z^* as one of the roots of $p(z)$ is small and vanishes exponentially as increases in precision make more floating-point numbers available.

Unfortunately, the situation is not quite so simple because floating-point arithmetic is approximate. So even if $p(z)$ is identically zero, its evaluation at a random point will seldom be exactly zero. Very small values of $|p(z^*)|$ will have to be accepted as evidence that $p(z)$ could be exactly zero. In the opposite direction, if $|p(z^*)|$ is large enough, we can conclude with relative certainty that $p(z)$ is not identically zero. In fact, it is possible to rigorously compute an upper bound on the evaluation error, for example, by using complex disk arithmetic [62]. In such a way, one can achieve rigorous proof that a polynomial is not identically zero using only floating-point arithmetic. Unfortunately, that approach cannot prove that a polynomial is zero.

A facility for increasing precision as necessary can make such decisions more robust. Consider a nonzero univariate polynomial $p(z)$ that has a root $\hat{z}$, and suppose u is the floating-point round-off factor, i.e., $u = 10^{-P}$ in P-digit arithmetic ($u \approx 10^{-16}$ in double precision). Then as u decreases, to first order the error in evaluating $p(z)$ is proportional to u: $e(z,u) := |\tilde{p}(z,u) - p(z)| = O(u)$, where $\tilde{p}(z,u)$ is the calculated floating-point value in precision u. Meanwhile, to first order the magnitude of the polynomial $|p(z)|$ near a *nonsingular* root $\hat{z}$ is proportional to the distance from the root: $|p(z)| \approx |p'(\hat{z})||z - \hat{z}|$. If $|\tilde{p}(z^*,u)| > e(z^*,u)$ for any evaluation point z^*, then it is clear that the polynomial p is nonzero. So for well-separated roots, the disk in $\mathbb{C}$ centered on $\hat{z}$, where the result is uncertain, has radius approximately $Cu/|p'(\hat{z})|$ for some finite constant C, and hence the area of the uncertain zone is $O(u^2)$. Increasing precision by one decimal digit reduces the probability of choosing a test point in the uncertain zone by a factor of 100! If $\hat{z}$ is a singular root of multiplicity μ, we have to work harder, because now the magnitude of the polynomial scales as $|z - \hat{z}|^{\mu}$, so the uncertain area scales as $O(u^{2/\mu})$. But even so, high enough precision still reduces the probability of uncertain results to any desired level.

On the flip side, if the polynomial $p(z)$ really is zero, then we expect the evaluation error $e(z,u)$, and hence $|\tilde{p}(z^*,u)|$, to fall in proportion to u as precision is increased. This contrasts with the behavior for nonzero $p(z)$, where $|\tilde{p}(z^*,u)|$ converges to a nonzero value as precision increases. Checking the trend in $|\tilde{p}(z^*, u)|$ for several different precisions significantly decreases the probability of making an erroneous decision as to whether p is zero.[10]

Worrying about falling short of probability-one assertions is not warrantless! In the basic linear homotopy (5.7) discussed in §5.1, a choice of a random complex number γ is made so that with probability one no paths of the homotopy cross. In the nine-point four-bar problem of §5.5.2, which consists of eight polynomials of degree seven, experiments show that for a random γ roughly 0.8% of paths get so close that a path tracker executed in only double precision is prone to jumping from one path to the other. Therefore, for the best reliability at a reasonable computational cost, it is valuable to switch computations to higher precision when such danger is detected, as Bertini does by default.

Precision questions

Adaptive multiprecision can go a long way in overcoming ill-conditioning and preventing failure of the probability-one algorithms used in numerical algebraic geometry. This raises some questions about how to deploy this capability judiciously:

1. What is the computational cost of higher precision?
2. How do we choose a "correct" level of precision?
3. Is it reasonable to use higher precision when the system we want to solve is only given to double precision?

Let's discuss these points one at a time.

3.2.2 ▪ The Cost of Higher Precision

As a famous application of the fast Fourier transform [108], there is an algorithm and an integer n_0 such that for n digit numbers and $n \geq n_0$, the cost of floating-point operations is $O(n \log(n))$. There are different methods that are best for $n < n_0$, and a good multiprecision package will make good choices of the algorithms used for different n. The change from n to $2n$ digits does not seem particularly severe until it is realized that with the GMP-based library MPFR [57], there is an order-of-magnitude hit when switching from hardware double precision to MPFR arbitrary precision; e.g., see [23, Table 4]. The MPFR package is extremely useful because it is accurate and works on any machine with a standard flavor of Linux. The disadvantage is that MPFR seems to make little use of the floating-point hardware. Consequently the cost of doing the same calculation in MPFR double precision differs from doing it in hardware double precision by an order of magnitude.

We hope this will change in the future, but even so, we often need to use hundreds of digits in computations. Even if we started with hardware double precision as a base, the extra cost of computation will be high. Yet it would be impossible to avoid path-crossing using only double precision. Moreover, without multiprecision, we would often be unable to track to the regions where the endgames of §3.3 work.

Our conclusion is that the only way to have accurate polynomial solving software is to use multiprecision. Furthermore, to control the cost of using multiprecision, we must resort to it only when necessary.

[10] For an identically zero $p(z)$, strict proportionality of $|\tilde{p}(z^*, u)|$ to u in a set of evaluations at different precisions is not ensured, as proportionality applies to the error bound, not to the actual function values one might chance upon inside that bound.

3.2.3 ▪ Choosing a Correct Level of Precision

When tracking a path, we set a tolerance, i.e., a maximum distance, within which we wish to track the path for all $t \in (0,1]$; see §E.4. We also set `FinalTol` (default 10^{-11}) for how closely we compute the endpoint; see §E.5.1. We track by making a predictor step followed by a small number of Newton corrector steps. Numerical analysis provides several ways of measuring errors, all of which have traditionally been used to change step sizes adaptively. These error estimates assume that all computations have a certain number of digits of accuracy.

Prediction and correction steps involve solving linear systems whose coefficients are polynomials. The condition number of the system magnifies any errors in the evaluation of the coefficients. But even evaluation of a polynomial to within a desired error may be impossible in double precision. For example, as the simple degree-10 polynomial in (1.5) shows, the evaluation of even a moderate-degree, three-term polynomial with moderate coefficients can give poor results in double precision. Evaluations of systems of polynomials can be even worse.

The details and theory behind adaptive multiprecision are not within the scope of this book. A main determinant in the need for additional digits of accuracy is the condition number. To make estimates for the number of digits needed, we need to know maximum degrees of polynomials in the system and maximum absolute values of coefficients of monomials in the polynomials. A careful analysis of this, along with the development of sophisticated rules for deciding when to increase or decrease precision, may be found in [23, 24]. The main idea is that Bertini watches the condition number of the Jacobian matrix and other, similar information as it tracks each path, increasing and decreasing precision as necessary. See §E.2.2 for all the settings connected with multiprecision.

3.2.4 ▪ Collateral Requirements of Higher Precision

Using many-digit arithmetic on a system whose parameters are given only to double precision will produce a highly accurate solution to a perturbation of the intended problem. Besides perturbing the solution points, this can sometimes lead to inaccurate conclusions about the structure of the solution set, even something as basic as the number of isolated solutions. For example, consider the inverse position problem for six-revolute serial-link robots. One formulation of this, which we call the IPP system, is available at `bertini.nd.edu/ExampleSystems.html`. If the coefficients of that system are given to just a few digits, certain singular solutions that are "at infinity" will appear as a cluster of finite, nonsingular isolated solutions.

The key point is that when all parameters are included, systems arising in applications usually are defined by equations with exact (often small integer) coefficients. Approximate floating-point coefficients arise when a point on a parameter space is chosen. For example, an angle θ^* might be chosen approximately and might enter the system as coefficients built out of $\cos(\theta^*)$ and $\sin(\theta^*)$. Evaluating the cosine and sine to 16 digits before posing the problem and then using more than that number of digits when solving the problem will lead to a failure of the relation $\cos^2(\theta^*)+\sin^2(\theta^*)=1$. This can easily lead to errors, such as in the inexact IPP problem, where a multiple point at infinity appears in the solution set as several finite nonsingular points. This is reminiscent of the pejorative manifolds of Kahan for one-variable polynomials (see [186] for a good discussion of this). For this reason, it is necessary to stay on the appropriate parameter space as the number of digits increases. For the user, this comes down to specifying coefficients and constants, which themselves need to satisfy relations, to enough digits so that those relations are still true to the maximum number of digits used. We provide a simple example of this in §7.8.

3.3 ▪ Endgames for Singular Solutions

In this section we discuss the final phase of tracking a solution path $p(t) \in \mathbb{C}^N$, i.e., $H(p(t),t) = 0$, going from some $t \approx 0$ to the endpoint $\lim_{t\to 0} p(t)$ at the end of the path.[11] Of course, a homotopy $H(z,t) = 0$ often has many solution paths, but we consider these one at a time in the endgame. The original articles on *endgames*, where the word was coined, are [135, 136, 137]. The book [161] gives a thorough presentation of these algorithms, while the parallel version of the Cauchy endgame, first presented in [21], is more recent. See also [102, 163]. We will not repeat the details here but instead settle for some basic notions.

Fix a homotopy $H(z,t)$. The whole point of endgames is to follow the very end of the path, from $t = \epsilon \approx 0$ to $t = 0$. These specialized methods are necessary near singularities, as the numerical linear algebra needed for path tracking will rapidly become very ill-conditioned as $t \to 0$ if the endpoint $\lim_{t\to 0} z(t)$ is singular. Endgames are designed to approximate the endpoint of the path without stepping too close to the potential singularity at $t = 0$. We say "potential" because we must work under the assumption that the endpoint of any path could be singular, as discussed in the previous chapter.

Remark 3.2. Everything we say works for a system of complex analytic functions that are polynomial in the z variables and defined on $\mathbb{C}^N$ for some open disk Δ around the origin in $\mathbb{C}$, i.e., a system of polynomials in the z variables with coefficients that are complex analytic functions of t in an open disk Δ around $0 \in \mathbb{C}$.

To see why we need endgames, consider the simple homotopy $z^6 - t = 0$. If we wanted the solution to within 10^{-8} of the root, we would need to track t to 10^{-48}. This would be slow and, in analogous situations in higher dimensions, would require very high precision considering that the condition number of the Jacobian grows large in the vicinity of a singularity. It can also happen that the endpoint belongs to a positive-dimensional solution set, in which case Newton's method applied at the endpoint can wander freely on that set instead of giving the true path endpoint.

Bertini has three endgames, controlled by the Bertini setting `EndgameNum`. See Appendix E.5 for all the endgame settings. Bertini treats the tracking before the endgame and the tracking during the endgame as different: the associated settings such as `TrackTolBeforeEG` and `TrackTolDuringEG` can be found in Appendix E.

The default (`EndgameNum:1`) is the power-series endgame, which works by building an approximation of the Puiseux expansion of $p(t)$ around 0.

The second and third approaches (`EndgameNum:2` and `EndgameNum:3`, respectively) are the Cauchy endgame and a parallel version of the Cauchy endgame. The parallel Cauchy endgame has the advantage over the other endgames that it reduces computation significantly if there are many solutions with high *cycle number* (also called the *winding number*). These methods work by using the Cauchy integral formula to express the limit of a path $p(t)$ as $t \to 0$ in terms of an integral of $p(t)$ continued around a path $\epsilon e^{i\theta}$ until $p(t)$ returns to its initial value. A novice Bertini user need not know the details of these methods, so the remainder of this chapter is optional.

[11] As described in the previous chapter, some paths could diverge to ∞. This is handled by working over projective space, the subject of the next chapter. For now, we are focusing on finite endpoints.

3.3.1 ▪ The Endgame Setup*

All the endgames have the same setup. Assume that we have a path for a good homotopy $H(z,t) = 0$, i.e., let $p(t) : (0,1] \to \mathbb{C}^N$ be a function such that $H(p(t),t) = 0$. We assume we have tracked the path to $t = \epsilon$, where $\epsilon > 0$ is some small number. This can be adjusted with the Bertini configuration `EndgameBdry`, the default of which is 0.1. The endgames work under the technical assumptions that the irreducible analytic set $Z \subset \mathbb{C}^N \times \{t \in \mathbb{C} \mid |t| < \epsilon\}$ on which $(p(t),t)$ lies for $t \in (0,\epsilon)$ is the analytic image of a disk $\Delta_r(0) = \{s \in \mathbb{C} \mid |s| < r\}$ of some radius r,

$$\psi : \Delta_r(0) \to Z,$$

with

1. $\psi(0) = (z^*, 0) := \lim_{t\to 0}(p(t),t)$ and
2. $d\psi$ is not zero anywhere in $\Delta_r(0)$ except possibly at $s = 0$.

Note that z^* is the endpoint of the path $p(t)$, and it is what we wish to compute in the endgame.

Under these assumptions, using s as the coordinate in the disk $\Delta_r(0)$, we can assume that the last coordinate of $\psi(s)$, which sends $s \to t$, is given by $t = s^c$ for some positive integer c. This integer is the classical *cycle number*, also called the *winding number*. Accordingly, letting $\phi(s)$ be the first N coordinates of $\psi(s)$, we have that $\psi(s)$ is of the form $\psi(s) = (\phi(s), s^c)$, and $z^* = \phi(0)$. After this transformation, we have swapped the problem of finding $p(0)$ for the problem of finding $\phi(0)$, with the crucial improvement that $\phi(s)$ is analytic on the disk.

Bottom line. Suppose we have tracked the path to the point z_ϵ at $t = \epsilon$. If the cycle number, c, is known, then the transformation $t = s^c$ gives a new system, say, $h(z,s) := H(z, s^c)$, such that the function $\phi(s)$ implicitly defined by $h(\phi(s),s) = 0$ starting from $(\phi(s),s) = (z_\epsilon, \sqrt[c]{\epsilon})$ is a smooth analytic function of s in the disk $\Delta_r(0)$ for some radius $r > 0$. Hence the limit as $s \to 0$ can be accurately estimated by sampling this function near zero, such as by fitting a truncated power series for $\phi(s)$ near $s = 0$ or by performing a Cauchy integral around a loop that encircles the origin, as described below. In general, we must estimate $\phi(0)$ from samples of $\phi(s)$ for nonzero s, because $\partial H/\partial z$ may vanish at $t = 0$, thereby causing Newton's method at the endpoint to fail.

Before we go further, let's look at the reasonableness of these assumptions. From the uniformization theorem for one-dimensional analytic sets, we know there exists an $\epsilon > 0$ such that the above assertions are true. The upper limit on ϵ is the magnitude of t at the nearest branch point. (See §2.1.2.) If we have tracked to some $t = \epsilon$ for which the assertions are true, we say we are in the *endgame operating zone*; see Figure 3.1. If the endpoint is singular, the condition number of the Jacobian matrix, cond(JH), goes to infinity at the endpoint, so for any given precision the endgame operating zone has an inner boundary, beyond which the ill-conditioning prevents numerical computation. If the endpoint is nonsingular, this inner boundary can be abolished with sufficient precision, and even for singular endpoints the ill-conditioned region can be shrunk as small as needed by invoking higher precision. Bertini will increase precision automatically as necessary to track the path in the endgame operating zone so long as we do not attempt to track all the way to $t = 0$.

Rigorously checking that we are in the endgame operating zone would be extremely expensive. Bertini uses heuristics to make a decision if this is true and, if not, it goes to a smaller ϵ. The heuristics check whether the data collected in the endgame match

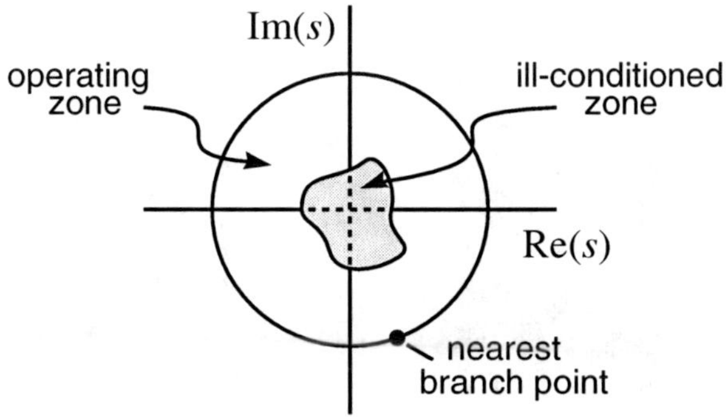

Figure 3.1. *Endgame operating zone.*

adequately to the endgame setup; that is, can we find a c so that $\phi(s)$ looks like an analytic function? If the data adequately match a low-order power-series approximation, then even if we are not yet in the operating zone, it is a sign that the path is heading toward a very tight cluster of roots. In such a case, the endgame will produce an estimate of the centroid of the cluster, and this will still be a good estimate of the true endpoint. While this general picture appears sound, a full analysis of the relationship between the heuristic measures and bounds on the endpoint error is still open for study.

In the next two subsections, we illustrate endgames using their performance on a simple example that has two paths converging to the same double root. As in §2.1, we use the homotopy

$$H(z,t)=\gamma t\, g(z)+(1-t)f(z),$$

this time with

$$g(z)=z^3-1, \qquad f(z)=z^3+2z^2+z, \qquad \gamma=0.5-0.9i.$$

The target polynomial, $f(z)$, has a nonsingular root at $z=0$ and a double root at $z=-1$. The paths of the homotopy are shown in Figure 3.2. Notice that the pair of roots meeting at $z=-1$ come in tangent to the $t=0$ plane.

The endgame setup says nothing about how big the operating zone is, just that a disk of finite radius exists. We do not normally compute the radius r that limits the endgame operating zone, but we may do so for this simple example. The disk $\Delta_r(0)$ can be extended out by continuation until it just meets the nearest branch point. Using the same procedure as in §2.1.2, one may find that the ramification points (where H and $\partial H/\partial z$ vanish simultaneously) for our new illustrative example are at the z-values

$$z\approx\{-0.6395+1.3747i,-1.0000+0.0000i,0.0589+0.4593i,\\ -0.5080-0.4080i,-0.3642-0.0108i\}$$

with corresponding t-values

$$t\approx\{2.7203+2.9627i,0.0000+0.0000i,-1.3030-0.2953i,\\ -0.3927-0.0441i,-0.0550-0.1344i\}.$$

Of course, the one at $t=0$ is the double root we are seeking. The next-closest branch point to the origin, $t=-0.0550-0.1344i$, has approximate magnitude 0.145. This point limits the endgame operating zone to $|t|<a$, $a\approx 0.145$, or since $t=s^2$, $r\approx\sqrt{0.145}$.

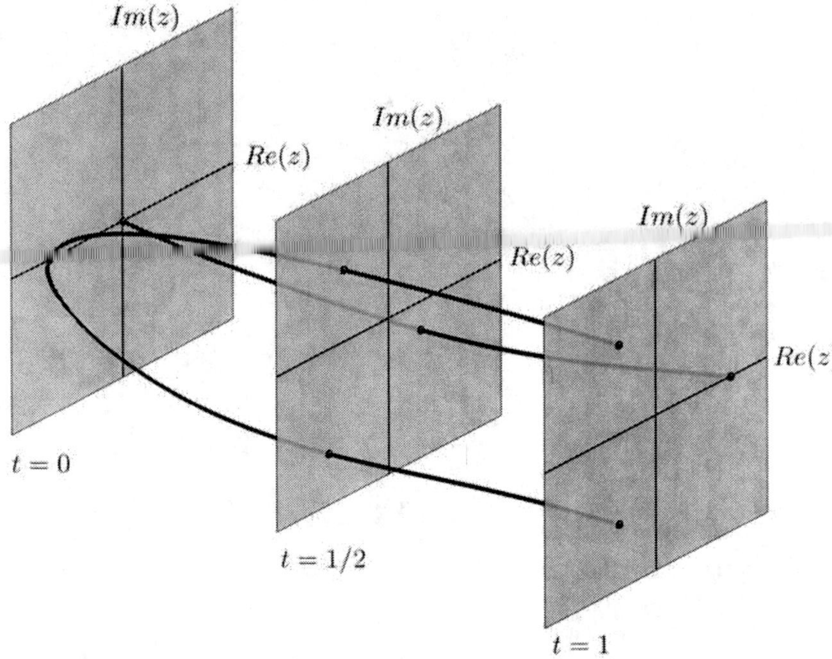

Figure 3.2. *A homotopy that has two paths converging to a singular point at* $t = 0$.

3.3.2 ▪ Fractional Power-Series Endgame*

One way to estimate z^* is to approximate $\phi(s)$ in the neighborhood of $s = 0$ by a truncated power series. That is, the power-series endgame is based on the fact that any of the coordinate functions $z_j = \phi_j(s)$ is a power series $z_j^* + s^k(a_{j,0} + a_{j,1}s + \cdots)$, where $s = \sqrt[c]{t}$ for some positive integer c. Choosing successive points, e.g., $t = \epsilon, \epsilon\lambda, \epsilon\lambda^2$ for some $0 < \lambda < 1$, we can try to detect c, as it is closely connected to the curvature of the path.[12] To do this, we follow several steps:

1. We make several guesses at c (typically $c = 1, \ldots, 6$).[13]

2. For each c, approximate the first few terms of the power series for each $\phi_j(s)$ as a best fit to the previously computed points along the path.

3. Use these approximations to extrapolate $z(t)$ to a lower value of t, say, $t = t'$.

4. March forward along the path to get the actual point on the path at $t = t'$.

5. Choose the value of c whose extrapolant is nearest the point on the path at $t = t'$.

6. Use this value of c to estimate the endpoint of the path at $t = 0$ as the constant term of the power series.

[12]The number λ is given by the Bertini setting `SampleFactor`, which by default is 0.5.

[13]Bertini's `MaxCycleNum` setting controls this.

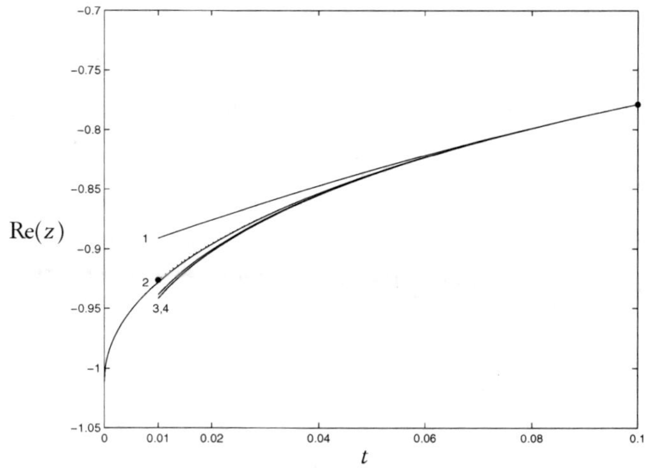

Figure 3.3. *One step of the power-series endgame.*

We say that the endgame has converged (and we have thus adequately computed the endpoint of the path) when two estimates of the endpoint agree to the desired accuracy, `FinalTol`. We repeat this procedure until convergence by setting $t' := \lambda t'$ and going back to step 1.

In Bertini, the power-series estimate is based on the values of z and its derivative dz/dt at $n =$ `NumSamplePoints` points, giving a prediction to $s = 0$ of order $2n-1$ with a truncation error of $O(|s|^{2n})$. But since $t = s^c$, this means the truncation error is $O(|t|^{2n/c})$.

One step of this procedure is illustrated in Figure 3.3. The path of the example homotopy is tracked to $t = 0.1$ and $t = 0.01$, and the corresponding values of z are marked with dots in the figure. To make errors more apparent in the drawing, we use a second-order prediction in s, instead of Bertini's default of third order. From the value of z and its first two derivatives at $t = 0.1$, fractional power-series predictions are made to $t = 0.01$. (Although we need only to compute the predictions at $t = 0.01$, we plot the predicted curve on the entire interval $t \in [0.01, 0.1]$.) The $c = 2$ prediction lies very close to the tracked curve, barely visible as a dashed line, and in particular, it is the closest of the predictions to the value of z at $t = 0.01$. Accordingly, we use $c = 2$ to extrapolate to $t = 0$ to estimate the endpoint. In this case, the extrapolation from $t = 0.1$ to $t = 0$ gives $z_1 \approx -1.0115 - 0.0092i$, while the extrapolation from $t = 0.01$ (not shown) gives $z_2 \approx -1.0011 + 0.0004i$. Since these successive extrapolations agree to only about 0.01, the procedure will be repeated, this time starting from $t = 0.01$.

Just using double precision, we would be severely limited in how far we can go toward $t = 0$, which is an important reason to use multiprecision. The minimum ϵ is set by `NbhdRadius` with Bertini's default 10^{-100}. If this is reached without a clear decision, Bertini declares failure for the path. If we want to estimate a root within 10^{-T} with a given c, we need on the order of cT digits.

By homogenizing to work in projective space (see §4.3) and choosing a random patch (see §4.4), the decision of whether a path is going to ∞ comes down to checking whether its endpoint, $\lim_{t \to 0} p(t)$, satisfies the equation corresponding to the hyperplane at ∞. In our experience, points at ∞ often have c much larger than 10. Depending on the security settings (see §E.6), a lot of computation can sometimes be saved by allowing Bertini to stop early on paths whose endpoints appear to be heading to infinity.

3.3.3 ▪ Cauchy Methods*

Assume that we have a path for a good homotopy $H(z,t)=0$. As described in the endgame setup, §3.3.1, after the transformation $t=s^c$, where c is the cycle number of the endpoint, we have the related homotopy $h(z,s):=H(z,s^c)$ with a solution path $\phi(s):(0,1]\to\mathbb{C}^N$ that is an analytic function such that $h(\phi(s),s)=0$. We assume we have tracked the path to $t=\epsilon$, where $\epsilon>0$ is some small number. The key to the Cauchy endgame is a classical fact discovered by its namesake, the Cauchy integral formula. Given a function $f(s)$ holomorphic (i.e., analytic) in a disk D with center s^* and boundary ΔD, one version of this theorem is that

$$f(s^*)=\frac{1}{2\pi i}\int_{\Delta D}\frac{f(s)}{s-s^*}dz.$$

In this endgame, our disk D is a circle of radius $\epsilon^{1/c}$ centered at $s^*=0$, i.e., $\epsilon^{1/c}e^{i\theta}$ for $\theta\in[0,2\pi]$, and the function value that we wish to compute is the value of the path at the origin, i.e., $f(0)=\phi(0)$.

Since we are using numerical methods, it is difficult to compute the integral exactly. However, we can use numerical integration to estimate the integral. Beginning at $t=\epsilon$, we follow a polygonal path around $t=0$ with vertices evenly spaced on the circle $\epsilon e^{i\theta}$. The number of vertices of this polygon is equal to $1+$ `NumSamplePoints`, so that the default polygon is an equilateral triangle. By collecting the path values at these vertices, we use the trapezoid integration rule with the Cauchy integral formula to estimate the path value at $t=s=0$. In fact, we must track around the circle (polygon, technically) c times, where c is the same cycle number described in the fractional power-series endgame above. In this case, we can detect c by storing the starting value of the path at $t=\epsilon$ and walking around the circle repeatedly until we arrive back at the stored starting value. In this way, the cycle number (or winding number) c is the number of times we must wind around $t=0$ before the path closes up.

Although the Cauchy integral is to be done in the transformed s coordinates, there is a simple way to calculate it without ever computing a value of s. This is because our sample points in s are at the vertices of a regular polygon centered on the origin. In this case, it is easy to see that the trapezoid integration rule applied to the Cauchy integral yields just the mean of the z-values at the vertices. It also can be shown that this averaging rule applied to the vertices of a regular N-gon gives an $(N-1)$th order approximation of the value at the origin, i.e., the truncation error is $O(|s|^N)$. Since $t=s^c$, the truncation error in t is $O(|t|^{N/c})$. But since we go around c times in t, $N=nc$, where $n=1+$ `NumSamplePoints`, giving a truncation error of $O(|t|^n)$.

Figure 3.4 illustrates this procedure for the running example, where $c=2$. The value of c is revealed when the path returns to its original point after two trips of t around a triangle inscribed in the circle centered on the origin. The figure shows $\mathrm{Re}(p(t))$ versus t with the path in t drawn below for reference. The illustration is made for a starting point at $t=0.1$, which is inside the radius of convergence. The average of the six vertex points is $z^*\approx-0.9968-0.0009i$. Repeating the procedure starting at $t=0.01$ gives

$$z^*\approx-0.99999702-0.00000058i,$$

which is indeed converging to $z=1$ with error proportional to $|t|^3$.

Convergence for this endgame is similar to that of the fractional power-series endgame above. Until convergence, we repeat the procedure by replacing ϵ with $\lambda\epsilon$, estimating the

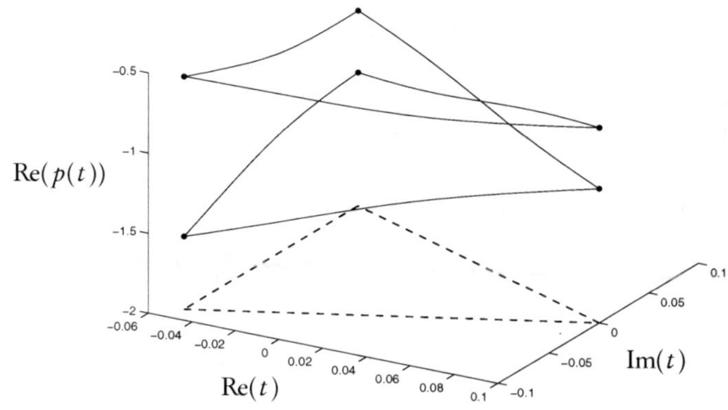

Figure 3.4. *Cauchy endgame: the path $(p(t), t)$ closes up after tracking t around a triangle twice.*

endpoint, and checking for convergence.[14] Convergence occurs when two estimates of the endpoint agree to the desired level of accuracy in `FinalTol`.

Since approximating the Cauchy integral can be relatively expensive, Bertini includes two heuristics for deciding when to actually start approximating these integrals, both appearing in [21]. The first is a cycle test based on four points collected in a geometric sequence, e.g., $z(t)$, $z(\lambda t)$, $z(\lambda^2 t)$, and $z(\lambda^3 t)$. For a general vector v and t in the endgame operating zone, there is $k \geq 1$ with $a_k \neq 0$ such that

$$v \cdot z(t) = v \cdot z(0) + v \cdot a_k t^{k/c} + \sum_{j=c+1}^{\infty} v \cdot a_j t^{j/c},$$

where c is the cycle number. When t is in the endgame operating zone, we expect

$$G(t) := \frac{\log \left| \frac{v\cdot z(\lambda t) - v\cdot z(\lambda^2 t)}{v\cdot z(t) - v\cdot z(\lambda t)} \right|}{\log \lambda} \approx \frac{k}{c}.$$

Therefore, the cycle test is passed if

$$\frac{3}{4} \leq \frac{G(t)}{G(\lambda t)} \leq \frac{4}{3}.$$

If the cycle test is passed, the second heuristic, discussed below, is performed. As a fail-safe mechanism, Bertini includes the configuration `CycleTimeCutoff`, default being 10^{-8}, which deems the cycle test to be passed when collecting sample points at values of t smaller than this value.

The second is a test based on collecting points on $z(te^{i\theta})$ for $\theta \in [0, 2\pi]$. Let M and m be the maximum and minimum norms of the 1+`NumSamplePoints` collected. This test is passed when

$$\text{either } M - m < \texttt{FinalTol} \text{ or } \frac{m}{M} > \frac{1}{2}.$$

Similar to the fail-safe mechanism for the cycle test, Bertini includes the configuration `RatioTimeCutoff`, default being 10^{-14}, which deems this test to be passed when collecting sample points at values of t smaller than this value.

[14]As in the power-series endgame, λ is given by the Bertini setting `SampleFactor`.

The parallel version of the Cauchy endgame is based on the fact that we go around the circle c times and have found c distinct roots of $H(z,\epsilon)=0$. Tracking these backward, i.e., from $t=\epsilon$ to $t=1$, we can find the full set of c startpoints for these paths. This eliminates the cost of running the endgame for each of the c paths. In practice, we have often found that it is not worth carrying out this back tracking step unless $c \geq 4$. The Bertini setting `MinCycleTrackback` is the minimum c for which back tracking is performed, with the default being 4.

Chapter 4
Projective Space

Projective spaces are relevant to solving systems of polynomial equations in two ways. First, and most commonly, they provide a means of understanding how solutions of a polynomial system may exist "at infinity." When such solutions exist, some solution paths in a homotopy diverge to infinity, and projective spaces can be used to follow such paths securely. This helps make sure that large, but finite, solutions are not missed in the computation. Second, in some applications, groups of variables arise whose ratios matter but whose exact values are not of interest. As explained in §4.1, such a group of variables is said to live on projective space, or rather the variables are the homogeneous coordinates of a projective space.

When a problem has variables given in affine space, a process called *homogenization* can be used to transform them into homogeneous coordinates on a projective space. In this process, coordinates in affine (Euclidean) space are replaced by ratios of coordinates in projective space so that any large affine coordinate can be represented instead by a ratio with a small denominator. This stabilizes numerics and transforms infinitely long divergent paths in affine space into finite length paths in projective space. This allows such a path to be tracked numerically all the way to its endpoint. Moreover, in Bertini, the user can declare various groupings of the variables, thereby directing how Bertini uses projective spaces. This can change the number and character of solutions at infinity, which can have a big effect on how long it takes to solve a system. We will return to this idea in full in Chapter 5; here we sketch the underlying theory that makes homogenization so useful.

The idea of projective space is closely linked to two basic Bertini keywords:

- `hom_variable_group` declares a group of variables to be the homogeneous coordinates of a projective space; that is, a group of variables for which only their ratios matter.
- `variable_group` declares all other groups of variables.

Each `variable_group` will be transformed into its own projective space via a homogenization process. Except in the case of a user-defined homotopy, all unknowns must appear in one of these declarations.[15] The division of variables into groups affects the number of paths that Bertini uses and also affects the numerical treatment of paths that diverge to infinity. Under the default settings, the declaration of all variables in a single

[15] A third declaration type, `variable`, may appear in user-defined homotopies (see §7.5).

`variable_group` causes Bertini to form a total-degree homotopy (see §2.1.4), while multiple variable groups lead to multihomogeneous homotopies (see §4.5 and §5.1). This chapter concentrates on the meaning of projective space, how it resolves the problem of solutions at infinity, and how computations on projective space are implemented. A discussion of the related homotopies and how the declaration of variable groups affects the number of solution paths is postponed until the next chapter.

4.1 ▪ Naturally Arising Projective Spaces

Although in most applications the variables arise in affine space, that is, in $\mathbb{R}^n$ or $\mathbb{C}^n$, in some situations variables arise in projective space. When variables appear in affine space, as we shall see, it is useful to transform them into projective space. To prepare for that eventuality, we first discuss naturally arising projective spaces.

The ratios $1 : 1/2 : \sqrt{2}$ and $\sqrt{2} : \sqrt{2}/2 : 2$ are the same. So long as we care only about the ratios, it is natural to consider as the same the triplets $[1, 1/2, \sqrt{2}]$ and $[\sqrt{2}, \sqrt{2}/2, 2]$ and any other triplets proportional to them. This is the basic idea of a projective space.

A polynomial is said to be *homogeneous* if every monomial has the same degree. If only ratios matter in an answer, it must be that the system of equations that describe the problem is homogeneous, and vice versa. For example, the triplets above happen to be values that satisfy the system of polynomial equations

$$f(x,y,z) = \begin{bmatrix} f_1(x,y,z) \\ f_2(x,y,z) \end{bmatrix} = \begin{bmatrix} 4xy - z^2 \\ x - 2y \end{bmatrix} = 0.$$

In fact, $(x,y,z) = (\alpha, \alpha/2, \alpha\sqrt{2})$ satisfies $f(x,y,z) = 0$ for any $\alpha \in \mathbb{C}$. This is clear because for any $u \in \mathbb{C}^3$, $f_1(\alpha u) = \alpha^2 f_1(u)$ and $f_2(\alpha u) = \alpha f(u)$. In other words, f_1 and f_2 are *homogeneous* polynomials: every monomial of f_1 is degree 2, and every monomial of f_2 is degree 1.

We could be content to say that the solution set of f consists of lines through the origin in $\mathbb{C}^3$. Alternatively, we may say that because f consists of all homogeneous polynomials, we are interested only in the ratios of the variables, $x : y : z$. We consider each point $x : y : z$ to represent the unique line through the origin in $\mathbb{C}^3$ that passes through (x,y,z), namely, $\{(tx, ty, tz), t \in \mathbb{C}\}$, and disallow $0 : 0 : 0$ as meaningless, because it doesn't determine such a line. The set of all possible ratios $x : y : z$, for $(x,y,z) \in \mathbb{C}^3, (x,y,z) \neq 0$, and equivalently the set of all lines through the origin in $\mathbb{C}^3$, is a two-dimensional space. Mathematicians call this the *complex projective plane*, denoted as $\mathbb{P}^2$. (Some authors use $\mathbb{P}^2_{\mathbb{C}}$ instead and use $\mathbb{P}^2_{\mathbb{R}}$ for the real projective plane.) Also, rather than the colon notation, $x : y : z$, it is standard to denote the coordinates of a projective space with square brackets, $[x,y,z]$, a notation that we will adopt from here on. The triplet $[x,y,z]$ is said to be a set of *homogeneous coordinates* for $\mathbb{P}^2$. With this concept, we may say that the solution of $f = 0$ is a pair of points in $\mathbb{P}^2$:

$$\mathbf{V}(f) = \left\{ \left[1, 1/2, \sqrt{2}\right], \left[1, 1/2, -\sqrt{2}\right] \right\}.$$

Readers who have not seen this concept before may think it a bit strange, and yet it is almost certain that they have used it often. This is because one of the most common problems in linear algebra, arising in many practical applications, is the matrix eigenvalue-eigenvector problem. For an $N \times N$ matrix A, the eigenvalue-eigenvector problem is to find pairs of scalars λ and vectors v, called eigenvalues and eigenvectors, respectively, such that

$$Av = \lambda v. \tag{4.1}$$

(Equivalently, we may consider $(A-\lambda I)v = 0$.) Notice that (4.1) is homogeneous in v, so for any solution point (λ_0, v_0), $v_0 \neq 0$, there is an entire line of solutions $(\lambda_0, t v_0)$, $t \in \mathbb{C}$. Accordingly, we may cast the problem as finding all $(\lambda, v) \in \mathbb{C} \times \mathbb{P}^{N-1}$ that satisfy (4.1). With such a formulation, for general A, the problem has N distinct eigenvalue-eigenvector pairs. Standard eigenvalue software computes these, usually normalizing the length in the Euclidean norm of the reported eigenvectors to one, i.e., $v^H v = 1$, where v^H is the Hermitian transpose (conjugate transpose) of v. These packages report the eigenvector solutions in projective space: for each eigenvector direction, they give just one of the two vectors v_0 and $-v_0$ whose length is one.

Recall that the standard declaration type for complex variables in Bertini is `variable_group`. Variables declared that way live on complex Euclidean space. To declare a set of variables as homogeneous coordinates on a projective space, use instead the declaration `hom_variable_group`.

Example 4.1 (a simple eigenvalue problem). Eigenvalue problems are so common and so important that highly efficient, special-purpose software exists for solving them. Even though Bertini will not be competitive with these, since the problem is so familiar, let's use it to illustrate an example of a Bertini run that uses projective space. The input file to find the eigenvalues and eigenvectors of the matrix $\begin{bmatrix} 1 & 2 \\ 3 & 4 \end{bmatrix}$ is as follows, where `lambda` is used for λ and the coordinates of v are `v1,v2`:

```
% Eigenvalue2x2.input
% A simple eigenvalue/eigenvector problem
%   illustrating usage of hom_variable_group.
INPUT
 function f1,f2;
 variable_group lambda;
 hom_variable_group v1,v2;

 f1 = (1-lambda)*v1 + 2*v2;
 f2 = 3*v1 + (4-lambda)*v2;
END;
```

Bertini gives two solutions to this problem, indicating that they are both real. The file `main_data` contains the following solutions:

```
Accuracy estimate, internal coordinates (difference of last two
endpoint estimates):  2.904071858576524e-13
Accuracy estimate, user's coordinates (after dehomogenization, if
applicable):  4.634050781165563c-13
Cycle number: 1
-3.722813232690144e-01  1.040834085586084e-17
 1.000000000000000e+00  0.000000000000000e+00
-6.861406616345074e-01 -1.387778780781446e-17

Accuracy estimate, internal coordinates (difference of last two
endpoint estimates):  8.370208695162804e-13
Accuracy estimate, user's coordinates (after dehomogenization, if
applicable):  2.882335236869105e-11
Cycle number: 1
5.372281323269016e+00 -1.332267629550188e-15
4.574271077563377e-01  0.000000000000000e+00
1.000000000000000e+00  0.000000000000000e+00
```

The solutions are computed in complex numbers, but the magnitude of the largest imaginary part is `1.4e-17` for the first solution and `1.3e-15` for the second. These are declared to be real and so are also reported in the file `real_finite_solutions`, because the imaginary parts are smaller than `ImagThreshold`, which defaults to `1e-8`. Notice that for both solutions the size of the imaginary parts is four orders of magnitude smaller than the accuracy estimate in the user's coordinates. The accuracy estimate is intended to be conservative. See §4.6 for more details. ∎

Example 4.2 (generalized eigenvalue problems). A generalization of the standard eigenvalue problem involves two $N \times N$ matrices, say, A and B, for which we wish to find $([\mu,\lambda],[v_1,\dots,v_N]) \in \mathbb{P}^1 \times \mathbb{P}^{N-1}$ to satisfy the equation

$$\mu A v = \lambda B v. \tag{4.2}$$

This is the same as the standard eigenvalue problem when $B = I$. Notice that the variables for the generalized problem appear in two separate sets of homogeneous coordinates. If either of the matrices is nonsingular, the problem can be converted back to the standard one, so the most interesting situation is when both are singular. Then it is possible to have some roots with $[\mu,\lambda] = [0,1]$ and others with $[\mu,\lambda] = [1,0]$, so it is not advisable to try to reduce the number of variables by setting either μ or λ to 1. It is better to keep both in play and work on $\mathbb{P}^1$ instead.

For a particular example, let us use

$$A = \begin{bmatrix} 1 & 2 \\ 2 & 4 \end{bmatrix}, \qquad B = \begin{bmatrix} 4 & -2 \\ -2 & 1 \end{bmatrix}. \tag{4.3}$$

A Bertini file to solve this problem is

```
% GeneralizedEigenvalue.input
% Simple example of using two homogeneous groups.
% This is a 2x2 generalized eigenvalue problem.
INPUT
 function f1,f2;
 hom_variable_group v1,v2;
 hom_variable_group mu,lambda;

 f1 = mu*(1*v1 + 2*v2) - lambda*( 4*v1 - 2*v2);
 f2 = mu*(2*v1 + 4*v2) - lambda*(-2*v1 + 1*v2);
END;
```

Bertini reports two real solutions, which have generalized eigenvalue pairs that are to about 15 digits:

$$[\mu,\lambda] = [1,0] \quad \text{and} \quad [\mu,\lambda] = [0,1]. \quad \blacksquare$$

Example 4.3 (rigid-body transforms). A rigid body is a collection of points in $\mathbb{R}^3$ that retain the same geometric shape as they move. That is, the distance between any two points remains constant, and the collection of points is not allowed to miraculously snap to its mirror image either. Any location of a rigid body can be described by a position vector, $p \in \mathbb{R}^3$, and a 3×3 rotation matrix $R \in SO(3)$. The space of all possible rigid-body motions is six-dimensional, having three degrees of freedom of translation and three more of rotation, and is called "special Euclidean 3-space," denoted

$$SE(3) := \{(p,R) \mid p \in \mathbb{R}^3,\ R \in \mathbb{R}^{3\times 3},\ R^T R = I,\ \det R = 1\}.$$

If $v \in \mathbb{R}^3$ is the coordinate vector of a point of the body in its reference position, then when the body is in location (p,R), the point's new location is $p+Rv$. The representation of $SE(3)$ in terms of (p,R) is not homogeneous, as any scaling αR, $|\alpha| \neq 1$, gives a matrix that is not in $SO(3)$.

There is a homogeneous representation of $SE(3)$ in $\mathbb{P}^7$ called *Study coordinates*, named after E. Study [166, 167]. Let the coordinates be written as

$$s = [e_0, e_1, e_2, e_3, g_0, g_1, g_2, g_3].$$

Consider the two quaternions formed as

$$e = e_0\mathbf{1} + e_1\mathbf{i} + e_2\mathbf{j} + e_3\mathbf{k} \quad \text{and} \quad g = g_0\mathbf{1} + g_1\mathbf{i} + g_2\mathbf{j} + g_3\mathbf{k}$$

and let the coordinates of a point on a rigid body be the pure vector $v = x\mathbf{i} + y\mathbf{j} + z\mathbf{k}$. Let $*$ denote quaternion multiplication, and write quaternion conjugation as

$$e' = e_0\mathbf{1} - e_1\mathbf{i} - e_2\mathbf{j} - e_3\mathbf{k}.$$

Then, $s = [e, g]$ encodes a rigid-body displacement where the new location of v is given by

$$(g * e' + e * v * e')/(e * e'). \tag{4.4}$$

The result must again be a pure vector, which requires that the real part (the $\mathbf{1}$ component) of $g * e'$ be zero; that is, $[e, g]$ must satisfy the *Study quadric*:

$$Q(e,g) = g_0e_0 + g_1e_1 + g_2e_2 + g_3e_3 = 0. \tag{4.5}$$

The transformation given in (4.4) is well formed so long as $e * e' \neq 0$. It can be shown that the set

$$S := \{[e,g] \in \mathbb{P}^7 \mid Q(e,g) = 0, e * e' \neq 0\}$$

is isomorphic to $SE(3)$. This means that every $(p,R) \in SE(3)$ has a unique representation in S and vice versa. See, for example, the appendix to [184].

Study coordinates are useful in robot kinematics, especially for solving the forward kinematics of Stewart–Gough platforms (see [106, 179] and [161, §7.7]). See also §6.3. Note that while a roboticist would be interested only in real solutions, all the formulas are algebraic and so carry over to complex projective space. ∎

4.2 ▪ Paths to Infinity

In the common case of variables defined on affine space, one often must consider issues of infinity. We have already introduced the basic notion of solving a system by continuation, and we have observed that sometimes solution paths may diverge to infinity. Projective space gives a way of seeing exactly where these paths go and provides a way to stably compute their endpoints. Compared to simply judging that a solution path is growing large, the computation of a path's endpoint in projective space provides a means of more reliably determining whether a path is diverging. When the variables of the problem live in Euclidean space, not projective space, we may invoke a homogenization operation that embeds Euclidean n-space inside projective n-space, thus capturing the picture at infinity and transforming any infinite length paths to infinity into finite length projective paths. Any solution that does not diverge to infinity can then be dehomogenized to get its corresponding value in the original, affine variables.

Consider the example of intersecting two circles, each given in the form

$$(x-a)^2+(y-b)^2-r^2=0. \tag{4.6}$$

This is a circle with center (a,b) and radius r. Two general circles intersect in two points in $\mathbb{C}^2$, but a total-degree homotopy (see §2.1.4) will have four paths, the expected number of roots for the intersection of two general quadrics. There is only one possibility for the two extra solution paths: they must diverge to infinity. As x and y grow large, the terms of highest degree dominate, which in this case means both circles "look like" $x^2+y^2=0$ for large (x,y). Accordingly, one might expect that the divergent paths asymptotically approach the lines $x=iy$ and $x=-iy$ as the homotopy parameter t goes to zero. The following Bertini run demonstrates that this is exactly what happens:

```
% TwoCircles.input
% Intersection of two circles
CONFIG
 % Force accurate determination of diverging paths
 SecurityLevel:1;
END;
INPUT
 function circle1, circle2;
 variable_group x, y;

 circle1 = x^2 + y^2 - 1;
 circle2 = (x-1)^2 + (y+1)^2 - 1;
END;
```

For this input, Bertini gives the following four path endpoints in the file `main_data`:

```
 1.000000000000008e+00 -1.593071541056629e-14
-2.936262242267982e-14  4.333401705869432e-14

-7.602781852215701e-14 -1.336478563670085e-13
-9.999999999999067e-01 -5.141796311679362e-14

 2.029753622301697e+13 -7.485726784374431e+12
-7.485726784373927e+12 -2.029753622301844e+13

1.500714493237600e+15 -2.869409837128654e+15
2.869409837128651e+15  1.500714493237596e+15
```

One sees that the first two solutions are close approximations to the expected roots, $(1,0)$ and $(0,-1)$. The third and fourth endpoints are quite large and satisfy $x=iy$ and $x=-iy$, respectively, to about 13 digits of accuracy. These are diverging to infinity in the expected pattern!

Tracking a path in Euclidean space from starting points of magnitude 1 to endpoints of magnitude 10^{13} or 10^{15} could take many steps and be very costly. Yet, examination of the file `output` shows that Bertini took less than 20 steps on either of these, hardly more than it used on the two finite solutions. The next subsection reveals how this is accomplished so efficiently.

4.3 ▪ Taming Infinity: Homogenization-Dehomogenization

The two-circle example illustrates the behavior of all divergent homotopy paths: as t goes to zero, the variables asymptotically approach ratios determined by the highest degree terms in each of the polynomials. Thus, it is natural to consider replacing each variable with a ratio. Inside of Bertini's computational engine, the declaration

```
variable_group x,y;
```

invokes substitutions involving three new variables, say, X, Y, Z, as $x = X/Z$ and $y = Y/Z$. Accordingly, after clearing denominators, the circle equation (4.6) becomes

$$(X - aZ)^2 + (Y - bZ)^2 - r^2Z^2 = 0. \tag{4.7}$$

Notice that this equation is now homogeneous, since every term in it has degree 2. This process is called *homogenization*. With this transformation, the intersection of two circles, originally a problem on $\mathbb{C}^2$, becomes a problem on $\mathbb{P}^2$ with homogeneous coordinates $[X, Y, Z]$.

Clearly, for any finite solution point, say, $(x_0, y_0) \in \mathbb{C}^2$, of the original equations, there is a corresponding point in $\mathbb{P}^2$ that satisfies the homogenized equations, namely, $[x_0, y_0, 1] \in \mathbb{P}^2$. Moreover, for the two-circle problem, the paths that were previously diverging toward infinity in $\mathbb{C}^2$ are now approaching the points $[1, i, 0]$ and $[1, -i, 0]$ in $\mathbb{P}^2$, which also satisfy (4.7). A key observation is that the homogeneous coordinates for both the finite solution points and the solution points at infinity are all finite numbers, which are easy to handle in a computer program.

After computing answers in projective space, the user usually wishes to see the final solutions in the original coordinates. *Dehomogenization* is the simple process of dividing through by the homogenizing coordinate, Z in the example above, and throwing it away, thusly:

$$[X, Y, Z] = [X/Z, Y/Z, 1] \mapsto (X/Z, Y/Z) = (x, y). \tag{4.8}$$

Of course, dehomogenization is well defined only for finite roots, i.e., the ones for which $Z \neq 0$. In principle, a computer code could detect the $Z = 0$ condition and dehomogenize only points that are finite. Bertini does something equivalent, but instead of testing for small $|Z|$, it dehomogenizes every path endpoint and then sorts them by whether any coordinate in $\mathbb{C}^2$ has magnitude greater than `EndpointFiniteThreshold`. (This parameter defaults to size `1e5`, which may be too small for some problems. If answers larger than the default are of interest, the user should change `EndpointFiniteThreshold` in the `CONFIG` section of the input file.)

The foregoing description of Bertini's behavior assumes that the configuration setting `SecurityLevel:1` is used, as illustrated in the two-circle intersection example above. This setting directs Bertini to be very careful by following all homotopy paths all the way to $t = 0$, using a singular endgame (see §3.3), if necessary. In the default setting, `SecurityLevel:0`, Bertini aborts the endgame whenever the path appears to be approaching infinity. (This is invoked if any coordinate in the dehomogenizations of two consecutive estimates of the endpoint exceeds `SecurityMaxNorm`.) The lower security setting can save a lot of computation time for problems having many singular endpoints at infinity, a common occurrence. But the lower security also may miss a few finite roots that are the endpoints of paths that grow large before stabilizing or receding to smaller values.

In the default case that early termination of large paths is allowed, one might wonder what value there is to homogenizing the problem. After all, we introduced homogenization as a method of computing the endpoints of diverging paths. The payoff is that

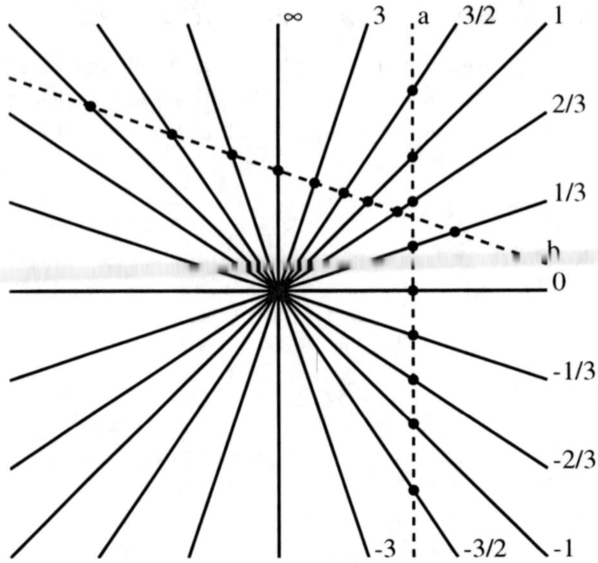

Figure 4.1. *The real projective line, $\mathbb{P}^1_{\mathbb{R}}$, with two patches.*

homogenization tends to make paths shorter, reducing the number of steps used to track the paths even before `SecurityMaxNorm` is exceeded.

To summarize, when only one variable group, say, $x_1,\ldots,x_N$, is declared, a polynomial $f(x_1,\ldots,x_N)$ of degree d is homogenized as

$$\hat{f}(X_0,X_1,\ldots,X_n) = X_0^d f(X_1/X_0,\ldots,X_n/X_0). \tag{4.9}$$

Any solution $[X_0,\ldots,X_n]$ to $\hat{f} = 0$ with $X_0 \neq 0$ is said to be finite, and it is dehomogenized as

$$(x_1,\ldots,x_N) = (X_1/X_0,\ldots,X_n/X_0). \tag{4.10}$$

4.4 ▪ Calculating Infinity: Projective Patches

The elements of projective space $\mathbb{P}^N$ are lines through the origin in $\mathbb{C}^{N+1}$. The equations do not determine unique homogeneous coordinates; just their ratios matter. Since the computation proceeds in $\mathbb{C}^{N+1}$, a question arises: how shall we pick out a unique point on the solution lines?

Although we generally work in complex space, it is illustrative to consider first the real projective line, $\mathbb{P}^1_{\mathbb{R}}$, as this is easier to visualize. Figure 4.1 shows the elements of $\mathbb{P}^1_{\mathbb{R}}$ as lines through the origin in $\mathbb{R}^2$, a few of which are drawn. The real numbers, $\mathbb{R}$, are in one-to-one correspondence with the slopes of lines in $\mathbb{R}^2$. The space $\mathbb{P}^1_{\mathbb{R}}$ is completed by adding a single vertical line, labeled ∞ in the figure. (Notice that $+\infty$ and $-\infty$ are equal in this compactification of the real line.) The equivalence of points on the real line to slopes of lines in the plane is already encoded into our treatment of homogeneous coordinates, where $[X_0,X_1]$ represents the line through the origin and point (X_0,X_1). The dehomogenization of $[X_0,X_1]$, $x = X_1/X_0$, is the slope of the line.

One simple approach to computing on projective space, the one used in Bertini, is to work on a random linear patch. This idea was first introduced in the context of polynomial continuation in [130]. For $[X_0, X_1, \ldots, X_N] \in \mathbb{P}^N$, the corresponding line in $\mathbb{C}^{N+1}$ is $(y_0, y_1, \ldots, y_N) = (\alpha X_0, \alpha X_1, \ldots, \alpha X_N)$, $\alpha \in \mathbb{C}$. For almost all such lines, the hyperplane

$$a_0 y_0 + a_1 y_1 + \cdots + a_N y_N = 1 \tag{4.11}$$

cuts the line in a unique point

$$(y_0, \ldots, y_N) = (X_0, \ldots, X_N)/(a_0 X_0 + \cdots + a_N X_N). \tag{4.12}$$

The exceptions are the points $[X_0, \ldots, X_N]$ such that

$$a_0 X_0 + \cdots + a_N X_N = 0. \tag{4.13}$$

If $[X_0, \ldots, X_N]$ are homogeneous coordinates obtained by homogenizing $f(x)$ for $x = (x_1, \ldots, x_N) \in \mathbb{C}^N$, the patch $X_0 = 1$ is equivalent to working in the original coordinates, and $X_0 = 0$ is the equation for the *hyperplane at infinity*. By homogenizing the system and picking a different patch, say, one where $a_0, \ldots, a_N$ are "general," points at infinity on the original patch act like finite points on the new patch, while the hyperplane (4.13) is now mapped to infinity.

The use of patches is also illustrated in Figure 4.1, where two patches are indicated by dashed lines. For each patch, the representation of an element in $\mathbb{P}^1_{\mathbb{R}}$—a line through the origin—is indicated by a dot where the patch intersects the element. The standard patch, $X_0 = 1$, labeled "a," gives $x = X_1/X_0 = X_1$. Accordingly, this patch includes all $x \in \mathbb{R}$ but misses ∞. A second patch, labeled "b," represents ∞ as a finite point but instead misses the element whose slope is equal to the patch's slope. Each possible patch gives a finite representation of all but one element of $\mathbb{P}^1_{\mathbb{R}}$. The essential fact is that for any finite set of points in $\mathbb{P}^1_{\mathbb{R}}$, a random patch will represent all the points in the set by finite numbers, even if one of the points in $\mathbb{P}^1_{\mathbb{R}}$ is ∞. The equivalent fact is true for patches on complex projective space in any dimension.

During a computation, it is feasible to switch patches whenever necessary to keep the solution path finite. For example, one could always make the homogeneous coordinate of largest magnitude equal to one. However, allowing switching of patches makes certain numerical processes more difficult. For example, multistep predictor-corrector path tracking algorithms do not expect the equations to suddenly change midway, so every patch switch would require some sort of remediation to reconstruct the history used to predict the next point on the path.

Clearly a computer code would be simpler if a single patch would suffice, and indeed, a random patch does suffice. Hence, this is the option used in Bertini. To see that this is valid, consider that in any homotopy we are seeking to find a finite number of solution points. Thus, with probability one, a random selection of $(a_0, \ldots, a_n)$ will not satisfy (4.13) for any of the points we seek to compute. Furthermore, since (4.13) is a complex equation, equivalent to two real equations, a finite number of one-real-dimensional homotopy paths also miss this exceptional hyperplane with probability one. With the use of multiple precision, even the occasional path that strays extremely near the transformed hyperplane at infinity can still be seen to be finite and can be tracked to its endpoint.

By selecting a random patch with coefficients of magnitude approximately 1, the homotopy paths tracked in $\mathbb{C}^{N+1}$ are normalized to approximate unit magnitude by the mapping (4.12). (In Bertini, the vector of patch coefficients, $(a_0, \ldots, a_N)$, is chosen to have unit length: $||a|| = 1$.) This keeps the homotopy paths short for the vast majority of cases, although a few that stray near the transformed hyperplane at infinity (4.13) can become long.

It can be important for a user to understand these inner workings to make choices in the trade-off between computational speed and security. For speed, Bertini defaults to `SecurityLevel:0`, which allows the endgame to abort early on large endpoint estimates. For some problems, one might expect large finite answers, in which case `SecurityMaxNorm` may need to be increased beyond its default value. Alternatively, one may select `SecurityLevel:1`, thereby disallowing early termination altogether. The advantage of allowing early termination lies mainly in the fact that often solutions at infinity are singular. Forcing the code to compute these accurately, while providing the highest level of security, also incurs a high computational cost.

4.5 ▪ Multihomogeneous Structures

Sometimes it is advantageous to partition the variables into several different groups and homogenize each group independently, forming a multihomogeneous structure. The choice of a multihomogeneous structure can affect the number of solution paths in the corresponding homotopy, a phenomenon that a Bertini user can use to decrease the computational work required to solve a system. The topic of homotopy types will be taken up more fully in the next chapter.

We already saw in the eigenvalue-eigenvector problems of Examples 4.1 and 4.2 that it was natural to put the variables into two groups: the eigenvector components in one group and the eigenvalue(s) in another. If these problems were treated by placing all variables in one group, then every polynomial in the system of equations would be treated as if it were a general quadratic instead of being bilinear. For an $N \times N$ matrix, this leads to a total degree of 2^N, whereas it is well known that the number of solutions is just N. By splitting the variables into two sets, the bilinear nature of the equations is recognized. It can be shown that a system of N bilinear equations on $\mathbb{P}^1 \times \mathbb{P}^{N-1}$ has at most N solutions. So the eigenvalue problems are just a special case of this general phenomenon.

In the eigenvalue problems, the eigenvector naturally lives on projective space, so it is clear that the eigenvector components should be declared as a separate homogeneous group. The choice of groupings is not always so clear. It is desirable to choose a grouping that reduces the computational work in solving the system, and in particular the choice generally affects the number of homotopy paths that must be tracked. Moreover, it often happens that many paths to infinity have endpoints that are singular roots of the homogenized system. Consequently, a side effect of reducing the number of homotopy paths that diverge to infinity is the elimination of singular endpoints that are difficult to compute to high numerical precision. If one is being careful, such as when using `SecurityLevel:1`, these singular endpoints can take a lot of CPU time, only to find that the endpoint is at infinity.

Example 4.4 (two hyperbolas). Consider the system

$$f(x,y)=\begin{bmatrix} xy-1 \\ xy+x-y-1 \end{bmatrix}=0, \tag{4.14}$$

which is the intersection of two hyperbolas whose asymptotes align with the x and y axes. By using the first equation to eliminate the term xy from the second, one sees that this system has exactly two finite roots: $(1,1)$ and $(-1,-1)$. Let's explore how different groupings of the variables affect the number of solutions after homogenization. With two variables, there are only two possible groupings: $\{(x,y)\}$ and $\{(x),(y)\}$.

Placing both x and y in the same group and homogenizing gives the system on $\mathbb{P}^2$ as

$$F_1 = \begin{bmatrix} XY - Z^2 \\ XY + XZ - YZ - Z^2 \end{bmatrix} = 0, \tag{4.15}$$

which has an additional two solutions at infinity, $[X,Y,Z] = [0,1,0]$ and $[1,0,0]$, for a total of four solutions. This means that in $\mathbb{P}^2$ the asymptotes of the hyperbolas meet at infinity once in the x direction and once in the y direction.

In contrast, 2-homogenizing by using the substitutions $x = X/U$, $y = Y/V$ gives, after clearing denominators, the system on $\mathbb{P}^1 \times \mathbb{P}^1$ as

$$F_2 = \begin{bmatrix} XY - UV \\ XY + XV - YU - UV \end{bmatrix} = 0. \tag{4.16}$$

This system has no solutions at infinity, which can be verified by checking each possibility. For example, if $U = 0$, then $X \neq 0$ (remember, $[U,X] = [0,0]$ is not allowed in projective space), so the first polynomial implies $Y = 0$. But substituting this into the second polynomial gives $XV = 0$, which has no solutions since $[V,Y] = [0,0]$ is not allowed either. Hence, this system has only the two finite solutions, now represented as $([U,X],[V,Y]) = ([1,1],[1,1])$ and $([1,-1],[1,-1])$. We have wrapped up infinity in a different way, and the asymptotes of the hyperbolas no longer meet. ∎

From the previous example, it should be clear how to multihomogenize a polynomial. Any group of N variables is replaced by ratios involving $N+1$ variables as $(x_1,\dots,x_N) = (X_1/X_0,\dots,X_N/X_0)$, $(y_1,\dots,y_N) = (Y_1/Y_0,\dots,Y_N/Y_0)$, etc. Afterward, one multiplies by the minimal powers of X_0, Y_0, etc., that clear the denominators. Doing the equivalent of this for each variable group gives a polynomial that is homogeneous in each group independently.

The possibility of using multihomogeneous structures to reduce the number of paths in a homotopy is discussed further in the next chapter.

4.6 ▪ Numerical Accuracy*

Suppose that an affine coordinate x is computed as $x = U/V$, where U and V are homogeneous coordinates on some projective patch. It is important to understand how the accuracy of computing U and V on the patch affects the accuracy of x. This is particularly true for understanding numerical accuracy in Bertini, since except for some user-defined homotopies, it computes all variables via projective patches.

Suppose then that U is computed as $u \in \mathbb{C}$ with absolute accuracy $r_u \in \mathbb{R}^+$, and similarly, V is computed as $v \in \mathbb{C}$ with absolute accuracy $r_v \in \mathbb{R}^+$. What is the accuracy of the value for x computed as $\hat{x} = u/v$?

A few notational conveniences make the discussion simpler:

- We shall use disks in the complex plane to model inaccuracy, so we define $\Delta_r(c)$, $c \in \mathbb{C}$, $r \in \mathbb{R}^+$ to be the complex disk with center c and nonzero radius r: $\Delta_r(c) := \{x \in \mathbb{C} \mid |x - c| \leq r\}$.

- For two subsets of the complex plane, $A, B \subset \mathbb{C}$, define their Minkowski product as the set
$$A \otimes B = \{ab \mid a \in A, b \in B\}.$$

- Similarly, define the Minkowski quotient of A and B as

$$A \oslash B = \{a/b \mid a \in A, b \in B\}.$$

With this notation, the question above can be restated as follows. If U could be anywhere in $\Delta_{r_u}(u)$ and V anywhere in $\Delta_{r_v}(v)$, then what is the maximum possible distance between u/v and U/V? That is, letting $Q := \Delta_{r_u}(u) \oslash \Delta_{r_v}(v)$, what is

$$\max_{q\in Q} |q-(u/v)|?$$

We state the answer as the following proposition.

Proposition 4.5. *Let* $Q = \Delta_{r_u}(u) \oslash \Delta_{r_v}(v)$ *and*

$$e = \max_{q\in Q} |q-(u/v)|.$$

If $|v| \le r_v$, *then* $e = \infty$; *otherwise*

$$e = \frac{1}{|v|^2 - r_v^2}\left(\frac{|u|}{|v|} r_v^2 + |v| r_u + |u| r_v + r_u r_v\right). \tag{4.17}$$

Proof. If $|v| \le r_v$, then $\Delta_{r_v}(v)$ includes the origin, and Q is infinite in extent. This establishes the first claim of the proposition. Otherwise, $\Delta_{r_v}(v)$ does not include the origin, and so Q is a finite region. In the case that $u = 0$, then $u/v = 0$, and e is the point of Q of maximum absolute value. The largest $|a|$ for $a \in \Delta_{r_u}(u)$ is r_u and the smallest $|b|$ for $b \in \Delta_{r_v}(v)$ is $|v| - r_v$, hence $e = r_u/(|v| - r_v)$, which agrees with (4.17). In the final case, we may rewrite

$$Q = (u/v)\Delta_{r_u/|u|}(1) \oslash \Delta_{r_v/|v|}(1).$$

It is well known that any inversion of a circle is a circle or a line, and in particular, the reciprocal of a circle that does not pass through the origin is a circle [38, Chapter 5]. More precisely, for $\rho < 1$,

$$1 \oslash \Delta_\rho(1) = (1/(1-\rho^2))\Delta_\rho(1),$$

so we have

$$Q = (u/v)\frac{|v|^2}{|v|^2 - r_v^2}\Delta_{r_u/|u|}(1) \otimes \Delta_{r_v/|v|}(1).$$

Let

$$E = Q - (u/v) = (u/v)\frac{|v|^2}{|v|^2 - r_v^2}\left(\Delta_{r_u/|u|}(1) \otimes \Delta_{r_v/|v|}(1) - (1 - r_v^2/|v|^2)\right).$$

Then, $e = \max_{y\in E}|y|$. The region $\Delta_{r_1}(1) \otimes \Delta_{r_2}(1)$ is bounded by a quartic curve known as a Cartesian oval [55]. It can be shown that the point on the curve farthest from any point in the segment $[0,1]$ is the point $1 + r_1 + r_2 + r_1 r_2$; hence for $\alpha \in [0,1]$ the largest distance from $1-\alpha$ is $\alpha + r_1 + r_2 + r_1 r_2$. Including the scale factor appearing in the definition of E, we have $e = |u/v|\frac{|v|^2}{|v|^2 - r_v^2}(\alpha + r_1 + r_2 + r_1 r_2)$, where $\alpha = r_v^2/|v|^2$, $r_1 = r_u/|u|$, and $r_2 = r_v/|v|$. A bit of simplification leads to (4.17). $\square$

The proposition gives an upper bound on the error. In general, this value is attained by only one point in $\Delta_{r_u}(u) \oslash \Delta_{r_v}(v)$, so it tends to be a pessimistic estimate of the true error. It is even more so if the error radii, r_u and r_v, are conservative estimates.

Once we have computed homogeneous coordinates U, V to known (or estimated) accuracy, Proposition 4.5 tells us how to estimate the accuracy of the dehomogenized affine coordinate $x = U/V$. When a projective patch is chosen at random, it tends to have a normalizing affect on the computed homogeneous coordinates: we expect the largest of U and V to be approximately unit magnitude. (See §4.7.) Suppose that U and V are both computed to the same absolute accuracy, $r_u = r_v = \epsilon \ll 1$. A few special cases illustrate the expected results for the accuracy of $x = U/V$:

- If $|x| \ll 1$, we expect $|u| \approx |x| \ll 1$ and $|v| \approx 1$. Then, Proposition 4.5 gives $e \approx \epsilon$, so ϵ is a good estimate of the absolute accuracy of x.
- If $|x| \approx 1$, we expect $|u| \approx 1$ and $|v| \approx 1$. Then the proposition gives $e \approx 2\epsilon$.
- If x is a large, finite root, $|x| \gg 1$, then we expect $|u| \approx 1$, $|v| \ll 1$. Assume that x is not too large, so $|x| \ll 1/\epsilon$, $|v| \approx 1/|x| \gg \epsilon$. Then, we get $e \approx |x|^2\epsilon$. This means that the relative accuracy of the large root is $e/|x| \approx |x|\epsilon$. Accordingly, if $|x|$ is on the order of 10^a and ϵ is on the order of 10^{-b}, $b > a$, then we expect about $b - a$ digits of accuracy in x.
- For $|x|$ on the order of $1/\epsilon$ or larger, we cannot expect any digits of accuracy. In fact, if $|v| \approx 1/|x|$ drops below ϵ, Proposition 4.5 indicates that the answer may be meaningless ($e = \infty$). When homogeneous coordinates are calculated to accuracy ϵ, we must expect that finite roots of magnitude $1/\epsilon$ or larger will be indistinguishable from true roots at infinity.

Accuracy in Bertini

In Bertini, except for the case of some user-defined homotopies, all affine variables are computed via homogeneous coordinates. Bertini attempts to compute the homogeneous coordinates to absolute accuracy `FinalTol`, and when adaptive precision is enabled, it allocates extra digits when needed to achieve this. Accordingly, the discussion above applies with $\epsilon \approx$ `FinalTol`. Here, we give additional details on how accuracy is treated in Bertini.

Bertini reports two accuracy estimates in the file `main_data`. One is labeled as

```
Accuracy estimate, internal coordinates (difference of last two
endpoint estimates):
```

This is measured using the coordinates of the underlying homotopy, either the variables for a user-defined homotopy in affine coordinates (see §7.5.1) or, in all other cases, the homogeneous coordinates of the homogenized system. The estimate is the magnitude of the largest difference in any coordinate over the last two estimates of the path endpoint at $t = 0$. (These are obtained using an *endgame*. See §3.3.) Let us call this estimate ϵ_1, which Bertini tries to make less than `FinalTol`. Because the endgame is expected to be converging on the solution, the error estimated is approximately the error in the penultimate estimate of the endpoint and is a pessimistic estimate of the error in the final endpoint estimate.

The second accuracy estimate is for the user's original coordinates, labeled as

```
Accuracy estimate, user's coordinates (after dehomogenization,
if applicable):
```

For a user-defined homotopy in affine coordinates, this is exactly the same as the first accuracy estimate. Otherwise, the second accuracy estimate is based on Proposition 4.5. In the case of a `hom_variable_group`, Bertini divides through by the coordinate of largest magnitude; in the case of `variable_group`, the affine coordinates are all found by division in the usual way. In either case, Proposition 4.5 applies, and Bertini uses $r_u = r_v = \epsilon_1$, where ϵ_1 is the accuracy estimate for the homogeneous coordinates, as above. In accordance with Proposition 4.5, solutions at infinity will usually be assigned an accuracy in the user's coordinates of "`infinity`." This is not a failure; it is an appropriate recognition that any finite value reported for the root, no matter how large, is still infinitely far from the true root. If a finite root is too large compared to the reciprocal of `FinalTol`, its error estimate may be very large or declared infinite. The user may need to rescale the problem or change the setting of `FinalTol` to obtain the desired accuracy in both small and large finite roots.

Because ϵ_1 is already a pessimistic error estimate for the internal coordinates, the estimate of the accuracy in the user's coordinates is doubly pessimistic. For example, rerunning the 2×2 eigenvalue problem of Example 4.1 with a smaller `FinalTol`, say, `1e-20`, reveals that the true errors in the solutions reported earlier are about 1000 times less than the reported accuracy estimate.

4.7 ▪ Normalization Effect of Homogenization*

In §4.6, we claimed that the projective transformation, consisting of homogenization followed by the selection of a random projective patch, has a normalizing effect on the homogeneous coordinates. This is desirable for two reasons:

- it tends to make the homotopy path leading to the solution shorter, and
- it tends to make numerical calculations more stable.

In this section, we take a closer look at the normalization claim by computing the probability distribution of the two-norm of the solution after transformation. In this section, all norms are the usual two-norm, $||y|| := \sqrt{y^H y}$, where we treat $y \in \mathbb{C}^M$ as a column vector, and y^H is its Hermitian (complex-conjugate) transpose.

The projective transformation is described in §4.3 and §4.4. Let us suppose that the random patch (4.11) is chosen with a coefficient vector that is unit length in the two-norm; that is, $a = (a_0, \ldots, a_N) \in \mathbb{C}^N$ satisfies

$$a^H a = \mathrm{conj}(a_0)a_0 + \cdots + \mathrm{conj}(a_N)a_N = 1. \tag{4.18}$$

We may call this a unitary projective patch. Since each coefficient has a real and an imaginary part, (4.18) describes the unit hypersphere in $\mathbb{R}^{2N+2}$, which we denote as S_{2N+2}, using the geometer's convention.[16]

Suppose we have $x \in \mathbb{P}^N$ given by homogeneous coordinates $[x_0, \ldots, x_N] \neq 0$. This is equivalent to the line in $\mathbb{C}^{N+1}$ given by $\{y \mid y = \alpha x, \alpha \in \mathbb{C}\}$. The projective patch $a^T y = 1$ intersects this line at the point $y = x/(a^T x)$. By imposing the patch equation, we select a unique representation of x along the line $y = \alpha x$. We see that $||y||$ is independent of $||x||$ but dependent on the direction of a. The most favorable directions are $a = \pm\mathrm{conj}(x)/||x||$, in which case $||y|| = 1$. The most unfavorable directions are the ones orthogonal to x, namely, a such that $a^T x = 0$. For these, x lies in the hyperplane at infinity for the patch,

[16]Topologists denote this as S_{2N+1} instead, as it is a $(2N+1)$-dimensional hypersurface.

and $||y|| = \infty$. For a fixed x, the plane $a^T x = 0$ intersects the hypersphere $a^H a = 1$ in a $2N$-dimensional sphere, the "great hypersphere orthogonal to x."

Suppose x is fixed and a is chosen uniformly random on S_{2N+2}. Then, there is a zero probability that a will fall on the great hypersphere orthogonal to x, but it might happen to be close enough to make $||y||$ large. This concern raises the following question:

> For a random unitary projective patch, as above, what is the probability that $||y|| \geq \rho$ for $\rho \in \mathbb{R}^+$?

We know that $\text{Prob}(||y|| \geq 1) = 1$, since 1 is the smallest possible value. We shall not provide a full answer to the question posed but instead estimate the probability for large ρ. We shall show that

$$\text{Prob}(||y|| \geq \rho) \approx N/\rho^2, \qquad \rho \gg 1. \tag{4.19}$$

To simplify the derivation, we note that since a is uniformly random over all unitary vectors, the probability is independent of x and we may assume $x = [1, 0, \ldots, 0]$. So $\text{Prob}(||y|| \geq \rho) = \text{Prob}(|a_0| \leq 1/\rho)$, which, writing out the real and imaginary parts of a_0, becomes $\text{Prob}((\Re(a_0)^2 + \Im(a_0)^2)^{1/2} \leq 1/\rho)$. For the circle $|a_0| = r$, the real and imaginary parts of the remaining elements of a, $(a_1, \ldots, a_N)$, range over the hypersphere in $\mathbb{R}^{2N}$ of radius $\sqrt{1-r^2}$. For small r, this is approximately the unit hypersphere S_{2N}. Hence, as a_0 ranges over the entire disk of radius $1/\rho$, $\rho \gg 1$, to good approximation the remaining elements of a independently range over S_{2N}. The percentage of S_{2N+2} that falls into this region is the probability we seek, so

$$\text{Prob}(||y|| \geq \rho) \approx \frac{(\pi/\rho^2)\text{Area}(S_{2N})}{\text{Area}(S_{2N+2})}. \tag{4.20}$$

It happens that the areas of unit hyperspheres obey the recursion

$$\text{Area}(S_{k+2}) = \frac{2\pi}{k}\text{Area}(S_k),$$

which when substituted into (4.20) gives (4.19).

The result shows that using a random patch, the probability of a root in $\mathbb{P}^1$ being represented by homogeneous coordinates with norm as large as 1000 is about 1 in a million. For larger projective spaces, say, $\mathbb{P}^N$, the probability rises proportional to N, a small effect. There are, however, two effects that can substantially increase the probability of encountering large homogeneous coordinates. First, a problem could have many solutions. Second, in a homotopy algorithm, some paths could stray near the hyperplane at infinity somewhere mid path, between $t = 1$ and $t = 0$. However, in Bertini, adaptive precision is available to handle these occurrences, and since they are relatively rare, this maintains the reliability of the algorithms at a relatively small increase in the total computational cost. Because of this, Bertini chooses one random patch at the outset of each run and does not change it.

An alternative would be to adjust the patch whenever the transformed coordinates y grow large. As mentioned above, the optimal choice would be $a = \text{conj}(x)/|x|$ so that $||y|| = 1$: the transformed coordinates have two-norm equal to one. However, multistep prediction methods for path tracking and most endgame processes need to maintain the same patch over successive steps, so any change of patch causes a restart to the process. For simplicity of the underlying computer program, it is advantageous to eschew patch changes altogether.

Chapter 5

Types of Homotopies

There are several types of homotopy functions that can be usefully employed to solve polynomial systems. The most basic among these is the *total-degree homotopy*, discussed in §2.1.4. Depending on the structure of the polynomial system at hand, other homotopy types are often more efficient. In some cases, the difference in the number of solution paths to be tracked can be extreme.

The most obvious structural characteristic of a polynomial system is the set of monomials that appear in each polynomial. As discussed in §1.1.1, for a degree d and a number of variables n, there are $\binom{d+n}{n}$ possible monomials of degree at most d. Polynomials that omit most of these monomials are said to be *sparse*. For the family of all polynomial systems that have a given monomial structure, there exists a remarkable theory that gives the maximum number of isolated roots that any member of the family can have. From this theory, one may construct a *polyhedral homotopy* for which the number of solution paths is exactly this maximal number. In general, finding the exact maximal root count for a given sparse structure and setting up a compatible homotopy is a combinatorial process with a high computational complexity.

Fortunately, there is a middle ground between total-degree homotopy and polyhedral homotopy where some sparsity can be exploited without excessive combinatorics. In this middle ground lie *multihomogeneous homotopies* and *linear product homotopies*. Although Bertini implements only multihomogeneous homotopies and its special case, a total-degree homotopy, for completeness we describe all these: multihomogeneous in §5.1, linear product in §5.2, and polyhedral in §5.3. We point out resources where implementations of all the sparse homotopies can be found.

The approaches based on sparsity do not capture all the structure that may be present in a system. In particular, there may be interrelationships between coefficients in the system; that is, the coefficients might lie in some algebraic set for which the number of isolated roots is smaller than the sparse system theory predicts. There is very little general theory available for detecting this sort of structure a priori, and yet there are methods for exposing the structure in the process of solving the system.

Unlike the traditional homotopies, which find all isolated solutions in one shot, *equation-by-equation* approaches work through the polynomials in a sequential fashion, potentially discovering structurally induced simplifications as they go. We will treat one of these in detail here: the *regeneration* procedure in §5.4.

In applications, systems often arise in families where certain physical values—the length of a robot link, the temperature of a isothermal chemical reaction, etc.—are assumed

Table 5.1. *Illustration of the growth of Bell(N) as N grows.*

N	1	2	3	4	5	6	7	8
Bell(N)	1	2	5	15	52	203	877	4140
N	9	10	11	12				
Bell(N)	21,147	115,975	678,570	4,213,597				
N	25							
Bell(N)	4,638,590,332,229,999,353							

known but one might wish to solve such a system several times for different values of these. In the next chapter, we will meet the *parameter homotopy*, which enables efficient solution of the second case, the third case, and so on, if one can only manage to solve a first general example from a family. But consider this: the number of paths to track in a total-degree homotopy is equal to the product of the degrees of the equations, so a system of N quadratics in N unknowns has 2^N paths in the total-degree homotopy. Due to this exponential growth, the solution of a first example may become too computationally expensive to execute even once. Hence, it can be important to have alternatives to the total-degree approach. We begin with the easiest generalization: the multihomogeneous homotopy.

5.1 ▪ Multihomogeneous Homotopy

The multihomogeneous homotopy is easy to apply. The user need only partition the list of variables into several groups. In Bertini, this is accomplished with `variable_group` and `hom_variable_group` declaration statements. The latter is relevant only when the group of variables appears homogeneously in the polynomials, such as the examples we saw in §4.1. Putting all variables into a single group yields the total-degree homotopy. Dividing the variables into groups may significantly increase or decrease the number of homotopy paths, depending on whether the groups are compatible with the sparse monomial structure of the target system, if indeed such sparsity is present.

Bertini does not automatically choose variable groups: this is the user's responsibility. The objective in picking groups is to minimize the root count of the start system of the homotopy, which is the same as the number of paths in the homotopy. Unfortunately, there does not yet exist a truly efficient algorithm for finding optimal groupings, and the combinatorics are such that an exhaustive examination of all possible groupings becomes impractical as the number of variables grows much larger than 10 [178]. The number of partitions of N items is the Nth Bell number. The rapid growth rate of the Bell numbers is indicated in Table 5.1.

When an exhaustive search for the optimal subdivision is intractable, keeping a few rules of thumb in mind and using familiarity with the derivation of the polynomial system, one can often arrive in short order at a subdivision that is good enough, if not optimal, allowing the homotopy to complete in an acceptable time. Moreover, there exist heuristic methods for arriving at good multihomogeneous groupings [170].

A subdivision of the variables is advantageous only if it lowers the degree of the polynomials when considered with respect to the groups taken independently. For example, the cubic $x^2z+xyz+1$ is quadratic in (x,y) and linear in z, so a partition of the variables as $\{(x,y),(z)\}$ *could* be advantageous. If all three polynomials in a system of three polynomials in $\mathbb{C}[x,y,z]$ had this same degree structure, then it would certainly be an advantage to partition the variables in this manner. The situation becomes more complicated when a subdivision reduces degrees in some polynomials of the system but increases them in others.

5.1.1 ▪ Multihomogeneous Bézout Number

The number of paths in a multihomogeneous homotopy, also called the *multihomogeneous Bézout number*[17] of the system, is closely tied to multihomogeneous projective spaces mentioned in §4.5. To decide which grouping of the variables to use, one may easily compute the multihomogeneous Bézout number of each grouping under consideration and pick the smallest one. As mentioned, the total number of possible groupings grows quickly with the number of variables, so Bertini does not automate an exhaustive search. But with a little practice, a user can identify and check the most promising candidates.

Instead of introducing a detailed notation for expressing the multihomogeneous Bézout number, let us demonstrate the idea through examples. Let's start with the system

$$xy - 1 = 0, \quad x^2 - 1 = 0. \tag{5.1}$$

There are two possible groupings, $\{(x,y)\}$ and $\{(x),(y)\}$, each with its own Bézout number. For each grouping, we make a multidegree table with one column per group and one row per equation, as follows. For the grouping $\{(x,y)\}$, the table is

	(x,y)
$xy-1$	2
x^2-1	2

since both equations have degree 2, while for $\{(x),(y)\}$, we have

	(x)	(y)
$xy-1$	1	1
x^2-1	2	0

indicating the degrees of the polynomials with respect to each variable group. In the latter case, we may say the polynomial $xy-1$ has *multidegree* $(1,1)$, or, more simply, the polynomial is type $(1,1)$, while x^2-1 is type $(2,0)$.

The rules for computing the Bézout number for a given degree table are as follows. As a preparatory step, replace the variable groups with the dimension of the subspace they live in. That is, $(x,y) \in \mathbb{C}^2$, so the first table becomes

	2
$xy-1$	2
x^2-1	2

while since $x \in \mathbb{C}$ and $y \in \mathbb{C}$, the second one becomes

	1	1
$xy-1$	1	1
x^2-1	2	0

[17] Sometimes we call the Bézout number simply the root count.

Before evaluating the multihomogeneous Bézout number for these and other examples, let us describe the general methodology. First, lay out the multidegree table for an m-homogeneous system as

$$\begin{array}{c|ccc} & k_1 & \cdots & k_m \\ \hline f_1 & d_{1,1} & \cdots & d_{1,m} \\ \vdots & \vdots & & \vdots \\ f_N & d_{N,1} & \cdots & d_{N,m} \end{array}$$

So k_j is the dimension of the jth group of variables ($\mathbb{C}^{k_j}$ or $\mathbb{P}^{k_j}$, as appropriate), and $d_{i,j}$ is the degree of the ith polynomial f_i with respect to the jth variable group. With this notation, the rules for evaluating the m-homogeneous Bézout number are as follows:

1. Find all ways of choosing N of the $d_{i,j}$ such that there is one chosen in each row and k_j chosen in each column j.
2. For each such choice, multiply the chosen degrees.
3. Sum these products over all the choices.

Let us return to the example above. In the 1-homogeneous case of $\{(x,y)\}$, there is only one combination of the multidegrees that is possible, so the Bézout number is $2 \cdot 2 = 4$, the total degree of the system. In the 2-homogeneous case of $\{(x),(y)\}$, there are two valid combinations of multidegrees,

$$\begin{array}{r|cc} & 1 & 1 \\ \hline xy-1 & 1 & 1 \\ x^2-1 & 2 & 0 \\ & 2 & 0 \end{array}$$

so summing the products, one gets a 2-homogeneous Bézout number of $1 \cdot 0 + 1 \cdot 2 = 2$.

For a second simple example, consider a 4×4 generalized eigenvalue problem, formulated on $[\lambda, \mu] \times v \in \mathbb{P}^1 \times \mathbb{P}^3$, as

$$(A\lambda + B\mu)v = 0, \quad A, B \in \mathbb{C}^{4\times 4}.$$

The corresponding multidegree table is

$[\lambda,\mu]$	$[v_0,v_1,v_2,v_3]$
1	1
1	1
1	1
1	1

or simply

1	3
1	1
1	1
1	1
1	1

where we replace variable names with the dimensions of the subspace they live on. (When using homogeneous coordinates, the dimension of the projective space is one less than the number of coordinates that appear. See Chapter 4 for details.) There are four ways to choose one element from the first column of degrees, which forces the choice of the remaining three elements from the second column, giving a Bézout number of $4(1 \cdot 1 \cdot 1 \cdot 1) = 4$, as expected, much lower than the 2^4 total degree of the system. It is easy to see that this approach generalizes to give a Bézout number of N for the $N \times N$ eigenvalue problem.

When several equations have the same multidegree, as in the eigenvalue problem, it can be convenient to compress the table by introducing a left column containing the number of repeats. In this format, the 4×4 eigenvalue problem's degree table becomes simply

	1	3
4×	1	1

which contains the same information as the longer version.

An equivalent way of evaluating the multihomogeneous Bézout number uses a polynomial expression to capture the combinatorics. Again, let us illustrate first on the system (5.1). We introduce a dummy variable associated to each group, say, α for (x) and β for (y), and form a new last column:

	1	1	$\alpha^1\beta^1$
$xy-1$	1	1	$(1\alpha+1\beta)$
x^2-1	2	0	$(2\alpha+0\beta)$

Here, the first row gives exponents of the monomial atop the column, whereas the multidegree columns give coefficients in the sums in their right-hand entries. From these expressions, the Bézout number is the coefficient of $\alpha^1\beta^1$ in the polynomial $(1\alpha+1\beta)(2\alpha+0\beta)$, which is 2, as we obtained before.

The rule is that the Bézout number is the coefficient of the monomial atop the column in the product of all the polynomial factors below it.

Applying the same method to the 4×4 eigenvalue problem gives

	1	3	$\alpha\beta^3$
4×	1	1	$(\alpha+\beta)^4$

By the binomial theorem, the coefficient of $\alpha\beta^3$ in $(\alpha+\beta)^4$ is $\binom{4}{1}=4$, the same Bézout number as we obtained before.

Example 5.1. Let's also revisit the intersection of two hyperbolas from Example 4.4, that is,

$$f(x,y)=\begin{bmatrix} xy-1 \\ xy+x-y-1 \end{bmatrix}=0.$$

Using either the method for directly summing terms from the degree table or the equivalent combinatorial polynomial, one sees that the 2-homogeneous Bézout number for this system is 2. ∎

Let's examine a more substantial example from robot kinematics.

Example 5.2. Consider the inverse kinematics of a general six-revolute (6R) serial-link robot as formulated in [181] and also treated in [161, §9.4]. The symbols appearing in the problem are $\mathbf{z}_1,\dots,\mathbf{z}_6 \in \mathbb{R}^3$ (unit vectors along the joint axes), a_1,a_2,a_3,a_4,a_5 (associated to link lengths), d_2,d_3,d_4,d_5 (link offset distances), c_1,c_2,c_3,c_4,c_5 (cosines of twist angles), and $\mathbf{p}\in\mathbb{R}^3$ (position of the endpoint). These quantities are known to roboticists as the Denavit–Hartenburg parameters of the robot.[18] In the problem, $\mathbf{z}_2,\mathbf{z}_3,\mathbf{z}_4,\mathbf{z}_5$ are unknown, and the rest are given. The solutions reveal all the ways the robot can reach a desired position and orientation.

[18] For roboticists: to be exact, $a_i=\hat{a}_i/s_i$, where $\hat{a}_i$ is the D-H link length and s_i is the sine of the twist angle.

The equations written using vector dot products and vector cross products are

$$\mathbf{z}_i \cdot \mathbf{z}_i - 1 = 0, \qquad i = 2,3,4,5, \tag{5.2}$$

$$\mathbf{z}_i \cdot \mathbf{z}_{i+1} - c_i = 0, \qquad i = 1,2,3,4,5, \tag{5.3}$$

$$\begin{aligned} a_1\mathbf{z}_1 \times \mathbf{z}_2 + d_2\mathbf{z}_2 + a_2\mathbf{z}_2 \times \mathbf{z}_3 + d_3\mathbf{z}_3 + a_3\mathbf{z}_3 \times \mathbf{z}_4 \\ + d_4\mathbf{z}_4 + a_4\mathbf{z}_4 \times \mathbf{z}_5 + d_5\mathbf{z}_5 + a_5\mathbf{z}_5 \times \mathbf{z}_6 - \mathbf{p} = 0. \end{aligned} \tag{5.4}$$

The last of these is a vector equation, so it has three scalar parts. Altogether, there are 12 equations in the 12 unknowns. Two of the polynomials, $\mathbf{z}_1 \cdot \mathbf{z}_2 - c_1$ and $\mathbf{z}_5 \cdot \mathbf{z}_6 - c_5$, are linear (because $\mathbf{z}_1$ and $\mathbf{z}_6$ are given), while the rest are all quadratic, so the total degree of the system is $2^{10} = 1024$.

There are more than 4 million possible groupings of 12 variables (see Bell(12) above), so an exhaustive search through these would be arduous. But it is clear from the vectorial form of the equations that any grouping that breaks up the three entries in $\mathbf{z}_2$ or any of the other joint vectors would be useless, since doing so would just increase the number of variable groups without reducing degrees. Keeping the entries of each unit vector together, we need to consider at most Bell(4) $= 15$ possible groupings. In fact, all we wish to do is break quadratic terms of the form xy into bilinear terms $(x)(y)$, which is accomplished only with the groupings $\{(\mathbf{z}_2,\mathbf{z}_4),(\mathbf{z}_3,\mathbf{z}_5)\}$ and $\{(\mathbf{z}_1),(\mathbf{z}_2),(\mathbf{z}_3),(\mathbf{z}_4)\}$. Let us first consider the former. To fit it more compactly on the page, this time we turn the degree table on its side:

	(5.2)				(5.3)					(5.4)		
$\#(\mathbf{z}_2,\mathbf{z}_4)=6$	2	0	2	0	1	1	1	1	0	1	1	1
$\#(\mathbf{z}_3,\mathbf{z}_5)=6$	0	2	0	2	0	1	1	1	1	1	1	1

In the six columns that contain an entry of 0, only one choice contributes to the Bézout number. We indicate this with boxes,

	(5.2)				(5.3)					(5.4)		
$\#(\mathbf{z}_2,\mathbf{z}_4)=6$	$\boxed{2}$	0	$\boxed{2}$	0	$\boxed{1}$	1	1	1	0	1	1	1
$\#(\mathbf{z}_3,\mathbf{z}_5)=6$	0	$\boxed{2}$	0	$\boxed{2}$	0	1	1	1	$\boxed{1}$	1	1	1

and note that there remain three choices in each row and all the remaining columns are type $(1,1)$. From this, one finds the root count to be $2^4\binom{6}{3} = 320$.

The other subdivision of the variables, $\{(\mathbf{z}_2),(\mathbf{z}_3),(\mathbf{z}_4),(\mathbf{z}_5)\}$, is disadvantageous because it turns the last equations, (5.4), into type (1,1,1,1). General polynomials of this type could have terms involving all four joint axis vectors, when the actual equations at hand involve only two at a time. As an exercise, the reader may wish to verify that the 4-homogeneous Bézout number is 576.

Regardless of which variable grouping is used, the number of finite solutions of the 6R problem for general parameters is only 16. As we shall see in the next chapter, this means that after solving one general example, subsequent 6R examples can be solved much more efficiently with a 16-path parameter homotopy. ∎

5.1.2 ▪ Degeneration to Linear Products

While one can use variable groupings advantageously without understanding the theory behind the multihomogeneous Bézout number, the underlying idea of *degeneration* illuminates many of the alternative homotopies that are also useful. A degeneration argument not only shows the proper multihomogeneous root count but it also provides a start system for a multihomogeneous homotopy to find all isolated roots.

Consider again the degree tables above. To the right of the degree table, we make a new version in which each entry on the left becomes an equal number of linear factors on the right. Moreover, the variables that appear in the linear factors are exactly the variable group from the table heading. It is easier to do than say, so let's revisit the basic system (5.1), whose augmented degree table now becomes

	(x)	(y)	(x)	(y)
$xy-1$	1	1	$(\square+\square x)$	$(\square+\square y)$
x^2-1	2	0	$(\square+\square x)(\square+\square x)$	1

The idea is that we insert a random coefficient from $\mathbb{C}$ into each box, $\square$. Then, the right-hand version of the table becomes a system of polynomials in $\mathbb{C}[x,y]$ that we call g. By this construction, the system g has the same degree table as the original system f. A particular instance of g, not using random numbers, is

$$\begin{aligned} (x-1)(y-2) &= 0, \\ (x-3)(x-4) &= 0. \end{aligned} \tag{5.5}$$

This system has two solutions, $(x,y) = (3,2)$ and $(x,y) = (4,2)$, the same number as the 2-homogeneous Bézout number for (5.1). The calculation of the roots follows the evaluation of the Bézout number from the degree table quite directly:

(x)	(y)	
$(x-1)$	$(y-2)$	$=0$
$(x-3)(x-4)$	1	$=0$
2 solutions	no solution	

It is clear that f and g have the same degree structure, and the set of all polynomial systems in $\mathbb{C}[x,y]$ that have the same multidegree table as f and g forms a family. In particular, all type $(1,1)$ polynomials can be written in the form

$$c_0 + c_1 x + c_2 y + c_3 xy,$$

and all type $(2,0)$ polynomials can be written in the form

$$c_4 + c_5 x + c_6 x^2,$$

so the family sharing the same degree table with f and g is parameterized by the seven coefficients $c_0,\dots,c_6$. For generic members of the family, i.e., for all choices of $c_0,\dots,c_6$ not on a particular proper algebraic subset of $\mathbb{C}^7$, the polynomial system of equations

$$F(x,y;c_0,\dots,c_6) = \begin{bmatrix} c_0 + c_1 x + c_2 y + c_3 xy \\ c_4 + c_5 x + c_6 x^2 \end{bmatrix} = 0$$

has two isolated roots in $\mathbb{C}^2$.

In the family F, there are some polynomial systems in which each equation can be factored into a product of linear expressions; in fact, g is one of these. Interpreting this geometrically, we may say that as we move continuously through the parameter space from a general member of F to one of these factorable members, the curve defined by each equation in $F=0$ *degenerates* into a union of lines. Computing the solutions to g is the same as finding the intersections of various pairs of lines.

Here is the crucial fact: For general choices of the coefficients appearing in the linear factors that form g, the system has the same number of roots as the most general systems in the family. But the factored systems are easy to solve, and for general coefficients all the roots are nonsingular! This makes g an ideal start system for a homotopy

$$H(x,y,t)=\gamma t\, g(x,y)+(1-t)f(x,y)=0, \tag{5.6}$$

where γ is a random, complex number. As t goes from 1 to 0, we follow the solution paths of H from the starting solutions of g to the solutions of the target system f. In essence, this homotopy reverses the degeneration to linear products described above in order to solve the system f, which usually cannot be factorized.

Of course, this would not be interesting if it were limited to the simple system at hand; the idea is useful because it holds in general. That is, for any square polynomial system, $f:\mathbb{C}^N\to\mathbb{C}^N$, when one forms a degree table and constructs g from matching linear products, for generic choices of the coefficients in the linear forms, $g=0$ has all isolated nonsingular solutions and the number of roots of g is the same as the number of roots for a general member of the family defined by the degree table.

5.1.3 ▪ Solving the System

Putting everything together, we obtain a multihomogeneous homotopy for solving polynomial systems. This is the procedure that Bertini follows in its default solving mode. The procedure linearly interpolates between the start system g and the target system f using the homotopy

$$H(x,t)=\gamma t\, g(x)+(1-t)f(x)=0. \tag{5.7}$$

In more detail, the procedure is as follows:

Given:

Polynomial system $f(x):\mathbb{C}^N\to\mathbb{C}^N$.
A partition of the variables $x_1,\dots,x_N$ into groups.

Procedure:

Form the multidegree table of f with respect to the variable groups.
Convert each row of the table into a matching product of linear forms.
Choose random, complex coefficients for the linear forms.
Call the system so formed as g.
Solve g to get the start solutions.
Form $H(x,t)$ as in (5.7).
Multihomogenize H and append patch equations (see §4.3).
Follow the solution paths as t goes from 1 to 0 in (5.7).
Use an endgame to compute the endpoint of each solution path at $t=0$.
Dehomogenize the endpoints.

Output:

The set of endpoints includes all isolated solutions of $f(x)=0$.

The number of paths in the homotopy is equal to the multihomogeneous Bézout number of the system.

In general, the target system f might not be a general member of the family of systems defined by its degree table. In fact, since the Bézout number usually changes for different variable groupings, the actual number of finite isolated roots of f is at most the smallest among these. Yet any of the grouping choices will give rise to a start system that can be used to find all the roots of f. For any homotopy having more paths than f has isolated solutions, the extra solution paths either diverge to infinity as $t \to 0$ or land on positive-dimensional solution components (if these exist).

Example 5.3. The following input file gives a two-path homotopy for the simple example we have been using: $\{xy-1, x^2-1\}=0$.

```
% TwoQuadrics_2hom.input
INPUT
 function f1, f2;
 variable_group x;
 variable_group y;
 f1 = x*y - 1;
 f2 = x^2 - 1;
END;
```

Note that if we modify this to place both variables in one group with the declaration

```
 variable_group x, y;
```

then one obtains a total-degree homotopy having four paths, two of which end at the roots of the system, while the other two diverge to infinity. ∎

5.1.4 ▪ Justification for the Multihomogeneous Homotopy*

One may well wonder why this is called a *multihomogeneous* homotopy, as neither f nor g is homogeneous. The reason is that in order to tame any paths diverging to infinity, we multihomogenize the systems and work on a projective patch for each variable group. (See Chapter 4.) Also, although we have not made any rigorous arguments here, the simplest justification for the fact that the number of roots of g is the same as the generic root count for the family defined by the multidegree table of f is one based on a multihomogeneous formulation. The 2-homogenized version of our running example is

	$[U,X]$	$[V,Y]$	$[U,X]$	$[V,Y]$
$XY-UV$	1	1	$(\square U+\square X)$	$(\square V+\square Y)$
X^2-U^2	2	0	$(\square U+\square X)(\square V+\square Y)$	1

Using the right-hand columns to form a 2-homogenized start system with random parameters in the empty boxes—call it $G([U,X],[V,Y])$—one may check that for each root of g in $\mathbb{C}\times\mathbb{C}$ there is a corresponding root of G in $\mathbb{P}^1\times\mathbb{P}^1$. Moreover, there are no roots of G at infinity. It is this last fact that lets us know that nothing was lost in the degeneration from general systems to g in factored form—no solutions diverged to infinity—so we get a full root count when we reverse the degeneration in the homotopy (5.6) and its generalization to N variables. For a rigorous justification of the multihomogeneous homotopy, see [132].

In the multihomogeneous homotopy, variables are divided into groups, and these groups are applied uniformly to all the polynomials in the system to determine the degree table from which the start system and homotopy are derived. As we saw in the previous section, the key feature of the multihomogeneous homotopy is that each polynomial in its start system is a product of linear forms. This makes the start system easy to solve. The validity of the multihomogeneous homotopy is due to the fact that degeneration from a general member of the family (defined by the degree table) to the most general ones that can be composed as products of linear forms does not lose any solution paths—no root diverges to infinity—so the process can be reversed in a homotopy to solve nonfactored systems.

5.2 ▪ Linear Product Homotopy*

The justification just given for the validity of the multihomogeneous homotopy begs the question of whether there might be other factored systems that could serve as a start system, possibly having a lower root count. It turns out that there are indeed other possibilities based on linear products. These are not provided in Bertini, but we discuss them as a natural step in the progression to the regeneration homotopy of §5.4, which is provided in Bertini.

The simplest generalization of the multihomogeneous homotopy is to allow the variable groups to change from one equation to the next. Then, each polynomial still has a well-defined multidegree and one can match this with a product of linear forms in the variable groups. We may call this a *partitioned linear product* homotopy, as each equation has its own partitioning of the variables into groups.

Since the variable groups change from equation to equation, there is no natural way to multihomogenize a partitioned linear product homotopy. Instead, to tame divergent solution paths, one may as well use a single homogenizing coordinate to cast the problem from $\mathbb{C}^N$ to $\mathbb{P}^N$.

Even more flexible are *general linear product* homotopies. Here, the rule is that the linear forms in the start system g must match the monomials appearing in the target system f in a particular way. As usual, instead of introducing new notation to present the general case, we illustrate the technique through an example. Suppose that one of the polynomials in f is formed using a subset of the monomials

$$1, x, y, x^2, xy, y^2, x^2y, xy^2. \tag{5.8}$$

Then a suitable matching g has the following form, a product of three linear forms, where as before each $\square$ is a placeholder for a generic complex number,

$$(\square + \square x)(\square + \square y)(\square + \square x + \square y). \tag{5.9}$$

If this expression is expanded, all the monomials in (5.8) appear. Compared to a general cubic in two variables, a polynomial formed with these monomials is missing x^3 and y^3, so a 1-homogeneous treatment (i.e., group $\{(x, y)\}$)

$$(\square + \square x + \square y)(\square + \square x + \square y)(\square + \square x + \square y) \tag{5.10}$$

does not model this sparseness. Likewise, in a 2-homogeneous treatment, $\{(x), (y)\}$, the expression

$$(\square + \square x)(\square + \square x)(\square + \square y)(\square + \square y) \tag{5.11}$$

includes the monomial x^2y^2 that is not present in (5.8). So (5.9) is the only product of linear forms that exactly models a polynomial having the monomials (5.8).

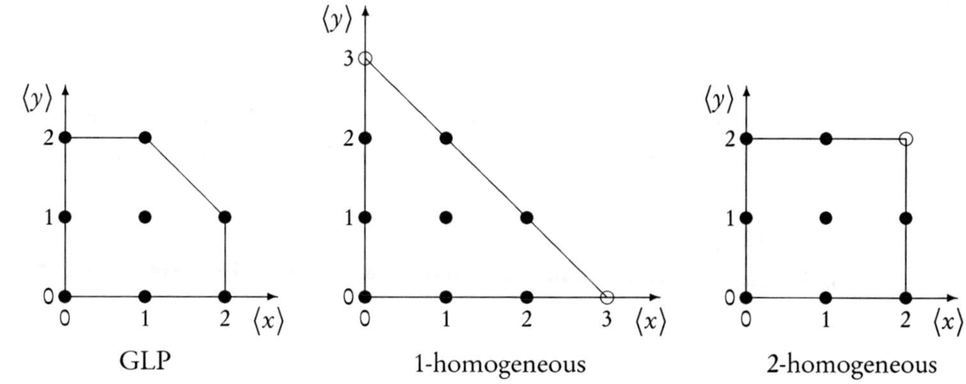

Figure 5.1. *Newton polygons for* (5.9), (5.10), (5.11).

The situation can be shown graphically, as follows. Given a polynomial in N variables, one may list the monomials and mark each one in an N-dimensional lattice according to the degrees of the variables. So x^2y becomes the point $(2,1)$, etc. The *Newton polytope* (or Newton polygon for $N = 2$, Newton polyhedron for $N = 3$) is the convex hull of these points. Accordingly, Figure 5.1 shows Newton polygons for the polynomial forms of (5.9), (5.10), (5.11), with open circles for the monomials that are not in (5.8). Only the general linear product matches the monomials (5.8) perfectly, so we would expect that degeneration to make the most efficient (least number of paths) homotopy for solving a system with a polynomial having just those monomials.

The factored form of (5.9) is allowed as a starting polynomial to match any target polynomial that has only the monomials (5.8). If this sort of relationship holds between each polynomial in the start system g and its corresponding entry in f, then the same linear homotopy as (5.6) still works for finding all the isolated roots of f. Theory justifying the general linear product homotopy can be found in [173]. (General linear products are also a special case of the polynomial-product homotopies developed in [138].) Essentially, we have the same situation as in the multihomogeneous homotopy: the roots for the start system are found by considering every possible choice of one linear factor from each equation. Some combinations are inconsistent, while others yield a nonsingular root. Either way there are no roots at infinity when the coefficients of the linear forms are generic, so the degeneration to a general linear product is without loss and can be reversed in the homotopy.

Example 5.4. Let us solve the system

$$f = \begin{bmatrix} 1 + 1.5y + 2x^2y + 5xy^2 \\ 3 + 0.4x + x^2 + 2.5y^2 + 7x^2y \end{bmatrix} = 0.$$

The total degree is $3 \cdot 3 = 9$, and the 2-homogeneous Bézout number is the coefficient of $\alpha\beta$ in $(2\alpha + 2\beta)^2$, which is 8. Using the general linear product form shown in (5.9) for both polynomials in the start system, say,

$$g = \begin{bmatrix} (x-1)(y-1)(x+y-1) \\ (x-2)(y-2)(x-y-2) \end{bmatrix} = 0,$$

we have seven start solutions, namely,

$$(x,y) = \{(1,2),(1,-1),(2,1),(3,1),(2,-1),(-1,2),(3/2,-1/2)\}.$$

(There are nine possible choices of one factor from each equation, but choosing the first factor in both is incompatible, as is choosing the second factor in both.) Accordingly, a general linear product homotopy can solve this problem by following only seven paths. As it turns out, f does in fact have seven roots. A 1-homogeneous or 2-homogeneous homotopy will also solve the problem, but with 2 or 1, respectively, extra path(s) diverging to infinity. ∎

In the example, both polynomials of f have been matched with the same form of general linear product, but this does not have to be the case. It also may happen that no general linear product homotopy perfectly matches a given f, so some extra solution paths have to be tracked.

Bertini does not currently offer partitioned linear products or general linear products; however, these are available in the packages POLSYS_PLP [185] and POLSYS_GLP [169], respectively. In Bertini, one would need to set up a user-defined homotopy, constructing and solving the start system by some other means. However, linear products turn up again in a moment in a homotopy that Bertini does provide: regeneration (§5.4).

5.3 ▪ Polyhedral Homotopy*

Although Bertini does not include an implementation of polyhedral homotopy methods, it is worth being aware of such methods as they are available elsewhere and can be effective, particularly for small systems or those with special structure.

Bernstein's theorem [33] relates the number of roots of a polynomial system to the Newton polytopes of its polynomials. Let $\mathrm{Vol}(P)$ mean the volume of polytope P. Let $P_1 + P_2$, the sum of two polytopes, mean the convex hull of the Minkowski sum of the polytopes, meaning

$$P_1 + P_2 = \mathrm{ConvexHull}(\{a + b \mid a \in P_1, b \in P_2\}).$$

Also, let $\mathbb{C}^* = \mathbb{C} \setminus 0$ (pronounced "Cee-Star"), the complex plane omitting the origin. Then, the number of isolated roots in $(\mathbb{C}^*)^N$ of a system $f = \{f_1, \ldots, f_N\} = 0$ of N polynomials in N variables with corresponding polytopes $\{P_1, \ldots, P_N\}$ is at most the coefficient of $\lambda_1 \cdots \lambda_N$ in the polynomial $\mathrm{Vol}(\lambda_1 P_1 + \cdots + \lambda_N P_N)$. This combinatorial quantity is called the *mixed volume* of the polynomials. Although it may not be obvious, when the vertices of the polytopes have integer coordinates, as they do for all Newton polytopes, the mixed volume is also an integer.

Bernstein's theorem is tight in the sense that if the polynomials have general coefficients, then the number of isolated roots is exactly the mixed volume (and all roots are nonsingular). This holds even if only the monomials corresponding to the vertices of the polytopes are general; monomials in the interior of any polytope can be omitted without changing the mixed volume. Of course, when we wish to solve a particular system, it may be that its coefficients are not general, but the mixed volume still gives an upper bound on the number of isolated roots.

A polyhedral homotopy is one that uses mixed volume computations to construct a start system so that the homotopy has exactly that number of solution paths to be tracked. Such a homotopy finds all the isolated roots in $(\mathbb{C}^*)^N$ of the system. Modified polyhedral homotopies find all isolated roots in $\mathbb{C}^N$, which can be larger in number.

The definition of the mixed volume should remind one of the combinatorial formula for the multihomogeneous Bézout number. In fact, it can be shown that the mixed volume of the start system for a multihomogeneous homotopy is exactly the same as its Bézout number. (The same is true for root counts coming from general linear products.)

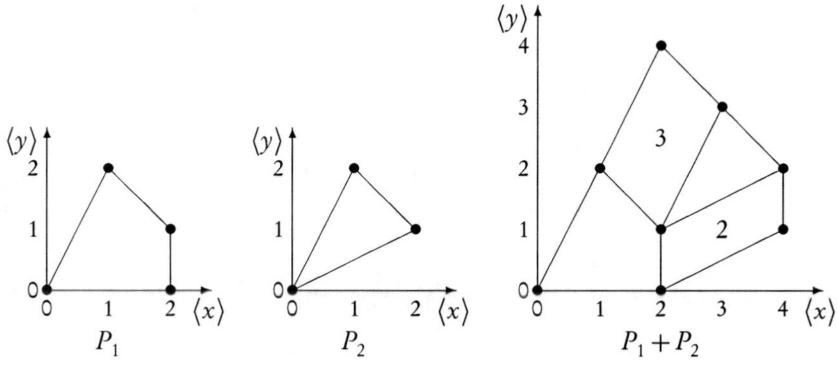

Figure 5.2. *Mixed volume for* (5.12).

In this sense, polyhedral homotopies generalize the linear product homotopies, and the number of paths in a polyhedral homotopy will never be greater than the best possible multihomogeneous or general linear product homotopy.

The exact methods of how to construct a polyhedral homotopy are outside the scope of this book. The foundational ideas can be found in [100, 174], the extension to $\mathbb{C}^N$ in [101], and good overviews in [119, 120]. The construction requires finding not just the total mixed volume but identifying all the "mixed cells" in $\lambda_1 P_1 + \cdots + \lambda_N P_N$, a hard combinatorial problem.

Example 5.5. For illustrative purposes, let us look at a system in just two variables:

$$f = \begin{bmatrix} f_1 \\ f_2 \end{bmatrix} = \begin{bmatrix} 1+3x^2+2x^2y+5xy^2 \\ 2-x^2y+1.6xy^2 \end{bmatrix} = 0. \tag{5.12}$$

In two variables, the mixed volume can be evaluated from the alternative formula

$$\mathrm{Vol}(P_1+P_2) - \mathrm{Vol}(P_1) - \mathrm{Vol}(P_2).$$

(Related alternating sign formulas apply in higher dimensions.) The polygons are shown in Figure 5.2, in which one sees that $P_1 + P_2$ has two mixed cells, contributing areas 2 and 3 to the mixed volume, summing to 5. Accordingly, a polyhedral homotopy can be constructed to solve f using five paths. Note that the general linear product form that best captures the sparsity of f_1 and f_2 is just the same as already considered in (5.9), and as in Example 5.4, the number of paths to solve f is nine for a total-degree homotopy, eight for a 2-homogeneous homotopy, and seven for the general linear product. Hence, the polyhedral homotopy will have the fewest paths, and indeed, f has exactly five solutions, as can be verified in Bertini by running a total-degree or 2-homogeneous homotopy. ∎

Bertini does not offer polyhedral homotopies. Some packages that do are PHCpack [172] and HOM4PS2 [114].

5.4 ▪ Regeneration

While Bertini does not offer linear product homotopies, the theory behind linear products enables a more sophisticated approach known as regeneration, which Bertini does provide. This methodology proceeds in stages, introducing one equation (or group of equations) at a time. If the equations are very general, this is not advantageous, but in the common case that the equations have special structure, the technique can pay handsomely. This is because it automatically reveals the special structure of the equations as it proceeds, reducing computation in the subsequent stages. This is made especially effective by the fact that the early stages are less costly than the later ones.

Important note. The standard implementation of regeneration in Bertini is designed to find only the *nonsingular* isolated solutions of the system. Singular isolated solutions may be missed, but in many applications, only the nonsingular solutions are of interest. It is possible to configure Bertini to find all isolated solutions, including any singular ones, at the cost of a longer runtime (in most cases). This is discussed in more detail below.

To invoke regeneration in Bertini, all one needs to do is include the appropriate configuration setting:

```
UseRegeneration:1;
```

Adding this line converts a total-degree or multihomogeneous homotopy to a multistage regeneration homotopy instead.

While one can execute a regeneration homotopy without understanding its internal workings, a basic understanding will give a user a better sense of when it may be useful as well as an understanding of the progress reports that Bertini writes to the screen.

The regeneration method is built using linear product homotopies. As usual, we start with a square polynomial system of N polynomials in N variables, $f = \{f_1,\dots,f_N\}$. Suppose that we also establish a matching set of linear product polynomials, say, $g = \{g_1,\dots,g_N\}$. Using the linear product approach, we could find all the isolated solutions of f in one shot using the basic linear homotopy (5.7), but regeneration follows a different approach.

In addition to the g_i, let us pick linear functions ℓ_i, $i = 1,\dots,N$. These could be completely general linear equations, that is,

$$\ell_i(x) = \Box x_1 + \cdots + \Box x_N + \Box,$$

but at least all the variables that appear in f_i must also appear in ℓ_i. Bertini uses the maximally sparse second option.

Suppose that somehow—we'll see how in a moment—we have managed to find all isolated solutions to the system

$$G_i = \{f_1,\dots,f_{i-1}, g_i, \ell_{i+1},\dots,\ell_N\} = 0. \tag{5.13}$$

Then, by the general linear product theory, it is clear that we could use this as the start system to solve

$$F_i = \{f_1,\dots,f_{i-1}, f_i, \ell_{i+1},\dots,\ell_N\} = 0, \tag{5.14}$$

where the only difference is the ith polynomial. The homotopy to use is just

$$h_i(x,t) = \{f_1,\dots,f_{i-1},\quad \gamma t g_i + (1-t) f_i,\quad \ell_{i+1},\dots,\ell_N\} = 0. \tag{5.15}$$

This is the key move that lets us introduce one f_i at a time, progressing from $i = 1$ to $i = N$, and thereby solving $f = 0$.

So how can we solve (5.13)? Recall that g_i is a product of linear factors, say,

$$g_i(x) = \ell_{i,1}(x)\ell_{i,2}(x)\cdots\ell_{i,k_i}(x),$$

where the number of linear factors must be at least as large as the degree of f_i. To solve (5.13), it suffices to independently solve each of the systems

$$G_{i,j} = \{f_1,\dots,f_{i-1},\quad \ell_{i,j},\quad \ell_{i+1},\dots,\ell_N\} = 0, \qquad j = 1,\dots,k_i. \tag{5.16}$$

But these all resemble the system F_{i-1} in (5.14). Supposing that we have the isolated solutions of $F_{i-1} = 0$, we can find all the isolated solutions of $G_{i,j}$ using the homotopy

$$\hat{h}_{i,j} = \{f_1, \ldots, f_{i-1}, \quad t\ell_i + (1-t)\ell_{i,j}, \quad \ell_{i+1}, \ldots, \ell_N\} = 0, \qquad j = 1, \ldots, k_i. \quad (5.17)$$

(This is justified as it is a form of parameter homotopy, as discussed later in Chapter 6.) The union of the finite nonsingular solutions found from these homotopies is the full solution list for $G_i = 0$.

A regeneration run uses all these pieces to move sequentially from G_1 to F_N in several stages. System G_1 has all linear factors, so it is easy to solve. Then, homotopy h_1 solves F_1. From there, we bootstrap to solving F_2 first using the homotopies $\hat{h}_{2,j}$, $j = 1, \ldots, k_2$, to assemble a full solution list for G_2 and then using h_2 to solve F_2. This process repeats in similar fashion until the target system F_N has been solved, which is of course exactly the target system f.

During this bootstrapping process, only the nonsingular endpoints at each stage are carried forward to the next stage. The endpoints of the paths from the nonsingular isolated solutions of one stage are guaranteed, with probability one, to include all nonsingular isolated solutions of that stage's target system. In this way, the method finds all nonsingular isolated solutions of the final target. (We discuss options concerning singular solutions below.) At any stage, any solution paths that diverge to infinity or end at singularities are dropped away, thereby reducing the number of paths, and therefore the computational cost, of subsequent stages.

The regeneration method is so named because the homotopies $\hat{h}_{i,j}$ do more than reverse a degeneration as we did in all the other linear homotopies. Collectively, they rebuild G_i from the simpler F_{i-1}, replacing the single linear expression ℓ_i in F_{i-1} by the product of linear factors g_i in G_i.

Early levels of regeneration proceed quickly for two reasons. First, they tend to have fewer paths as the total degree of F_i grows as i increases. Second, the early stages include linear equations that do not change during the homotopies h_i and $\hat{h}_{i,j}$. Instead of explicitly carrying a full set of variables through the homotopy, it is possible to compute a set of basis vectors for the solution set of these linear equations and work with a reduced set of variables. (In this maneuver, it is said that the linear equations are treated *intrinsically* instead of *extrinsically*.) The combination of inexpensive early levels with the possibility of dropping away some endpoints at the completion of each level gives regeneration its upside potential.

Each time a level is completed, that is, F_i is solved for some i, it is always possible that some paths have diverged to infinity, ended at singularities, or landed on positive-dimensional components. Since we are seeking only the finite isolated solutions of F_N, these endpoints can be dropped away. This is where regeneration can save a lot of work; each endpoint that is omitted at an early stage would have otherwise engendered a multitude of descendants in later stages.

The downside of regeneration occurs when little or no simplification happens, that is, when very few endpoints can be eliminated before the last level. Consider a square system of N quadratics. All the isolated solutions can be found using 2^N paths in a total-degree homotopy. If the system is completely general so that no paths are eliminated before level N, the regeneration method will track two paths at level 1, then four paths at level 2, and so on, for a total of $2^1 + 2^2 + \cdots + 2^N = 2^{N+1} - 2$ paths total (not counting the paths of the $\hat{h}_{i,j}$ homotopies). That approximately doubles the total number of paths compared to the total-degree homotopy. But remember that these paths are not all equal;

the early levels are cheaper. So the regeneration method has a good chance of paying off if the target system has a substantially lower number of solutions than its total degree. This happens often in large, sparse systems.

For small systems, there is not as much chance for simplifications found in early stages to pay off in later stages. Accordingly, one cannot expect regeneration to offer gains in efficiency for such systems, and multihomogeneous homotopy should be used instead.

Order of the equations

Since regeneration discovers structure as it introduces polynomials one at a time, the order in which the polynomials are considered can affect the amount of computation that is required. When regeneration is invoked, the order of the polynomials that Bertini uses is governed by the following rules. Let N be the number of unknowns, let n be the number of equations, and let r be the rank of the system.[19]

- If $r < N$, Bertini stops immediately, as the system cannot have isolated roots. See Part II of this book for details.
- If the system is square ($N = n$), the polynomials are considered in the same order in which they are declared using `function` statements.
- If $n > N$ in a 1-homogeneous formulation, the polynomials are sorted by descending degree (ties broken by declaration order) before squaring up using a randomization of the form $F = \begin{bmatrix} I & A \end{bmatrix} f$. Then, F is solved starting from the bottom. This avoids raising degrees in F and solves from lowest degree to highest.
- If $n > N$ in an m-homogeneous formulation with $m > 1$, Bertini will immediately exit, as Bertini does not randomize multihomogeneous polynomial systems. The user must switch to a 1-homogeneous formulation by declaring all variables in one `variable_group` to proceed.

If the system involves linear polynomials, it is often advantageous to work intrinsically on the defining linear space. Although Bertini can work intrinsically on linear spaces it has randomly generated, on user-specified linear spaces it can work only extrinsically. However, by using the straight-line form of the systems, it is often trivial for the user to change the system from an extrinsic formulation to an intrinsic formulation, and vice versa (see Appendix F for further details).

Also, there is no hard and fast rule for determining the best order of presenting the polynomials to regeneration. Generally, it is best to delay high-degree equations until last. It is also helpful to delay introducing new variables as long as possible to allow structure in subsets of the variables to be revealed early.

Implementation details

The details of how Bertini chooses linear factors for each $g_i(x)$ affects the number of homotopy paths.

For a homotopy with just one `variable_group` (or `hom_variable_group`), each linear factor $\ell_{i,j}(x)$, $j = 1, \ldots, \deg f_i$, is chosen as general in the variables that appear in f_i. Thus, some of the sparsity of f_i is inherited by each $\ell_{i,j}$. Moreover, the linear $\ell_i(x)$ that is used in $F_1, \ldots, F_{i-1}$ is chosen to be the same as the first linear factor, $\ell_{i,1}(x)$. This eliminates the need for the $j = 1$ case of (5.17), which saves some computation.

[19] The rank of a polynomial system is the rank of its Jacobian matrix at a generic point of $\mathbb{C}^N$. See §9.1.

For a homotopy in which several variable groups have been declared, Bertini chooses $g_i(x)$ as in a multihomogeneous homotopy, using the variable groups declared by the user but limiting the variables that appear in the linear factors to those that appear in f_i.

Screen output

During a regeneration run, Bertini prints progress reports to the terminal screen. This output allows the user to monitor Bertini's progress and may provide insight into how to subsequently present a similar system to Bertini in a way that might reduce computation.

Let us examine the screen output for a simple example. Consider the intersection of two spheres and a cone, with an input file as follows:

```
% TwoSpheres_OneCone.input
CONFIG
 UseRegeneration:1;
END;
INPUT
 function sphere1, sphere2, cone;
 variable_group x,y,z;
 random cx1,cy1,cz1,r1, cx2,cy2,cz2,r2, cx3,cy3,r3;

 sphere1 = (x-cx1)^2 + (y-cy1)^2 + (z-cz1)^2 - r1^2;
 sphere2 = (x-cx2)^2 + (y-cy2)^2 + (z-cz2)^2 - r2^2;
 cone    = (x-z*cx3)^2 + (y-z*cy3)^2 - (z*r3)^2;
END;
```

One could put in explicit numbers for `cx1, cy1,` etc., but here we let Bertini pick random complex values using the declaration `random`.[20] This example is too small for regeneration to offer much help, but it is instructive for following the action step by step.

Let us intersperse clips from the screen output with explanations. First, Bertini says

```
Tracking regeneration level 0 of 3: 2 paths to track.
Tracking path 0 of 2
```

These are the two paths of $h_1(x,t) = 0$ that solve $F_1(x) = \{f_1(x), \ell_2(x), \ell_3(x)\} = 0$, the intersection of sphere1 with two random planes.

Since $\ell_{2,1}$ is equal to ℓ_2, the next task is to solve

$$G_2(x) = \{f_1(x), \ell_2(x) \cdot \ell_{2,2}(x), \ell_3(x)\} = 0,$$

which requires following two paths in a homotopy like (5.17) that moves $\ell_2(x)$ to $\ell_{2,2}(x)$, generating the following screen output:

```
Preparing regeneration level 1 of 3: 2 witness points to move.
Moving 0 of 2
```

Since $G_2(x) = 0$ has four solutions, $h_2(x,t) = 0$ is a four-path homotopy that solves $F_2(x) = \{f_1(x), f_2(x), \ell_3(x)\} = 0$, the intersection of the two spheres with a single random plane:

```
Tracking regeneration level 1 of 3: 4 paths to track.
Tracking path 0 of 4
```

Since the intersection of two spheres is a circle, F_2 has just two roots.

[20]The values selected are placed in the file `random_values`.

We are now prepared to solve

$$G_3(x) = \{\{f_1(x), f_2(x), \ell_3(x) \cdot \ell_{3,2}(x)\} = 0,$$

which, similar to solving G_2, requires moving $\ell_3(x)$ to $\ell_{3,2}(x)$:

```
Preparing regeneration level 2 of 3: 2 witness points to move.
Moving 0 of 2
```

There are four solutions in all for G_3: two for $\{f_1(x), f_2(x), \ell_3(x)\} \equiv F_2(x)$ and two for $\{f_1(x), f_2(x), \ell_{3,2}(x)\}$.

This prepares Bertini for the final homotopy that solves $F_3(x) \equiv f(x) = 0$:

```
Tracking regeneration level 2 of 3: 4 paths to track.
Tracking path 0 of 4
```

All of this is summarized in a table. Note that at level 1, two roots went to infinity, since the intersection of two spheres is a circle, which is only degree 2, not 4. This has reduced the number of paths in the final level 2 homotopy from the total degree of 8 to just 4. The table reports only the number of paths in the homotopies $h_i(x,t) = 0$, which solve $F_i(x) = 0$.

```
*************** Regeneration summary ***************
NOTE: nonsingular vs singular is based on rank deficiency and
      identical endpoints
                 nonsing      total       singular    infinite
|level| paths | endpoints | discarded | endpoints | endpoints |
----------------------------------------------------------------
| 0    | 2    | 2         | 0         | 0         | 0         |
| 1    | 4    | 2         | 2         | 0         | 2         |
| 2    | 4    | 4         | 0         | 0         | 0         |
----------------------------------------------------------------
|total| 10
****************************************************
```

Finally, we obtain the usual summary of the number of roots found, in this case four nonreal, nonsingular ones:

```
Nonsingular Finite Solution Summary
NOTE: nonsingular vs singular is based on rank deficiency and
      identical endpoints

              Number of        Number of
            | real solns | nonreal solns | Total |
--------------------------------------------------
Nonsingular |     0      |       4       |   4   |
--------------------------------------------------
Total       |     0      |       4       |   4   |
```

Note that to get the reduction to just four paths in the final stage, it is necessary to intersect the two spheres first. Otherwise, with cone declared first or second, no simplification occurs until half of eight paths go to infinity in the final stage.

Finding all isolated solutions with regeneration

Standard regeneration as described above is guaranteed only to find *nonsingular* isolated roots. Consider the system $\{x^2, y^2\} = 0$, which has one root: $(x,y) = (0,0)$. This root is isolated, but it is singular; in fact, it is multiplicity 4. A total-degree homotopy will have four paths, all leading to this root at the origin, but regeneration will report only that the system has no nonsingular isolated roots.

A more general regeneration approach, as described in [93], can find singular as well as nonsingular roots. However, finding singular roots introduces a significant complication into the procedure: singular endpoints at early stages of regeneration must be carried forward to the later stages. Since path tracking works efficiently only for nonsingular homotopy paths, this means that the singular paths need to be *deflated* (see §11.6). Also, the general approach has to distinguish between endpoints that are isolated singular and those that lie on higher-dimensional sets. Only the isolated ones have well-defined homotopy paths in subsequent stages of regeneration, and only these can lead to isolated solutions of the final target system. While deflation and local dimension sorting (see §11.5) are both feasible, to keep the regeneration procedure simple Bertini does not currently perform this operation. Since in many contexts only the nonsingular solutions are interesting, the approach still remains broadly applicable.

But what can the user do if singular solutions are of interest? One option is to stick with a multihomogeneous homotopy, the default in Bertini. But if the system is large and has structure not easily captured by multihomogeneity, there are still ways to take advantage of regeneration while getting both singular and nonsingular isolated roots.

Cascade approach for finding all isolated solutions

The first approach uses the *regenerative cascade* of §9.4. This method generates points on all the irreducible solution components of the system at any dimension. That includes, of course, zero-dimensional components, the isolated points. To invoke this without running a complete numerical irreducible decomposition, the user may include the configurations

```
CONFIG
 TrackType:1;                 % Find points on all components
 UseRegeneration:1;           % Do it with the regenerative cascade
 % Do not decompose higher-dimensional components
 WitnessSupersetOnly:1;
END;
```

This approach comes with some caveats, because the regenerative cascade comes with some side effects. First, it is not compatible with multiple variable groups. Second, it begins with a full randomization of the equations. Both of these can mean an increase in the number of homotopy paths to track, and sometimes this increase can be substantial.

Perturbation approach for finding all isolated solutions

A second approach to finding all isolated solutions of a square system,

$$f(z) = \{f_1(z), \dots, f_N(z)\} = 0, \qquad z \in \mathbb{C}^N,$$

by regeneration proceeds in two stages. First, one could use regeneration to solve a perturbed system:

$$g(z) := f(z) + c = \{f_1(z) + c_1, \dots, f_N(z) + c_N\} = 0, \tag{5.18}$$

where $c = (c_1, \ldots, c_N)$ is chosen at random in $\mathbb{C}^N$. With probability one, all the finite roots of $g(z)$ will be isolated and nonsingular, which can be seen by applying the simple Bertini theorem [161, Theorem A.8.7, p. 330] to the system $\{f(z), 1\}$. Then, use the solutions of (5.18) as startpoints for the homotopy

$$h(z,t) := f(z) + tc = \{f_1(z) + tc_1, \ldots, f_N(z) + tc_N\} = 0. \tag{5.19}$$

As t goes from 1 to 0, the paths starting at the finite roots of $g(z)$ are nonsingular for $t \in (0,1]$ and their endpoints include all the isolated roots of $f(z)$. This follows from the parameter homotopy theory covered in Chapter 6. This two-step approach resembles the "cheater's homotopy" of [121], which also regularizes solutions by adding random constants. The method is studied in detail in [13]. While the maneuver of adding random constants seems innocuous, and often is, it can sometimes have a profound effect on the system. In particular, if $f(z) = 0$ has positive-dimensional components, the regularization by random constants turns these into (possibly many) finite nonsingular solutions.

5.5 ▪ Examples

Let's use multihomogeneous and regeneration homotopies on two examples.

5.5.1 ▪ Lagrange Points

In a prizewinning paper published in 1772, Joseph-Louis Lagrange found exact solutions to the three-body problem of gravitation [110]. These solutions, initially regarded as an academic exercise, have been observed in nature and can serve as advantageous locations for space probes.

Although in general orbits are elliptical, we consider the special case of two celestial bodies in circular orbits about their common center of mass. In our solar system, this is a good first approximation for the motion of the Earth–Moon system, the Earth–Sun system, and many other orbiting pairs. Lagrange points are where a third, much smaller body can be in equilibrium while remaining in a fixed position relative to the two larger ones. As the system of three bodies rotates together, the resultant of the gravitational forces from the large bodies exactly matches the small body's centripetal acceleration, keeping it also in circular orbit. A similar balance applies at the large bodies, but we assume that the small one exerts a negligible gravitational force on them.

The arrangement of the bodies and their coordinates are illustrated in Figure 5.3. Let the large bodies have masses m_1 and m_2 and suppose they are distance r_1 and r_2 from the center of rotation O of the system. The small body m_3, with $m_3 \ll m_1$ and $m_3 \ll m_2$, is located in the orbital plane at position $\vec{p} = \left[\begin{smallmatrix} x \\ y \end{smallmatrix}\right]$. Let ω be the rotation rate of the system and let d_{ij} be the distance between masses i and j. Let G be the universal gravitational constant. The gravitational force between two bodies has magnitude $F = m_i m_j G/d_{ij}^2$, directed toward the opposing mass, and the centripetal acceleration of a mass at $\vec{p}$ from the center of rotation is $-\omega^2 \vec{p}$. Accordingly, the balance equations $\vec{f} - m\vec{a} = 0$ for the three masses are

$$-m_1 m_2 G/d_{12}^2 + m_1 \omega^2 r_1 = 0 \qquad \text{(mass 1)}, \tag{5.20}$$

$$m_1 m_2 G/d_{12}^2 - m_2 \omega^2 r_2 = 0 \qquad \text{(mass 2)}, \tag{5.21}$$

$$\frac{m_1 m_3 G}{d_{13}^2} \frac{(r_1 \mathbf{x} - \vec{p})}{d_{13}} + \frac{m_2 m_3 G}{d_{23}^2} \frac{(-r_2 \mathbf{x} - \vec{p})}{d_{23}} + m_3 \omega^2 \vec{p} = 0 \qquad \text{(mass 3)}. \tag{5.22}$$

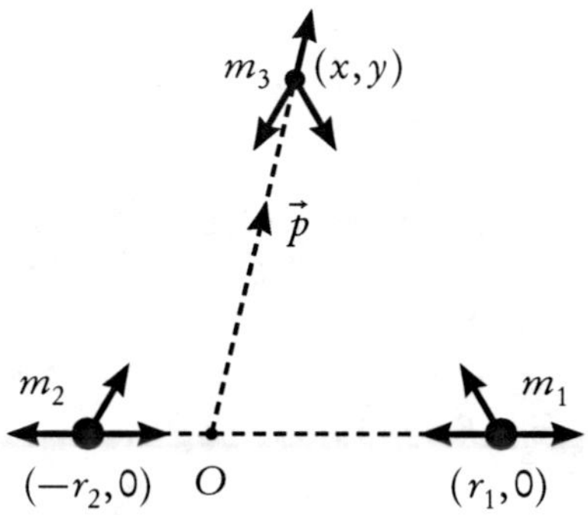

Figure 5.3. *Lagrange point coordinates. Solid arrows indicate gravitational and centrifugal forces. When m_3 is small, its effect on m_1 and m_2 can be ignored.*

Suppose we know m_1, m_2, and distance d_{12}. The unknowns are r_1, r_2, ω, x, y with $r_1 + r_2 = d_{12}$. Equilibrium does not depend on m_3, since it cancels out of (5.22). To make the system polynomial, we must move distances out of the denominators and define them in terms of r_1, r_2, x, y using the Pythagorean theorem. We can also simplify by introducing nondimensionalized variables, $\mu = m_1/m_2$, $\rho_1 = r_1/d_{12}$, $\rho_2 = r_2/d_{12}$, $\tilde{x} = x/d_{12}$, $\tilde{y} = y/d_{12}$, $\delta_{13} = d_{13}/d_{12}$, $\delta_{13} = d_{13}/d_{12}$, and $w = \omega^2 d_{12}^3/(Gm_2)$, so that the system becomes

$$\rho_1 + \rho_2 = 1, \tag{5.23}$$

$$1 - w\rho_1 = 0, \qquad \mu - w\rho_2 = 0, \tag{5.24}$$

$$(\rho_1 - \tilde{x})^2 + \tilde{y}^2 - \delta_{13}^2 = 0, \qquad (\rho_2 + \tilde{x})^2 + \tilde{y}^2 - \delta_{23}^2 = 0, \tag{5.25}$$

$$\mu\delta_{23}^3(\rho_1 - \tilde{x}) + \delta_{13}^3(-\rho_2 - \tilde{x}) + w\tilde{x}\delta_{12}^3\delta_{13}^3 = 0, \tag{5.26}$$

$$(-\mu\delta_{23}^3 - m_2\delta_{13}^3 + w\delta_{12}^3\delta_{13}^3)\tilde{y} = 0. \tag{5.27}$$

This system has one parameter, μ, the ratio of the two large masses.

It is clear from (5.27) that the problem can be split into two cases: $\tilde{y} = 0$ and $\tilde{y} \neq 0$. While this would be a commendable way to proceed, taking advantage of simplifications in (5.25) when $\tilde{y} = 0$, it is not necessary: we can proceed with the equations in their raw form. One can use (5.23) to eliminate ρ_2 from the remaining equations, leaving a system of six polynomials in unknowns $w, \rho_1, \delta_{13}, \delta_{23}, \tilde{x}, \tilde{y}$. The total degree of the system is $2^4 8^2 = 1024$. A presentation of this system as a Bertini input file is as follows:

```
% LagrangePoints.input
% Equilibria for a 3rd small body rotating with two large
%       ones in circular orbit
INPUT
  function  fma1,fma2,dist13,dist23,fma3x,fma3y;
  % definition: w = omega^2 d12^3/(G m2)
  % The remaining variables are nondimensionalized as
  %   ratio to d12.
  variable_group w;
  variable_group r1;
  variable_group x,y,d13,d23;
  constant mu;
```

```
    % choose value for the mass ratio
    mu = 9;

    % the following eliminates r2
    r2 = 1-r1;
    % f=ma on mass 1
    fma1 = w*r1 - 1;
    % f=ma on mass 2
    fma2 = w*r2 - mu;

    % distance m1 to m3
    dist13 = (x-r1)^2 + y^2 - d13^2;
    % distance m2 to m3
    dist23 = (x+r2)^2 + y^2 - d23^2;

    % f=ma on m3
    a = w*d13^3*d23^3; b1 = mu*d23^3; b2 = d13^3;
    fma3x = a*x + b1*(r1-x) + b2*(-r2-x);
    fma3y = a*y + b1*(-y)    + b2*(-y);
  END;
```

A 3-homogeneous treatment with the groups $\{(w),(r_1),(d_{13},d_{23},x,y)\}$ gives a Bézout number of 392. For this formulation, Bertini gives the following solution counts, which we show as they print to the screen:

```
                Number of          Number of
              | real solns  | nonreal solns  | Total  |
-------------------------------------------------------
Nonsingular   |      8      |       24       |   32   |
Singular      |      4      |        4       |    8   |
-------------------------------------------------------
Total         |     12      |       28       |   40   |

Finite Multiplicity Summary
  Multi-     Number of         Number of
  plicity  | real solns  | nonreal solns  |
-------------------------------------------
     1     |      8      |       24       |
     2     |      4      |        0       |
     3     |      0      |        2       |
     9     |      0      |        2       |
```

Using the same variable groups but invoking regeneration, the path count in the final level of regeneration is 196. Since this is the most expensive stage, this represents a significant savings compared to the 3-homogeneous run. The summary of the stages of regeneration is shown below:

level	paths	nonsing endpoints	total discarded	singular endpoints	infinite endpoints
0	2	2	0	0	0
1	2	1	1	0	1
2	2	2	0	0	0
3	4	4	0	0	0
4	28	28	0	0	0
5	196	32	164	32	132
total	234				

At the end of level 1, when (5.24) have been solved, there is only one nonsingular endpoint. This is the key result that reduces the number of paths in the succeeding levels. If one solved the level 1 subsystem by hand at the beginning, the remaining equations could be solved using a 196-path multihomogeneous homotopy. But the point is that regeneration often takes advantage of such simplifications without the user having recognized it. Of course, the simplification gained here is most effective when (5.24) are placed at the top when ordering the equations.

If one runs regeneration with all the variables in one group, instead of the 1024 paths of a total-degree homotopy, one gets just 224 paths in the final stage:

level	paths	nonsing endpoints	total discarded	singular endpoints	infinite endpoints
0	2	2	0	0	0
1	4	1	3	0	3
2	2	2	0	0	0
3	4	4	0	0	0
4	32	28	4	0	4
5	224	32	192	32	160
total	268				

As this illustrates, regeneration is usually much less sensitive to the choice of variable groups than the multihomogeneous approach. All three of these runs take roughly the same amount of time.

It should be acknowledged that the multihomogeneous run does provide more information than the regeneration runs: its homotopy should find all singular isolated roots. Because of this, a multihomogeneous run does some analysis on the singular endpoints and, as shown above, lists their apparent multiplicities. These are based on grouping endpoints that cluster tightly together. Note that although regeneration is not guaranteed to find singular isolated roots, in these runs it does give 32 singular endpoints. This is a raw count of the endpoints: they have not been clustered together. When the multiplicities of the eight singular endpoints found by the 3-homogeneous run are taken into account, the tally also comes to 32, so there is no discrepancy. In either case, there is no guarantee that the singular endpoints are indeed isolated—some may be points on higher-dimensional sets. But in the case of the multihomogeneous homotopy, the isolated singular solutions are guaranteed to be included among the endpoints.

One final note on running this system: The degree 8 equations (5.26), (5.27) present some challenge to the numerical processes. Using the default settings, the regeneration

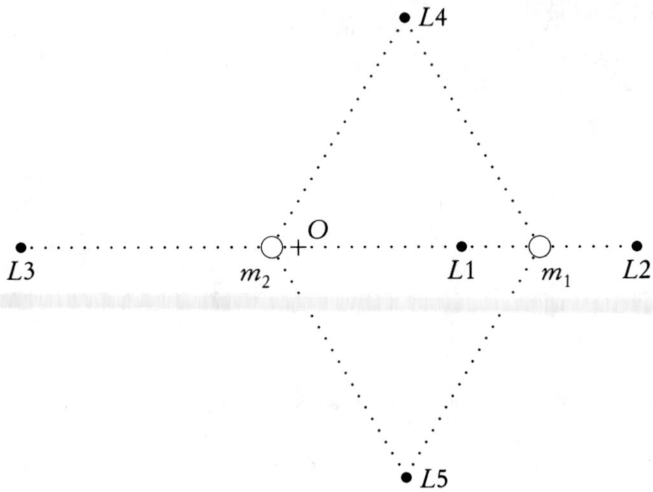

Figure 5.4. *Lagrange points. There are five equilibrium locations for the small third body.*

run with one variable group shows some variability in the summary table, although the 32 nonsingular roots are found consistently. For higher-degree equations, tighter tracking tolerances (see §E.4) are strongly encouraged, which reduces the variability. Another option, applicable in many situations, is to run the system allowing Bertini to select the parameters of the problem at random. That is, if one replaces

```
constant mu;
mu = 1/9;
```

with

```
random mu;
```

more consistent and reliable results will be obtained. As described in Chapter 6, one may then obtain solutions for a desired set of parameters by using a parameter homotopy that tracks just the 32 nonsingular solutions as μ is moved from a random value to a desired one. To solve the Earth–Sun system, where $\mu \approx 333,000$, this kind of approach is required for good results.

This section would not be complete without a brief discussion of the solutions obtained. Notice above that eight real nonsingular solutions were found. Among these are three in which one or both of the distances d_{13} and d_{23} are negative, so these are meaningless. The remaining five valid Lagrange points are known in astrophysics circles as $L1,\ldots,L5$, as labeled in Figure 5.4. Points $L1$, $L2$, and $L3$, which lie on the line through the two large masses, are unstable equilibria, while the off-axis points, $L4$ and $L5$, are stable if the ratio of the masses is large enough [43]. By convention, $L4$ precedes the orbit of the smaller of m_1 and m_2, while $L5$ trails behind. In the results computed by Bertini, one finds that for $L4$ and $L5$, $d_{13} = d_{23} = d_{13}$ to high precision, matching the result found by Lagrange that the corresponding exact solutions form equilateral triangles.

In addition to the nonsingular solutions, the 3-homogeneous homotopy found four real isolated singular solutions, each of multiplicity 2. These turn out to coincide with the locations of the large masses. There are two of each of these since the distance to the other mass can be considered either positive or negative in the algebraic formulation.

At any rate, as is so often the case with singular solutions, these are all uninteresting from a practical standpoint.

The nonsingular equilibrium points $L1,\ldots,L5$ have implications in the real world. Objects have been observed in the off-axis locations for various orbiting pairs. The first one, found in 1906, was the asteroid Achilles in the $L4$ location of Jupiter's orbit around the Sun. The result also applies to orbits of moons around planets: dust has been observed in these locations in the Earth–Moon system. Due to orbital instability, natural objects are not to be found in $L1$, $L2$, or $L3$. Even so, although a satellite must make periodic corrections to stay in $L1$ or $L2$, these are convenient locations for space observatories. For the Earth–Sun system, the Solar and Heliospheric Observatory satellite resides at $L1$, and $L2$ is both the current home of the Planck satellite (studying the cosmic microwave background radiation) and the future target of the James Webb space telescope [43]. $L3$, which always remains hidden on the other side of the Sun from Earth, is not an advantageous target.

5.5.2 ▪ Nine-Point Four-Bar Design

Our next problem is a mechanical design problem. The objective is to create a four-bar linkage that draws a desired path in the plane. A classical approach to such problems is to specify a number of points in the plane through which a point of the linkage must pass. The points are called "precision points," since the curve is required to pass exactly through them.

A four-bar linkage is a hinged quadrilateral. With one link held stationary, the points of the adjacent links trace circles, while the points of the opposing "coupler link" trace sextic curves. We would like one of those sextic curves to interpolate a set of precision points. The point whose motion forms this curve is called the "coupler point."

The maximum number of precision points that can be specified for a four-bar linkage is nine, which constitutes the nine-point path synthesis problem first proposed by Alt in 1923 [9]. A quick count of the design freedoms of the four-bar is as follows:

- four lengths of the sides of the quadrilateral,
- two translations and one rotation to locate the stationary link in the plane, and
- two coordinates to specify the coupler point.

As this tallies to nine, there are just enough design parameters to match nine precision points.

The formulation of the equations we shall use for the four-bar coupler curve is adapted from [182], which was based on an earlier derivation by Roth and Freudenstein [149]. The main difference in [182] is that isotropic coordinates are used. In isotropic coordinates, a point (x,y) in the plane is represented instead by $(z,\bar{z})=(x+iy,x-iy)$. When x and y are real, z and $\bar{z}$ are complex conjugates. Isotropic coordinates have the following two advantages over the usual Cartesian coordinates:

- Squared distances become bilinear: $x^2+y^2=z\bar{z}$.
- Rotations in Cartesian coordinates

$$(cx-sy,sx+cy) \text{ subject to } c^2+s^2=1$$

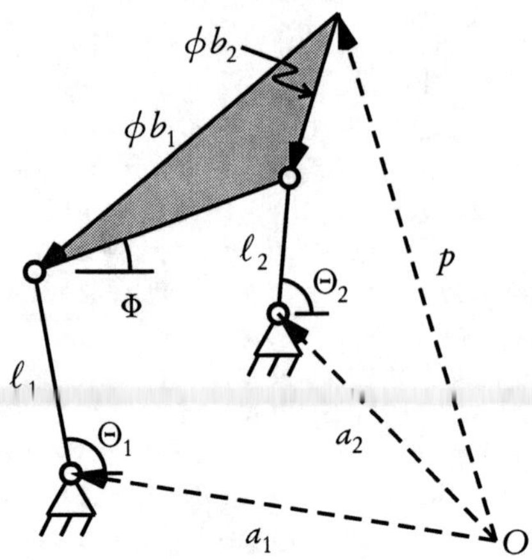

Figure 5.5. *Vector diagram of a four-bar linkage.*

take the simpler form

$$(\theta z, \bar{\theta}\bar{z}) \text{ subject to } \theta\bar{\theta} = 1.$$

These are equivalent for real rotation angle Θ by taking $c = \cos\Theta$, $s = \sin\Theta$, $\theta = e^{i\Theta}$, and $\bar{\theta} = e^{-i\Theta}$.

The use of isotropic coordinates is advantageous when applying multihomogeneous homotopy, since it converts many of the quadratic expressions that appear in planar kinematics into bilinear expressions. In contrast, since regeneration discovers structure during computation instead of analyzing a multidegree table, its efficiency is less affected by the choice of coordinates.

In Figure 5.5, where we label a vector as p, it is implied that its isotropic coordinates are $(p, \bar{p})$, and similarly for the other vectors. The parameters of the linkage,

$$(a_1, \bar{a}_1), (a_2, \bar{a}_2), (b_1, \bar{b}_1), (b_2, \bar{b}_2), \ell_1, \ell_2,$$

are constant, while the angles Θ_1, Θ_2, and Φ vary along a one-dimensional curve, sweeping out a coupler curve path, $(p, \bar{p})$, in the process. Using isotropic coordinates $(\phi, \bar{\phi})$, $\phi\bar{\phi} = 1$, to represent the rotation Φ, the sides of the coupler triangle rotate to $(\phi b_1, \bar{\phi}\bar{b}_1)$ and $(\phi b_2, \bar{\phi}\bar{b}_2)$. Accordingly, the squared lengths of the sides of the four-bar can be written as

$$\ell_1^2 = (p + \phi b_1 - a_1)(\bar{p} + \bar{\phi}\bar{b}_1 - \bar{a}_1), \tag{5.28}$$

$$\ell_2^2 = (p + \phi b_2 - a_2)(\bar{p} + \bar{\phi}\bar{b}_2 - \bar{a}_2). \tag{5.29}$$

For a given four-bar, these two equations, along with

$$\phi\bar{\phi} = 1, \tag{5.30}$$

constitute a system of three equations in four unknowns, $(p, \bar{p}, \phi, \bar{\phi})$, that describes the four-bar motion and its coupler curve.

When the right-hand sides of (5.28), (5.29) are expanded, (5.30) can be used to write them as linear expressions in $(\phi, \bar{\phi})$. Then, Cramer's rule can be applied to solve for these and substitute them into (5.30), with the result that we have a single coupler curve equation written in terms of several 2×2 determinants as

$$f_{cc}(p,\bar{p};q) = |v\ \bar{u}||v\ u| + |u\ \bar{u}|^2 = 0, \tag{5.31}$$

where

$$v = \begin{bmatrix} (p-a_1)(\bar{p}-\bar{a}_1) + b_1\bar{b}_1 - \ell_1^2 \\ (p-a_2)(\bar{p}-\bar{a}_2) + b_2\bar{b}_2 - \ell_2^2 \end{bmatrix}, \quad u = \begin{bmatrix} b_1(\bar{p}-\bar{a}_1) \\ b_2(\bar{p}-\bar{a}_2) \end{bmatrix}, \quad \bar{u} = \begin{bmatrix} \bar{b}_1(p-a_1) \\ \bar{b}_2(p-a_2) \end{bmatrix} \tag{5.32}$$

and where q is the set of linkage parameters:

$$q = (a_1, a_2, b_1, b_2, \bar{a}_1, \bar{a}_2, \bar{b}_1, \bar{b}_2, \ell_1, \ell_2).$$

In the nine-point problem, we are given nine precision points $(p_j, \bar{p}_j)$, $j = 0, \dots, 8$, and we must find the parameters of the linkage so that the coupler curve interpolates them. Without loss of generality, we can translate the origin of coordinates to make $(p_0, \bar{p}_0) = (0,0)$, and by matching $(b_1, \bar{b}_1), (b_2, \bar{b}_2)$ to the pose of the coupler when it visits this initial point, we can assume that $(\phi, \bar{\phi}) = (1,1)$. With these choices, (5.28), (5.29) evaluated at the initial point 0 become

$$\ell_1^2 = (b_1 - a_1)(\bar{b}_1 - \bar{a}_1), \tag{5.33}$$

$$\ell_2^2 = (b_2 - a_2)(\bar{b}_2 - \bar{a}_2). \tag{5.34}$$

Substituting these into the coupler curve polynomial f_{cc} above and evaluating it at each of the precision points, we have the system of equations

$$f_{cc}(p_j, \bar{p}_j, a_1, a_2, b_1, b_2, \bar{a}_1, \bar{a}_2, \bar{b}_1, \bar{b}_2) = 0, \qquad j = 1, \dots, 8. \tag{5.35}$$

Since the $(p_i, \bar{p}_i)$ are given, this is a system of eight polynomial equations in eight unknown linkage parameters.

Each of the elements in the column vectors $v, u, \bar{u}$ in (5.32) is quadratic, so f_{cc} appears to be degree 8. However, close examination reveals that the highest-degree terms cancel out, leaving equations of seventh degree. Consequently, the total degree of the system is $7^8 = 5{,}764{,}801$. This is a daunting number of paths to track, although it could be done on a sufficiently large parallel computer. However, there is no need for that.

Remark 5.6. To run this example in Bertini, it is useful to note that the equations can be given in straight-line form. That is, one does not need to expand out the expressions fully. Instead, subexpressions appearing in the derivation above can be formed, given names, and then combined to form the coupler curve polynomial. (A partial expansion is required to expose the cancellation of the eighth-degree terms so that Bertini can recognize that f_{cc} is of degree seven.) The straight-line form is easier to write and more efficient to evaluate than a fully expanded version. The reader is encouraged to visit the website for this book, `www.siam.org/books/se25`, to see how the straight-line form is used in the input file for this problem.

To take advantage of a multihomogeneous treatment, one might try the natural grouping of the variables,

$$\{(a_1, \bar{a}_1, b_1, \bar{b}_1), (a_2, \bar{a}_2, b_2, \bar{b}_2)\},$$

which leads to the following degree table:

	4	4
8×	4	4

This gives a 2-homogeneous Bézout number of $4^8\binom{8}{4} = 4,587,520$, which is only a little better than the total degree. One can get this same count in the original Roth–Freudenstein formulation in Cartesian coordinates. That formulation is equivalent to using substitutions like $(a_1, \bar{a}_1) = (a_{1x} + ia_{1y}, a_{1x} - ia_{1y})$, after which the 2-homogeneous grouping is

$$\{(a_{1x}, a_{1y}, b_{1x}, b_{1y}), (a_{2x}, a_{2y}, b_{2x}, b_{2y})\}.$$

A better alternative is to use a 4-homogeneous homotopy using the groupings

$$\{(a_1, b_1), (a_2, b_2), (\bar{a}_1, \bar{b}_1), (\bar{a}_2, \bar{b}_2)\}.$$

All eight equations have the same multidegrees, so the multidegree table can be written in the concise form

	2	2	2	2
8×	2	2	2	2

This gives a Bézout number of

$$2^8\binom{8}{2}\binom{6}{2}\binom{4}{2}\binom{2}{2} = 645,120.$$

By defining some extra variables related to the highest degree terms of the equations, it is also possible to formulate a system of 12 equations in 12 unknowns that has a 2-homogeneous Bézout number of 286,720 [182], but we do not take that up here.

Notice that the reduced 4-homogeneous Bézout number is directly enabled by the use of isotropic coordinates; in Cartesian coordinates, the multidegrees cannot be broken up so finely. While the multihomogeneous approach is clearly powerful, it can take considerable cleverness and experience to find a formulation that is amenable to it, if such a formulation even exists.

Regeneration does not require this level of cleverness. In fact, with regeneration, it is immaterial whether the system is formulated with Cartesian or isotropic coordinates, except for minor differences in the cost of evaluating the polynomials. A summary of the path counts for a regeneration run using isotropic coordinates in two groups is as follows:

level	paths	nonsing endpoints	total discarded	singular endpoints	infinite endpoints	higher dim'l
0	8	8	0	0	0	0
1	64	61	3	0	1	2
2	488	352	136	28	104	4
3	2816	1528	1288	346	936	6
4	11840	4688	7152	2100	5052	0
5	33064	9804	23260	7576	15684	0
6	55240	12176	43064	13304	29760	0
7	48704	8652	40052	11488	28904	0
total	152224					

For comparison, the summary using four groups is as follows:

level	paths	nonsing endpoints	total discarded	singular endpoints	infinite endpoints	higher dim'l
0	8	8	0	0	0	0
1	64	62	2	0	0	2
2	464	428	36	28	4	4
3	2904	2274	630	458	166	6
4	13552	7888	5664	3112	2552	0
5	38776	16672	22104	10560	11544	0
6	60464	18848	41616	18320	23296	0
7	37696	8652	29044	11488	17896	0
total	153928					

Since this system has various positive components and many points at infinity and consists of higher-degree polynomials with a large Bézout number, it is prudent to modify the default settings when running this system (more details are provided in Appendix F). One change is to tighten all the tracking tolerances (see Appendix E) used in regeneration to, say, 10^{-8} so that the tracker more closely follows each path. This is accomplished by including the following configurations:

```
TrackTolBeforeEG:1e-8;
TrackTolDuringEG:1e-8;
SliceTolBeforeEG:1e-8;
SliceTolDuringEG:1e-8;
```

Another change is to increase the size of the finite region under consideration for the endpoints. Although the size of the true finite region is dependent upon the randomly selected parameters, we found that a maximum absolute value of 10^7 in each coordinate is a suitable choice. This is accomplished by including the following configurations:

```
SecurityMaxNorm:1e7;
EndpointFiniteThreshold:1e7;
PathTruncationThreshold:1e7;
```

The last change is to provide the power-series endgame (see §3.3.2) with points that are spaced farther apart by decreasing `SampleFactor`. This increases the numerical stability of the interpolation techniques used in the endgame and is accomplished by including the following configuration:

```
SampleFactor:0.1;
```

With this setup, these two runs took roughly the same amount of time.

Although regeneration is nearly as efficient as the most finely tuned multihomogeneous formulation of the nine-point problem—with a lot less cleverness required—both approaches track far more paths than the number of final solutions. Moreover, the 8652 nonsingular isolated solutions appear in a six-way symmetry group: a two-way symmetry that swaps a_1 with a_2 and so on, combined with a three-way symmetry called "Roberts cognates" [147, 37, 182]. Accordingly, there are only 1442 distinct coupler curves that pass through the precision points. (For most choices of the points, very few if any of the solutions are real.) Obtaining a complete solution for the first example of the problem is a somewhat expensive task, but the relatively low final solution count means that with one such solution set in hand, any other nine-point problem can be solved efficiently using just 1442 paths in a parameter homotopy. That is the topic of the next chapter.

5.6 ▪ Bertini Guide: Multihomogeneous and Regeneration Homotopy

Multihomogeneous Homotopy Summary

Input files
  `input` = run configuration & system definition
Output files
  Endpoints include all **isolated solutions**, sorted in various files.
Key configurations:
  `FinalTol`
Key declarations:
  `function`
  `variable_group` or `hom_variable_group`
Notes:
  If # equations > # variables, only 1 variable group is allowed.
Execution:
  `>> bertini <input>`

Regeneration Homotopy Summary

Input files
  `input` = run configuration & system definition
Output files
  Endpoints include all **nonsingular isolated solutions** in file
    `nonsingular_solutions`.
Key configurations:
  `UseRegeneration:1;`
Key declarations:
  `function`
  `variable_group` or `hom_variable_group`
Notes:
  If # equations > # variables, only 1 variable group is allowed.
Execution:
  `>> bertini <input>`

Chapter 6
Parameter Homotopies

In the previous chapter, we discussed various homotopies that take advantage of the sparse structure of a polynomial system (the patterns of monomials that appear) to reduce computation by decreasing the number of homotopy paths below the total degree. However, polynomial systems that arise in applications often have additional structure beyond the sparsity of their monomials. Table 6.1 summarizes some examples we have seen already. These illustrate the fact that naturally occurring systems often have fewer solutions than is predicted by a multihomogeneous Bézout number or the polyhedral mixed volume. While regeneration can take advantage of some structure beyond sparsity, its main strength is that it captures sparse structure automatically, requiring neither extreme cleverness on the part of the user nor complicated (and often expensive) combinatorial algorithms.

For examples like those in Table 6.1, a *parameter homotopy* takes advantage of any observed reduction in the number of roots. No matter what homotopy is used to solve the first general example from a family of systems, a parameter homotopy solves subsequent members of the family using a number of paths equal to the exact root count. These efficient homotopies are the second phase of a two-phase methodology wherein a sparse homotopy or regeneration is used to solve a general example *ab initio*, and a parameter homotopy is used on all remaining members of the family. Hence, whatever the cost the one-time ab initio phase may be, it may be a negligible cost when amortized over a sufficient number of subsequent parameter runs, which are as efficient as possible in terms of the number of paths to track. In some applications, the target system is a single example, not a member of a family, so the ab initio phase is the whole problem with no opportunity to amortize its cost. But in many applications, the coefficients of the system depend on some physical quantities, such as the lengths of a robot's links or temperature-dependent reaction rates of chemical species. If one wishes to solve several examples of the same family, differing only in the parameter values, then a parameter homotopy can be an important tool.

After carefully stating how parameter homotopies work in §6.1, we will discuss Bertini's implementation of it in §6.2. The key steps are to declare parameters and invoke the proper configuration setting. Between the ab initio phase and subsequent parameter runs, the user can make certain adjustments, for example, to take advantage of any symmetries that the solution set obeys. The application of these techniques is illustrated by revisiting the Lagrange points problem and the nine-point four-bar design problem in §6.3. Those systems are square (same number of equations, n, as variables, N), but as

Table 6.1. *Root counts for various large problems.*

Problem	Section	Equations	Best multihomogeneous Bézout number	Actual isolated root count
Six-revolute (6R) inverse kinematics	Ex.5.2	(5.2)–(5.4)	320	16
Lagrange points	§5.5.1	(5.23)–(5.27)	392	32
Nine-point four-bar design	§5.5.2	(5.35)	645,120	8652

discussed in §6.4, parameter homotopy can also be used for nonsquare systems with $n > N$. Also, it is easy to take advantage of nested parameter spaces—parameterized subfamilies inside a more general family (§6.5). Several more advanced refinements to the parameter technique are then treated, including taking advantage of symmetries (§6.6), using parameter homotopies for systems that have nonpolynomial (but analytic) dependence on the parameters, such as trigonometric functions (§6.7.2), and a discussion of how parameter homotopy applies if the parameter space is not Euclidean (§6.7.3).

It may happen that one wishes to solve more than a few cases in the same family; in fact, one might want to solve a problem at every point in a gridded lattice in the parameter space. A special Bertini module, called `paramotopy`, provides this functionality. After learning about how parameter homotopies work in this chapter, the reader may consult Appendix H for details on running the `paramotopy` module.

6.1 ▪ Parameter Homotopy Basics

Let $f(z;q) = 0$ be a parameterized family of polynomial systems:

$$f(z;q) := \begin{bmatrix} f_1(z_1,\ldots,z_N;q_1,\ldots,q_K) \\ \vdots \\ f_N(z_1,\ldots,z_N;q_1,\ldots,q_K) \end{bmatrix} = 0. \tag{6.1}$$

That is, $f(z;q)$ is a system of N polynomials in the variables $z \in \mathbb{C}^N$ and parameters $q \in \mathbb{C}^K$. In particular, for a given choice of $q^* \in \mathbb{C}^K$, we have a *square* system of the sort we have been discussing in previous chapters.

There are many results in algebraic geometry that assert some good behavior of the solution set of a parameterized system as the parameters vary. The following summarizes the basic facts that we will be using for a parameter homotopy:

1. The number of nonsingular isolated solutions of $f(z;q) = 0$, call it $\mathcal{N}_{\text{nonsing}}$, is constant for general parameter values $q \in \mathbb{C}^K$.

2. The set of nongeneric parameters, where the number of nonsingular isolated solutions changes, is a proper algebraic subset of $\mathbb{C}^K$. This implies that the set of general parameters is (path) connected.

3. For nongeneric parameters, the number of isolated solutions can only decrease.

4. Suppose that $\phi(t) \in \mathbb{C}^K$, $t \in [0,1]$, is a continuous path that stays generic for $t \in (0,1]$. Then,
    (a) the nonsingular isolated solutions of $f(z;\phi(t)) = 0$ vary continuously for $t \in [0,1]$; and
    (b) the endpoints of the $\mathscr{N}_{\text{nonsing}}$ solution paths of the homotopy $f(z;\phi(t)) = 0$ include all of the nonsingular isolated solutions of $f(z;\phi(0)) = 0$.

These key facts and generalized versions of them appear in [133] and are further discussed in [161]. We shall discuss some of the most important refinements of the approach in subsequent sections of this chapter.

These facts give us the two-phase process we call "parameter homotopy."

Ab Initio phase: Pick a generic set of parameters $q_1 \in \mathbb{C}^K$ and find all the nonsingular isolated solutions S of $f(z;q_1) = 0$ by any means, such as a multihomogeneous homotopy or a regeneration homotopy.

Parameter homotopy phase: For any $q_0 \in \mathbb{C}^K$, choose a path $\phi(t) \subset \mathbb{C}^K$ that is generic for $t \in (0,1]$ with $\phi(1) = q_1$ and $\phi(0) = q_0$ and follow the $\mathscr{N}_{\text{nonsing}}$ nonsingular isolated solutions of the homotopy

$$h(z,t) := f(z;\phi(t)) = 0 \tag{6.2}$$

starting at S as t goes from 1 to 0. The endpoints of these paths include all nonsingular isolated solutions of $f(z;q_0) = 0$.

The ab initio phase can be done just once, after which the parameter homotopy phase can be repeated as often as desired. Sometimes these are called the "off-line" and "on-line" phases, as the more computationally expensive ab initio phase is a preparation step for the faster-running parameter homotopies that follow it.

This raises the following questions:

Q1 How do we ensure that q_1 is generic?

Q2 Given a target $q_0 \in \mathbb{C}^K$, how do we get a continuous path $\phi(t) \subset \mathbb{C}^K$ from q_1 to q_0 that is generic for $t \in (0,1]$?

The simplest answer to Q1 is

A1 Choose q_1 at random in $\mathbb{C}^K$.

This suffices with *probability one* because the set of exceptional parameters where the number of nonsingular solutions decreases, call it $Q^* \subsetneq \mathbb{C}^K$, is a proper algebraic subset of $\mathbb{C}^K$. It also suffices to choose q_1 at random in $\mathbb{R}^K$, but this has implications for Q2 (see A2′ below). Another answer to Q1 is

A1′ Use additional theoretical knowledge about $f(z;q)$ to establish $\mathscr{N}_{\text{nonsing}}$ and pick a q_1 such that $f(z;q_1) = 0$ has $\mathscr{N}_{\text{nonsing}}$ nonsingular solutions.

In most cases, we do not have the knowledge required by A1′, so we resort to A1. Bertini has an automatic means of implementing A1. We list A1′ because it is implicitly what justifies the sparse homotopies: multihomogeneous, linear product, and polyhedral. (More on that in a moment.) Answer A1′ also comes into play when $f(z;q) = 0$ is a subfamily of a larger parameterized family of systems, that is, when we have nested parameter spaces. (More on that in a moment also.)

The answer to Q2 depends on how Q1 is answered and on the properties of $f(z;q)$. The simplest case is

A2 Suppose q_1 is chosen at random in $\mathbb{C}^K$. Then with probability one, it suffices to choose $\phi(t)$ as

$$\phi(t) = t q_0 + (1-t) q_1. \tag{6.3}$$

When q_1 is held fixed, there may be some q_0 for which this linear path crosses the exceptional set, Q^*. But if we solve for a finite number of different values of q_0, there is a zero probability we will hit such a special case. Nevertheless, one can buy insurance against unlucky circumstances by comparing several runs, moving q_1 to some other random q_1' first and then targeting q_0 from there. Adaptive multiprecision path tracking makes the process robust even if $\phi(t)$ passes close to an exceptional case, but cautious analysts may wish to double check anyway.

When q_1 is chosen according to A1′, then answer A2 may not suffice. Instead, we may employ the general "gamma trick":

A2′ (general gamma trick) Choose γ at random in $\mathbb{C}$ or at random on the unit circle $\{\gamma \in \mathbb{C} \mid |\gamma| = 1\}$ and use

$$\phi(t) = \tau(t)q_1 + (1-\tau(t))q_0 \quad \text{with} \quad \tau(t) = \frac{\gamma t}{1+(\gamma-1)t}. \tag{6.4}$$

This path stays general for $t \in (0,1]$ with probability one. This result applies also for random $q_1 \in \mathbb{C}^K$ and random $q_1 \in \mathbb{R}^K$. Note that A2 does not generally suffice if q_1 is merely chosen random in $\mathbb{R}^K$, especially if q_0 is also real. This is because in a real homotopy, real solutions often meet in a singularity before branching out into complex solutions or vice versa. With the general gamma trick, $\phi(t)$ takes a general path through complex Euclidean space even if the endpoints are real.[21]

Finally, there is a circumstance when A2′ can be simplified:

A2″ (special gamma trick) Suppose that $f(z;q)$ is *linear* in the parameters q. Then, with the same conditions on γ as in A2′, it suffices with probability one to use

$$h(z,t) = \gamma t f(z;q_1) + (1-t)f(z;q_0) = 0. \tag{6.5}$$

One can see this by substituting from (6.4) into (6.2) and then using the linearity of $f(z;q)$ with respect to q to clear out the denominator.

We have already made use of the special gamma trick earlier. We first invoked it in the basic example of §2.1.2 to get the results illustrated in Figures 2.4 and 2.5. It was also used in the sparse solver approaches; for example, see the multihomogeneous homotopy (5.7), which is of similar form to (6.5). In the sparse homotopies, the parameters are the coefficients of the monomials that appear in the multihomogeneous structure, linear product structure, or polyhedral structure used to define the homotopy. So the parameter space is complex Euclidean and the polynomial family is linear in these parameters, and hence the special gamma trick applies. One advantage of A2″ over either A2 or A2′ is that derivatives of $h(z,t)$ with respect to t are simpler, which speeds computation.

6.2 ▪ Parameter Homotopy in Bertini

Bertini provides an easy way of executing the combination of the A1-type ab initio homotopy with the A2-type parameter homotopy. As a reminder, these apply when the parameter space is Euclidean, i.e., $\mathbb{C}^K$ for some $K \in \mathbb{Z}_{>0}$. The steps are as follows:

Ab initio Create an input file as in any multihomogeneous or regeneration run with two changes.

---

[21] To be precise, $\tau(t)$ is a circular arc from 1 to 0 in the complex plane, $\mathbb{C}$, and because the arc specified by (6.4) depends on the angle of γ, it misses the nongeneric points of $\mathbb{C}$ with probability one [161, Lemma 7.1.3]. See also Figure 2.4.

- First, in the CONFIG section, use the setting

```
ParameterHomotopy:1;
```

- Second, in the INPUT section, declare parameter names using a statement like

```
parameter p1, p2, p3;
```

where the variable names can be any alphanumeric string of the user's choosing (the first character must be a letter). Any number of parameters can be declared in one such statement and multiple parameter statements can be used in one input file—the result will be the same. Bertini will generate a random number for each declared parameter and store the value in `start_parameters`.

Startpoints Once the ab initio run is complete, one must set up the file of start solutions corresponding to the start parameters. The default name of the file is `start`. Often the correct action is to rename the file `nonsingular_solutions` produced by the ab initio run as `start`. Later, we will discuss later scenarios where one retains a subset of the nonsingular solutions, such as when symmetries are present.

Parameter run With `start_parameters` and `start` in place, one needs to establish the target parameters. These go in the file `final_parameters`. (See Example 6.1 and Appendix B for details.) In addition, one must let Bertini know that it is to perform a parameter homotopy run by changing the configuration setting to

```
ParameterHomotopy:2;
```

The startpoint file does not have to be called `start`. Instead it can be given on the command line:

```
>> bertini <inputFile> <startpointFile>
```

Simple example: A sextic

Example 6.1 (sextic ab initio). For a simple example, let's consider solving a sextic polynomial in one variable. A parameter homotopy does not offer much advantage for the this example, although it is easier to automate the solution of several examples by writing a new `final_parameters` file for each run than to edit an input file to change the coefficients. Nevertheless, the problem can be used to illustrate some of the advanced capabilities of parameter homotopy in later sections.

A complete Bertini input file for the ab initio run follows.

```
% sextic.input
CONFIG
 ParameterHomotopy:1;
END;
INPUT
 function f;
 variable_group x;
 parameter a0,a1,a2,a3,a4,a5,a6;
 f = a0+x*(a1+x*(a2+x*(a3+x*(a4+x*(a5+x*a6)))));
END;
```

Let us suppose that this input file is stored as `sextic.input`. Then, the ab initio phase is completed by launching Bertini as usual:

```
>> bertini sextic.input
```

This executes a six-path total-degree homotopy run. In addition to the usual output files for solutions, this run will generate a file called `start_parameters` that contains the seven parameter values for $a_0,\dots,a_6$ that Bertini chooses at random. ∎

To prepare to solve other sextic equations by parameter homotopy, we rename the file `nonsingular_solutions` produced by the ab initio run as `start`.

Example 6.2 (sextic parameter). To solve a sextic equation by parameter homotopy, begin by creating the file `final_parameters`, which could be

```
7

 1.1 0
 2.4 0
 0.8 0
 3.6 0
-0.52 0
-1.8 0
 4.4 0
```

where the first line indicates the number of parameters and the remaining lines give their real and imaginary parts, listed in the same order as they are declared. In this example, all the coefficients of the target sextic are real.

Next, to tell Bertini that this is a parameter run, the configuration setting in the input file (`sextic.input`) is changed to

```
CONFIG
 ParameterHomotopy:2;
END;
```

With this change, all the files for a parameter run are ready:

- input file (`sextic.input`),
- starting set of parameters (`start_parameters`),
- the corresponding startpoints (`start`), and
- the target parameters (`final_parameters`).

To solve the sextic for the new coefficients, just launch Bertini with the command

```
>> bertini sextic.input
```

Note that if we wish to solve other sextics, we merely change the coefficients in `final_parameters` and call Bertini again. ∎

Parameter efficiency settings*

When calling Bertini over and over to solve many examples in the same parameterized family, only the `final_parameters` file changes from run to run. Even so, unless it is told not to, Bertini reads and parses the input file each time. Usually this is a small overhead on the process, but not always, especially if the homotopy has only a few paths to track. To eliminate this overhead, take the following actions:

1. Run the ab initio homotopy as usual and prepare the start file.
2. In the first parameter run, include the configuration setting to tell Bertini to save the results of parsing the input:

```
DeleteTempFiles:0;
```

3. In the second and subsequent parameter runs, tell Bertini to use the previously parsed results:

```
DeleteTempFiles:0;
NeedToDiff:2;
```

The configuration `NeedToDiff:2` is special for parameter runs, as it tells Bertini to read `final_parameters` and substitute these into the homotopy. (In contrast, `NeedToDiff:1` is the default, which parses the input, while `NeedToDiff:0` runs exactly the same homotopy as the last run without changing anything, including the final parameters.)

6.3 ▪ Examples

To illustrate the parameter homotopy method in practice, we consider several examples. One of these, the forward kinematics of Stewart–Gough platform robots, is newly introduced here, while the others are return visits to problems that have already been introduced in Chapter 5: six-revolute (6R) robots, Lagrange points, and the nine-point problem. In all these, we will see that the parameter homotopy method brings substantial gains in efficiency compared to the ab initio runs. Also, we will see later how special cases of these problems submit naturally (and easily) to more advanced refinements of parameter homotopy.

Example 6.3 (Stewart–Gough robots). A Stewart–Gough platform, illustrated in Figure 6.1, is a kind of parallel-link robotic device in which a moving platform is supported on six struts ("legs") anchored to a stationary base. Each leg connects to the base and to the platform with a spherical joint (i.e., ball-and-socket equivalent) at each end.[22] The legs can be extended or retracted in length, typically by screw-type linear actuators or hydraulic cylinders. As the leg lengths change, the platform translates and changes orientation. Generically, each leg puts one constraint on the platform, so that when all six leg lengths are held constant, the platform is held rigidly in space. By coordinating the six leg lengths, a robot controller moves the platform through the six-dimensional space of translation and rotation in $\mathbb{R}^3$, known as $SE(3)$ ("special Euclidean 3-space").

In Example 4.3, we saw that $SE(3)$ is isomorphic to

$$S := \{[e,g] = [e_0,e_1,e_2,e_3,g_0,g_1,g_2,g_3] \in \mathbb{P}^7 \mid \\ Q(e,g) = 0, e * e' = e_0^2 + e_1^2 + e_2^2 + e_3^2 \neq 0\},$$

[22]One of the two can be changed to a universal joint with no change to the essential kinematics.

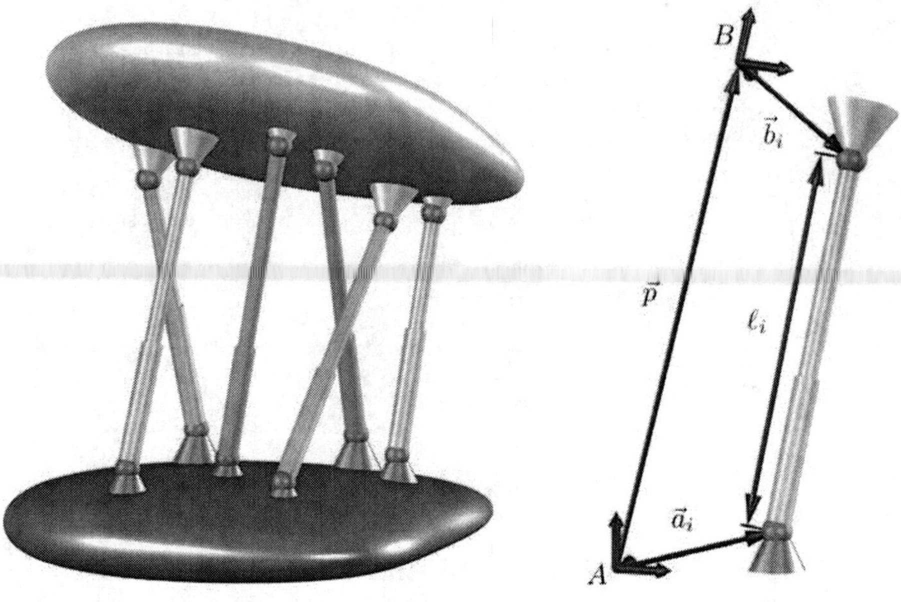

Figure 6.1. *Stewart–Gough platform and notation.*

where $Q(e,g)$ is the Study quadric

$$Q(e,g) = g_0e_0 + g_1e_1 + g_2e_2 + g_3e_3 = 0. \tag{6.6}$$

The kinematic equations for a Stewart–Gough platform are most concisely presented using quaternion notation. We treat e and g as quaternions (i.e., $e = e_0\mathbf{1}+e_1\mathbf{i}+e_2\mathbf{j}+e_3\mathbf{k}$ and similar for g), and vectors in 3-space are treated as quaternions having no 1 component (i.e., they have the form $\vec{v} = x\mathbf{i}+y\mathbf{j}+z\mathbf{k}$).

Let there be coordinate systems A and B in the base and platform, which coincide when $[e,g] = [\mathbf{1},0]$. After a displacement of the platform to $[e,g]$, the origin of B is located at $\vec{p} = g*e'/(e*e')$ with respect to A, and it also rotates according to the value of e. For leg i, $i = 1,\dots,6$, let $\vec{a}_i$ be the center of its spherical joint at the base and let $\vec{b}_i$ be the center of its universal joint at the platform. After displacement of the platform, $\vec{b}_i$ is relocated to

$$\vec{B}_i = (g*e' + e*\vec{b}_i*e')/(e*e'), \tag{6.7}$$

so the corresponding length ℓ_i of leg i is

$$\ell_i^2 = (\vec{B}_i - \vec{a}_i)*(\vec{B}_i - \vec{a}_i)', \quad i = 1,\dots,6. \tag{6.8}$$

The inverse kinematics problem of finding ℓ_i, $i = 1,\dots,6$, given $[e,g]$ is simple using (6.7), (6.8).

The forward kinematics problem of finding $[e,g]$ given the leg lengths requires solving a polynomial system. Since $e*e' \neq 0$ on S, (6.8) can be expanded and rearranged to obtain, for $i = 1,\dots,6$,

$$(\vec{a}_i * \vec{a}'_i + \vec{b}_i * \vec{b}'_i - \ell_i^2) e * e' - e * \vec{b}_i * e' * \vec{a}'_i - \vec{a}_i * e * \vec{b}'_i * e'$$
$$+ g * \vec{b}'_i * e' + e * \vec{b}_i * g' - g * e' * \vec{a}'_i - \vec{a}_i * e * g' + g * g' = 0. \quad (6.9)$$

The system to solve is (6.6), (6.9)) a system of seven quadrics on $\mathbb{P}^7$. Accordingly, the problem has total degree $2^7 = 128$.

One can algebraically reduce this system to a single degree 40 polynomial in one variable [103] or build a special 40-path homotopy [179], but parameter homotopy makes such measures unnecessary. Instead, one simply runs a 128-path total degree ab initio homotopy and then solves subsequent examples with parameter runs. The ab initio run on a general example gives 40 nonsingular solutions, so the parameter runs have 40 paths. The parameter space is Euclidean, $\mathbb{C}^{42}$, consisting of ℓ_i and the six coordinates of $\vec{b}_i, \vec{a}_i$, $i = 1, \ldots, 6$.

The family of general Stewart–Gough platforms includes many special subfamilies of interest to engineers, and we shall return to some of these later in the chapter. ∎

Example 6.4 (6R serial-link robots). We saw in Example 5.2 a formulation of the inverse kinematics problem for 6R robots that has total degree 1024 but a 2-homogeneous Bézout number of 320 and a nonsingular root count of 16. After a 320-path ab initio run, other instances of the problem can be solved with 16-path parameter runs. The parameter space is $\mathbb{C}^{23}$, consisting of $\mathbf{z}_1, \mathbf{z}_6 \in \mathbb{C}^3$ (vectors along the first and last joint axes), a_1, a_2, a_3, a_4, a_5 (associated to link lengths), d_2, d_3, d_4, d_5 (link offset distances), c_1, c_2, c_3, c_4, c_5 (cosines of twist angles), and $\mathbf{p} \in \mathbb{C}^3$ (position of the endpoint). ∎

Example 6.5 (Lagrange points revisited). The formulation for computing Lagrange points (§5.5.1) has a 3-homogeneous Bézout number of 392 and a nonsingular solution count of 32. So after an ab initio run, parameter runs have just 32 paths. The parameter space is $\mathbb{C}$, represented by μ. ∎

Example 6.6 (nine-point problem revisited). The parameter space of the nine-point problem (§5.5.2) is $\mathbb{C}^{16}$, the isotropic coordinates of points $P_1, \ldots, P_8$. (Point P_0 is taken as the origin.) While the ab initio run consists of a rather expensive 645,120 paths using a multihomogeneous homotopy, or roughly a quarter of this using a multihomogeneous regeneration approach, the number of nonsingular roots is just 8652, which is the number of paths in subsequent parameter runs. This reduction in the path count gives a considerable speedup in the execution, amplified by the fact that many troublesome paths having singular endpoints are out of the picture. We will see in a moment that even the 8652 figure can be further reduced sixfold using symmetry. ∎

6.4 ▪ Nonsquare Systems

Recall from §1.1.4 that the procedure for dealing with a nonsquare system $f(z) = 0$ having n equations in N variables, $n > N$, is to form a new system, say, $F(z) = 0$, of N random combinations of the original equations. Because of the randomization, some of the solutions of F might not satisfy f; we call these *nonsolutions*. After these are discarded, the remaining set of points includes all the isolated solutions of f.

When $f(z;q) = 0$ is a parameterized nonsquare system, $f : \mathbb{C}^N \times \mathbb{C}^K \to \mathbb{C}^n$ ($n > N$), the usual parameter homotopy technique works for all general $q \in \mathbb{C}^K$. That is, use randomization to produce a square family of systems $F(z;q) = 0$, then apply the usual procedure:

1. perform an ab initio run to solve $F(z;q_1)=0$ for random $q_1 \in \mathbb{C}^K$;
2. collect the nonsingular solutions, S, of $f(z;q_1)=0$ by removing nonsolutions; and
3. solve $f(z;q_0)=0$ for general $q_0 \in \mathbb{C}^K$, using the parameter homotopy

$$h(z,t) := F(z;tq_1 + (1-t)q_0) = 0, \tag{6.10}$$

tracking the paths that start at S.

The path endpoints of step 3 will all lie in $\mathbf{V}(f(z;q_0))$ and will include all its nonsingular isolated points.

The qualifier "general q_0" in step 3 is critical, because the technique can fail for nongeneric parameters. Consider the simple example

$$f(z;a) = \begin{bmatrix} z-a \\ z-1 \end{bmatrix}.$$

A valid randomization is $F(z;a) = z - a + \rho(z-1)$ for a random number ρ. This has the unique solution $z = (a+\rho)/(1+\rho)$, but that is a nonsolution. The three-step process above correctly tells us that for general a, $f(z;a) = 0$ has no solutions. But clearly $f(z;1) = 0$ has a nonsingular solution: $z = 1$.

This problem is easily remedied. If we wish to obtain all nonsingular solutions even for nongeneric parameters, then the correct procedure is as follows:

1. perform an ab initio run to solve $F(z;q_1)=0$ for random $q_1 \in \mathbb{C}^K$;
2. sort out the nonsingular solutions, S, of F; and
3. solve $f(z;q_0)=0$ for any[23] $q_0 \in \mathbb{C}^K$, using the parameter homotopy

$$h(z,t) := F(z;tq_1 + (1-t)q_0) = 0, \tag{6.11}$$

tracking the paths that start at S. Discard any nonsolutions among the endpoints of this homotopy.

Notice that we cannot discard nonsolutions from the ab initio run; we must track their paths in the parameter run in case they lead to solutions as $t \to 0$. Notice that in the simple example above, the point $z = (a+\rho)/(1+\rho)$, which is a nonsolution for generic a, becomes the desired solution, $z = 1$, when $a = 1$.

To facilitate this procedure, when $n > N$, Bertini writes out a file called `nonsolutions`. (By Bertini's theorem for systems (Theorem 9.3), these are always nonsingular.) When nonsolutions exist, general problems are solved more efficiently by using just the file of `nonsingular_solutions` as the startpoint set S, but to handle nongeneral problems securely, one must form S by merging both `nonsingular_solutions` and `nonsolutions` to form the startpoint file.

Another example of the phenomenon is the system

$$f(z;q) = \begin{bmatrix} z^2 - q^2 \\ z^3 - z - q^3 + q \end{bmatrix} = \begin{bmatrix} (z-q)(z+q) \\ (z-q)(z^2+qz+q^2-1) \end{bmatrix} = 0.$$

[23] To be precise, for any fixed q_1, there may exist some q_0 for which the homotopy fails by some path encountering a singularity. But if the random choice of q_1 is independent of q_0, the procedure succeeds with probability one.

For general q, this has a unique solution $z = q$, but for $q = \pm 1$, it has two nonsingular solutions $z = \pm 1$. A randomization F of f is a cubic equation having three nonsingular roots. Tracking these three roots as $q \to \pm 1$ gives the two endpoints $z = \pm 1$ and a nonsolution. If instead one were to discard nonsolutions immediately after the ab initio run, the parameter run to $q = 1$ would have only one path and it would only find the root $z = 1$.

The analyst has the option of discarding nonsolutions and running parameter homotopies starting from just the solutions from the ab initio run. As the examples show, this may result in missing some solutions at special parameter values, but one might view these as being unstable in that they disappear under perturbation of the parameters. Restricting one's attention to just the roots that are robust under perturbation may be justifiable in some applications. However, the user should be aware of the possible incompleteness of the solution set when taking this option.

6.5 ▪ Nested Parameter Spaces

Parameterized families of systems often have subfamilies that are special cases. An engineer might be interested in Stewart–Gough platforms in which some of the base or platform joint centers coincide. Or 6R arms with some joints intersecting might be useful; indeed, having joint axes 4, 5, 6 intersect to form a "spherical" wrist joint is common.

Suppose we have a subfamily of parameters $Q' \subset Q$, with both the parameter spaces Q and Q' being Euclidean. Then, we can apply the parameter homotopy methodology to either space. But if we have already solved an ab initio example from Q, then we do not need to start over from scratch when we wish to specialize to Q'. Since Q' lies inside Q, the ab initio run for Q' can use parameter homotopy!

Example 6.7 (6R spherical wrist robots). Referring to Example 5.2, recall that if $a_i = 0$, then joints i and $i+1$ intersect. To make joints 4, 5, 6 all intersect in a common point, we need $a_4 = a_5 = 0$, and additionally the offset along joint 5 must be zero: $d_5 = 0$. A robot with this geometric structure is said to have a spherical wrist. The parameter space for the inverse kinematic problem for this subfamily is a Euclidean space $\mathbb{C}^{20}$ obtained by setting $a_4 = a_5 = d_5 = 0$, which lies inside the $\mathbb{C}^{23}$ parameter space of general 6R problems.

Assuming that a parameter homotopy for the general case is already set up, we choose a random spherical wrist 6R problem and solve it, following 16 paths (not the 320 paths of a 2-homogeneous ab initio homotopy). One finds that this gives just eight nonsingular endpoints. These form a startpoint set for eight-path parameter runs to solve other spherical wrist 6R problems.

The inverse kinematics for spherical wrist robots can be broken into two subproblems: first find the poses of the proximal links that place the wrist center in the desired position and then find the rotations of the wrist joints to match the given orientation. These problems have four and two solutions, respectively, giving a total of $4 \cdot 2 = 8$ solutions. The solution of this special case was found by Pieper [144] in the early days of robotics (1968), and many robot arms are constructed this way. Control software for such arms takes advantage of this simplification, but we do not pursue this further here. ∎

Example 6.8 (coincident-joint Stewart–Gough platforms). General Stewart–Gough platforms have 40 forward kinematic solutions, but this number may drop if some of the joint centers coincide. A well-known special case is the *octahedral platform* for which the base and platform joints coincide in pairs to form triangles: $a_1 = a_2$, $a_3 = a_4$, $a_5 = a_6$ in the base and $b_2 = b_3$, $b_4 = b_5$, $b_6 = b_1$ in the platform. The forward kinematic problem for these devices has just 16 solutions. So if the general case has been solved once, the ab initio run for the octahedral case is a 40-path parameter run, while the subsequent parameter runs that stay within the octahedral family have just 16 paths. See [161, §7.7], [56] for more extensive compilations of coincident-joint Stewart–Gough platform families and their solution counts. ∎

Upper semicontinuity* Nested parameter spaces illustrate the general principle of *upper semicontinuity*. Let $f(z;q)$ be a square parameterized family of systems, i.e., $f: \mathbb{C}^N \times \mathbb{C}^K \to \mathbb{C}^N$. We have already said in items 1–3 at the beginning of §6.1 that for general $q \in \mathbb{C}^K$, the number of nonsingular solutions is constant, say, $\mathcal{N}_{\text{nonsing}}$. This count changes only on a proper algebraic subset of $\mathbb{C}^K$ and it can decrease only at nongeneric $q \in \mathbb{C}^K$. If Q is a complex Euclidean affine subspace of $\mathbb{C}^K$, then these same properties hold: there is a constant number of nonsingular solutions for generic points of Q that decreases (and never increases) on a proper algebraic subset of Q.

This is all we need for specializing parameter homotopies to find all nonsingular solutions on nested parameter spaces. However, it is useful to understand this as a special case of a more general principle of upper semicontinuity. Let $g(z;q) : \mathbb{C}^N \times \mathbb{C}^K \to \mathbb{C}^m$ be a polynomial system and let $\mathcal{N}_{\text{nonsing}}(g \neq 0)$ be the number of nonsingular roots of $f(z;q)$ for which $g(z;q) \neq 0$. This number is constant for generic $q \in Q = \mathbb{C}^K$. That is, the points in Q where the count changes is a proper algebraic subset of Q; let us call it $\hat{Q} \subsetneq Q$. For any nested complex Euclidean parameter space, $Q' \subsetneq Q$, we have that either $Q' \subset \hat{Q}$, in which case $\mathcal{N}_{\text{nonsing}}(g \neq 0)$ is smaller on Q' than on Q, or else $Q' \cap \hat{Q}$ is a proper algebraic subset of Q', in which case $\mathcal{N}_{\text{nonsing}}(g \neq 0)$ is generically the same on Q' and Q.

In a sense, $\mathcal{N}_{\text{nonsing}}$ is an instance of the more general principle in that a nonsingular solution path is lost in one of two ways: it lands on a singularity or it diverges to infinity. However, a root that lands on a singularity satisfies the discriminant equation $\det\left(\frac{\partial f}{\partial z}(z;q)\right) = 0$. So the discussion of the previous paragraph holds with $g(z) := \det\left(\frac{\partial f}{\partial z}(z;q)\right)$. Similarly, if we cast a problem into projective space, any solution path that diverges to infinity in $\mathbb{C}^N$ has landed on the hyperplane $z_0 = 0$ in projective space, where z_0 is the homogenizing coordinate. So the count of finite, nonsingular solutions is the same as the count of nonsingular solutions in projective space such that $g(z) := z_0 \neq 0$.

6.6 ▪ Symmetry

Symmetry is a common occurrence in nature and shows up in the solution sets of polynomial systems as well. Suppose that a homotopy $h(z,t) : \mathbb{C}^N \times \mathbb{C} \to \mathbb{C}^N$ respects a symmetry $\sigma(z) : \mathbb{C}^N \to \mathbb{C}^N$; that is, for any (z^*, t^*) such that $h(z^*, t^*) = 0$, it is also true that $h(\sigma(z^*), t^*) = 0$. Then, if z_1 and z_2 are startpoints with $z_2 = \sigma(z_1)$, there is no need to track both paths starting from these points. Instead, after tracking the path from z_1, one may recover the endpoint of the second path by applying the symmetry relation. In fact, if σ is a mapping with orbit k (meaning that k is the smallest integer such that $\sigma^k(z) = z$), then just one path in each symmetry orbit needs to be tracked, thereby reducing work by approximately a factor of k. (Not all paths in the orbit take the same amount of computation to track. If we knew which startpoint would give the cheapest path to track, we would choose it, but this is generally not knowable.) If more than one symmetry action exists, both can be used. For example, if σ_1 and σ_2 are independent symmetry relations with orbits of length k_1 and k_2, they yield a symmetric group of $k_1 k_2$ homotopy paths, only one of which needs to be tracked to find all their endpoints.

Sometimes these symmetries can be identified at the outset, before solving a system, while at other times the symmetry is noticed by examining the solutions of an ab initio run. If known at the outset, there is the possibility of building the symmetry into the ab initio homotopy itself, but this is often difficult: the usual multihomogeneous homotopy or its alternatives do not naturally obey the symmetry of the target system. However, at the parameter homotopy stage, one can take advantage of any symmetry that is obeyed on the whole parameter space. Since the ab initio run targets a generic example, any symmetries that appear in the solution set hold on the whole parameter space. (That is, since the ab initio run selects a target at random, there is a zero probability of choosing a target that displays nongeneric symmetry relations.)

Let's illustrate with a basic example and then two examples from kinematics.

Example 6.9 (symmetric sextics). Let us consider sextics in one variable as in Example 6.1 but restrict our attention to the palindromic subspace where $a_i = a_{6-i}$. In this case, it is easy to see that if z^* is a solution, then so is $1/z^*$. This is true also for the start system $g(z) = z^6 + 1$, so if $f(z;q)$ is a palindromic sextic, so is $h(z,t) = \gamma t g(z) + (1-t) f(z;q)$ at every t. So all six solutions of the problem can be found by following just three paths. If Bertini constructs and solves this ab initio homotopy, it will not know of the symmetry, and it will track six paths. But to set up parameter runs in the palindromic family, the user sorts the roots into reciprocal pairs and keeps just one of each in the `start` file. Then the parameter runs track just three paths. To recover the complete set of solutions, the user must apply the symmetry relation. ∎

Example 6.10 (planar Stewart–Gough platforms). Let's restrict the Stewart–Gough platforms of Example 6.3 to those whose base and platform joints each lie in a plane. In particular, let the z-coordinates of $\vec{a}_1, \dots, \vec{a}_6$ in the base and the z-coordinates of $\vec{b}_1, \dots, \vec{b}_6$ in the platform all be zero. Recall that the vectors $\vec{b}_i$ are relocated to $\vec{B}_i$ and the solutions of the problem are those rigid-body transforms that make the length of $\vec{B}_i - \vec{a}_i$ equal to ℓ_i. Clearly, since the $\vec{a}_i$ vectors all lie in the $z = 0$ plane, these lengths are preserved if the $\vec{B}_i$ vectors are sent to their mirror images through that plane. If the $\vec{b}_i$ are not all coplanar, then this mirror operation would not be a valid rigid-body transform, but it is for the coplanar case. The symmetry relation on $[e, g]$ that sends the vectors $\vec{B}_i$ to their mirror images while still being a proper rigid-body transform is

$$\sigma : [e_0, e_1, e_2, e_3, g_0, g_1, g_2, g_3] \mapsto [-e_0, e_1, e_2, -e_3, g_0, -g_1, -g_2, g_3].$$

This symmetry relation has an orbit of length 2 ($\sigma(\sigma(z)) = z$), so while general Stewart–Gough platforms have 40 solutions, a parameter homotopy for planar Stewart–Gough platforms can find the 20 symmetric pairs by tracking just 20 solution paths. ∎

Example 6.11 (nine-point path synthesis (cont.)). Unlike the previous two problems in which symmetries arise in special cases, every instance of the nine-point path synthesis problem for four-bar linkages has two symmetries built in. We already noted in §5.5.2 that solutions to this problem obey a two-way symmetry:

$$\sigma_1 : (a_1, b_1, a_2, b_2, \bar{a}_1, \bar{b}_1, \bar{a}_2, \bar{b}_2) \mapsto (a_2, b_2, a_1, b_1, \bar{a}_2, \bar{b}_2, \bar{a}_1, \bar{b}_1).$$

This amounts to simply interchanging the labels of the left and right sides of the four-bar in Figure 5.5.

There is a more subtle symmetry also at work, eponymously named "Roberts cognates" [147]. Roberts' theorem states that every four-bar coupler curve is triply generated; that is, for any given four-bar, there are two more that generate the same coupler curve. In our formulation of the problem, this symmetry is given by

$$r : (a_1, b_1, a_2, b_2) \mapsto \left(\frac{a_1 b_2 - a_2 b_1}{b_2 - b_1}, \frac{(a_1 - b_1) b_2}{b_2 - b_1}, a_1, a_1 - b_1 \right),$$

$$\sigma_2 : (a_1, b_1, a_2, b_2, \bar{a}_1, \bar{b}_1, \bar{a}_2, \bar{b}_2) \mapsto (r(a_1, b_1, a_2, b_2), r(\bar{a}_1, \bar{b}_1, \bar{a}_2, \bar{b}_2)).$$

Since σ_1 and σ_2 act independently, the 8652 solutions of a general nine-point problem appear in 1442 groups of six. Placing just one of each group in the `start` file produces a parameter homotopy that tracks just 1442 paths, a considerable reduction from the 645,120 paths in the 4-homogeneous ab initio run. ∎

6.7 ▪ Advanced Refinements

Nested parameter spaces and symmetries are the most important refinements to the parameter homotopy technique, but there are several more possibilities:

- inequalities: one may only desire solutions of $f(z;q) = 0$ for which $g(z;q) \neq 0$;
- nonpolynomial coefficients: one may wish to solve $f(z;q) = 0$, where f is polynomial in z but merely analytic in q; and
- non-Euclidean parameter spaces: the parameter space might be a nonlinear algebraic (or analytic) set.

We address each of these in succession.

6.7.1 ▪ Inequalities

It may happen that a solution set includes degenerate points, i.e., points that satisfy some extra condition which makes them uninteresting or undesirable. Suppose the degenerate condition is given by a polynomial system $g(z;q) = 0$. So for given target parameters q^*, we wish to find the set

$$\{z \in \mathbb{C}^N \mid f(z;q^*) = 0, g(z;q^*) \neq 0\}.$$

It turns out that such inequalities are easily accommodated in the parameter homotopy setting. The procedure is as follows:

- run an ab initio homotopy to solve $f(z;q_1) = 0$ for general parameters q_1;
- drop any solutions that satisfy the degeneracy conditions $g(z;q_1) = 0$; and
- run parameter homotopies $h(z,t) := f(z, tq_1 + (1-t)q_0) = 0$ from q_1 to any new target q_0 starting with the remaining solutions.

The justification for this is the principle of upper semicontinuity (at the end of §6.5 above).

For example, in a kinematics problem formulated in Study coordinates (e, g)—see §4.3—some solutions may have $e * e' = 0$. But the transform equation for Study coordinates acting on vector v is

$$(g * e' + e * v * e')/(e * e'),$$

which has no meaning when $e * e' = 0$. An ab initio run of the problem might yield such solutions. After all, in making the problem polynomial, one has to clear denominators by multiplying through by $e * e'$. The procedure above allows one to eliminate such solutions from the subsequent parameter homotopy runs. (Such points are usually singular as well, and if so, they can be cast aside on that basis also.)

6.7.2 ▪ Nonpolynomial Coefficient Functions

Suppose that $f(z;q)$ is polynomial in z but only analytic in q. That is, when the polynomial is expanded into terms in z, the coefficients may involve trigonometric, logarithmic, exponential, and other analytic functions of q. As long as q lives on complex Euclidean space, nothing changes in the foregoing theory. Bertini recognizes the most common analytic functions (exponential, cosine, sine, and powers) and checks the function definitions in the input file to make sure that the arguments of these functions are limited to numeric constants and symbols declared as constant, random, or parameter (or expressions formed from these). Full details on Bertini's allowed analytic functions can be found in Appendix B.

6.7.3 ▪ Non-Euclidean Parameter Spaces

In the foregoing, we have always restricted the parameter space Q and any nested parameter spaces to be Euclidean ($\mathbb{C}^K$). But this simplification is not crucial to the theory of parameter homotopy. The basic technique and the principle of upper semicontinuity hold as well if Q is an irreducible algebraic set or even an irreducible analytic set (one defined as the zero set of a system of analytic equations).

At present, Bertini does not support non-Euclidean parameter spaces, but this problem can be circumvented for many of the most useful non-Euclidean parameter spaces, as they can be reparameterized to appear as Euclidean spaces. For example, if parameters (q_1, q_2) live on the unit circle, $q_1^2 + q_2^2 = 1$, one can use the parameterization $q_1 = (1 - t^2)/(1 + t^2)$, $q_2 = 2t/(1 + t^2)$, which covers a dense open subset of the unit circle. (There is no t corresponding to $(q_1, q_2) = (-1, 0)$.) With these substitutions, $t \in \mathbb{C}$ becomes the new parameter space. In some cases, the reparameterization might involve analytic functions. For example, $SO(3)$, the space of 3×3 orthogonal matrices with determinant 1, can be most easily reparameterized using three Euler angles, in which case the entries of the matrix are trigonometric functions of the angles. The space of three independent angles is then $\mathbb{C}^3$, so one may then proceed as usual.

The reason Bertini does not support non-Euclidean parameter spaces is due to the difficulty of automatically establishing general paths on them. Whereas the straight-line segment between any two points of $\mathbb{C}^K$ is also in $\mathbb{C}^K$, this is not true on a nonlinear space.

6.7.4 ▪ Singular Isolated Solutions*

It may happen that a system has isolated solutions that are singular. Like the number of nonsingular solutions, the number of isolated singular solutions is constant on a dense Zariski open subset of parameter space (i.e., invariant on all but a proper algebraic subset), and for square systems, the principle of upper semicontinuity still applies in a modified form. The modification is that there are multiple counts that obey upper semicontinuity, each being the cumulative number of solutions up to a multiplicity limit. It is each of these counts that can only decrease as a parameter path lands on a nested parameter space. To be precise,

- let $\mathcal{N}_1 \equiv \mathcal{N}_{\text{nonsing}}$ be the number of nonsingular roots,
- let $\mathcal{N}_2$ be the number of isolated singular roots of multiplicity 2, and similarly
- let $\mathcal{N}_k$ be the number of isolated singular roots of multiplicity k.

Then, as we know, $\mathcal{N}_1$ can only decrease in approaching an exceptional set, but in doing so, some nonsingular roots may merge to form singular roots of higher multiplicity. For example, if two nonsingular roots merge to form a root of multiplicity 2, then $\mathcal{N}_1$ decreases by two while $\mathcal{N}_2$ increases by one. So the counts for higher multiplicity roots may increase, but only at the cost of more roots with lower multiplicity. So the upshot is that

- each count $\mathcal{N}_k$, $k = 1,2,\ldots$, is a constant on a dense Zariski open subset of Euclidean parameter space Q; and
- each of the sums $S_k := \sum_{\mu=1}^{k} \mathcal{N}_\mu$, $k = 1,2,\ldots$, can only decrease on approach to a nested Euclidean parameter space $Q' \subsetneq Q$.

The maximum multiplicity of any root is at most the total degree of the system, so there is a k' beyond which all the counts are zero: $\mathcal{N}_k = 0$, all $k > k'$. In fact, since an isolated solution of multiplicity μ contributes μ to the Bézout number, we have

$$\sum_{\mu=1}^{\infty} \mu \mathcal{N}_\mu \leq D = d_1 \cdots d_N,$$

where D is the total degree of the square system.

Given these facts, one may find all isolated solutions, nonsingular and singular, by tracking paths in a parameter homotopy just as we have been doing above for nonsingular solutions. And upon arriving at a general point of a nested parameter space, new solution counts can be established, and parameter homotopy restarts as usual inside this space. While at first sight of it this is a simple extension of the parameter theory for nonsingular roots, it presents several difficulties that we defer to later sections:

1. How do we tell which singular endpoints are isolated, that is, not on higher-dimensional solution sets?
2. How do we determine the multiplicity of an isolated root?
3. How do we track singular solutions along a path in parameter space since Newton's method may not converge or may converge slowly?

These questions lead us toward the idea of deflation, in which the singularity of a solution is analyzed and a new polynomial system is produced having the same root but now as a nonsingular solution. This deflated polynomial system can then be used to track the originally singular point. See §11.6 for details about multiplicity and deflation. Although the techniques are available and used in other contexts within Bertini, at present the capability to track singular isolated roots in a parameter homotopy is not offered.

6.8 ▪ Bertini Guide: Parameter Homotopy

All the variants of parameter homotopy follow the same three-step process:

- run an ab initio homotopy on a general example, which creates a random list of parameters in `start_parameters`;
- sort the solutions to create a file of startpoints, `start`; and
- run parameter homotopy on other instances of the problem whose parameters are listed in `final_parameters`.

The different variants taking advantage of nested parameter spaces, symmetries, etc., all differ only in which ab initio solutions are transferred into the `start` file.

Ab Initio Summary

Input files
 `input` = run configuration & system definition
Key configurations:
 `ParameterHomotopy:1;`
Key declarations:
 `parameter`
Notes:
 Use any multihomogeneous formulation and/or regeneration.
Execution:
 `>> bertini <input>`

Parameter Homotopy Summary

Input files
 `input` = run configuration & system definition
 `start` = list of startpoints
Key configurations:
 `ParameterHomotopy:2;`
Key declarations:
 `parameter`
Notes:
 Input file is exactly the same as the *ab initio* run
 except for the configuration setting.
Execution:
 `>> bertini <input> <start>`

Final note. For a parameter space Q' that is nested inside a larger one, Q, $(Q' \subsetneq Q)$, the ab initio step for Q' can be replaced by a parameter homotopy in Q that targets a general point of Q'.

Chapter 7

Advanced Topics about Isolated Solutions

In this chapter, we consider more advanced algorithms built from the basic building blocks presented in the previous chapters.

During a homotopy run, the endpoints are approximated by the endgame to the tolerance described by the configuration `FinalTol` (see §3.3). With appropriate settings, after the endgame has obtained an accurate approximation based on `FinalTol`, Newton's method will be used to approximate the nonsingular endpoints to arbitrarily many digits as shown in §7.1.

After a homotopy run, the set of endpoints can be sorted into sets including the real and nonreal points, singular and nonsingular points, and isolated and nonisolated points. The initial sorting of endpoints by Bertini depends on various user-specified (or default) configuration settings. Since these basic tests can yield incorrect results, Bertini allows for additional computations to be performed on the endpoints in order to classify them correctly. One option is to simply rerun Bertini's postprocessor with different configuration settings. Another option is to recompute selected endpoints to increased accuracy, using either Newton's method or an endgame, without needing to rerun the entire homotopy. Both options are accessible from Bertini's sharpening module, which we discuss in §7.2.

Independent of a homotopy run, Bertini can also evaluate a system and its Jacobian, perform a Newton iteration, and approximate the condition number of the Jacobian on a list of points using arbitrary fixed precision. Even though these options can be utilized without using Bertini to run a homotopy, they can be useful when trying to identify the endpoints of a homotopy of interest, e.g., to help find the solutions which satisfy some additional relations not built into the original system. These topics are presented in §7.3.

Due to the behavior of Newton's method near singular solutions, we discuss the use of parameter homotopies to sharpen solutions in §7.4.

User-defined homotopies first appeared in §2.1 for tracking around the unit circle. These homotopies, which are discussed in §7.5, provide functionality for general-purpose tracking routines.

In §7.6, we discuss how Bertini attempts to determine the rank of a matrix. For example, this is used in the sharpening module to decide which endpoints are singular. In particular, for singular solutions, the minimum singular value approaches zero as the error in approximating the point decreases. Therefore, by comparing the minimum singular value using different approximations, Bertini is able to reliably determine the singular endpoints.

In many applications, especially for polynomial systems arising in chemistry [3, 131], the coefficients for a polynomial system can be widely scaled. Since this can have a drastic impact on many aspects of numerically solving a polynomial system, function and variable scaling are discussed in §7.7.

The last section, §7.8, addresses rounding errors in polynomial construction.

7.1 ▪ Sharpening Using Newton's Method During a Homotopy Run

Due to the ubiquitous nature of Newton's method for solving systems of nonlinear equations, Bertini has three different ways of using Newton's method, aside from its usual use during path tracking. The first use, described here, performs Newton iterations on an endpoint during the course of a homotopy run after it has been approximated by an endgame. The second use, described in §7.2.2, is in the sharpening module, which allows further Newton iterations to be performed on an endpoint computed by a prior homotopy run. The final use, described in §7.3.2, performs a Newton iteration in arbitrary fixed precision on a given list of points for a given system.

During a homotopy run, an endgame is used to approximate the endpoints to a given tolerance, namely, `FinalTol`. Due to the quadratic convergence of Newton's method near nonsingular solutions, Newton's method is drastically more efficient than a standard endgame for computing a nonsingular endpoint to arbitrarily many digits. Therefore, Bertini provides the functionality to attempt to use Newton's method to increase the accuracy immediately after the endgame has finished, described by the configuration `SharpenDigits`.

As we saw in §1.1.2, Example 1.8, Newton's method applied to points near singular solutions can converge slowly or diverge. Thus, before Bertini attempts to sharpen endpoints using Newton's method, it analyzes the behavior of at most three Newton iterations, in varying precision if adaptive precision path tracking is being utilized. If these iterations demonstrate the potential for quadratic convergence to a solution, Bertini then performs the sharpening via repeated Newton iterations.

We illustrate the use of Newton's method during a homotopy run on

$$x^2+(y-1)^2=1, \quad y=2x^2. \tag{7.1}$$

This system describes the intersection of a circle and a parabola which has two nonsingular solutions $(\pm\sqrt{3}/2, 3/2)$ and a multiplicity 2 solution $(0,0)$.

Let's consider the following system:

```
% SharpenDigits.input
CONFIG
 % Solve using a total-degree homotopy
 SharpenDigits:20; % sharpen the endpoints to 20 digits
END;
INPUT
 function circle,parabola;
 variable_group x,y;

 circle = x^2 + (y-1)^2 - 1;
 parabola = y - 2*x^2;
END;
```

Running Bertini with this input will construct a total-degree homotopy and approximate the endpoints first using an endgame. After successful convergence of the endgame, the line

```
SharpenDigits:20; % sharpen the endpoints to 20 digits
```

instructs Bertini to sharpen the endpoints so that the resulting approximation is correct to 20 digits using Newton's method.

Before perusing the output, let's first consider the expected behavior of Bertini. We expect that the approximation generated by the endgame for the two nonsingular solutions to be close enough so that quadratic convergence of Newton's method is obtained. For the singular solution, we expect the analysis stage to fail. In particular, since the singular solution has multiplicity 2, we expect two failures. These failures are described to the user at the bottom of the screen via the following:

```
Number that failed to sharpen: 2
    Singular endpoints: 2
```

We can further investigate the outcome of the first stage analysis for testing for quadratic convergence in the file `output`. For example, the following is a summary for one of the paths that ended at the origin:

```
prec: 96 newton: 4.446718e-13 func: 7.863550e-26
prec: 128 newton: 2.172778e-13 func: 1.993445e-26
prec: 160 newton: 1.086389e-13 func: 4.983611e-27
min sv: 4.473054e-13
```

This summary shows that three Newton iterations using increasing precision resulted in poor convergence of both the Newton and the function residuals. Also, the estimated minimum singular value is approximately $4.5 \cdot 10^{-13}$.

We compare this with the summary of a path leading to a nonsingular solution:

```
prec: 96 newton: 1.724389e-15 func: 1.009750e-28
prec: 128 newton: 9.722574e-29 func: 4.702013e-38
prec: 160 newton: 2.297367e-38 func: 5.473822e-48
min sv: 2.264692e-01
```

This summary shows that three Newton iterations using increasing precision yielded a rapid decrease in both the Newton and the function residuals. Additionally, the estimated minimum singular value is approximately 0.23. We note that the limited precision prevents us from observing the quadratic convergence rate in the third iteration. Nonetheless, due to the rapid convergence, Bertini performs the second stage and yields the following approximation listed in `finite_solutions`:

```
0.866025403784438646763723170750e0 0.000000000000000000000000000000e0
0.150000000000000000000000000000e1 0.000000000000000000000000000000e0
```

To sharpen singular solutions, one may use endgame sharpening (see §7.2.2), parameter homotopy sharpening (see §7.4), or the positive-dimensional sharpening module which regularizes the singular solution using deflation (see §11.6).

7.2 ▪ Sharpening Module

In §7.1, Newton's method was applied during the running of a homotopy after the endgame. The *sharpening module* can be used to perform similar computations *after* running a homotopy. In particular, the sharpening module has two main parts: the postprocessor and automatic root sharpening. The postprocessor, discussed in §7.2.1, recreates the output files based on the new configurations. Root sharpening, which approximates the root to higher accuracy, can use either Newton's method or an endgame (see §3.3).

One invokes the sharpening module by using the line

```
SharpenOnly:1;
```

The required data from a previous run is stored in file `raw_data`. When using the endgame sharpening option, the file `midpath_data` is also needed since it contains the point on each path where the endgame was originally started.

7.2.1 ▪ Postprocessing

After running a homotopy, Bertini's postprocessor sorts the endpoints into real and nonreal points, singular and nonsingular points, and finite and infinite points when "infinity" is defined, e.g., when using `variable_group`. (Variables declared with `hom_variable_group` are coordinates of a projective space and so have no uniquely defined point at infinity.) An endpoint is considered real if the imaginary part of the point is smaller than `ImagThreshold`. An endpoint is considered nonsingular if all the other endpoints are outside the ball centered at the point of radius `FinalTol*TargetTolMultiplier` and the approximation of the condition number is smaller than `CondNumThreshold`. Finally, an endpoint is considered finite if the absolute value of each coordinate is smaller than `EndpointFiniteThreshold`.

Since these configurations may need to be adjusted depending on the polynomial system, Bertini's sharpening module can rerun the postprocessor using possibly different configurations specified in the `input` file. The only other file needed is `raw_data`, which is created during the course of a homotopy run. We demonstrate this idea on the example

$$y = x^{10}, \quad y = 10^{-10}, \tag{7.2}$$

which has 10 nonsingular isolated solutions, $(\omega^k/10, 10^{-10})$ for $k = 0, 1, \ldots, 9$, where $\omega = \exp(\pi\sqrt{-1}/5)$ is the 10th root of unity. The condition numbers for these nonsingular solutions are approximately $2 \cdot 10^8$, which arise from the degree of the first equation and the relatively small value of y in the second. Running Bertini with the input

```
% HighCondNum.input
 function f,g;
 variable_group x,y;

 f = y - x^10;
 g = y - 1e-10;
END;
```

produces the following onscreen summary, suggesting all 10 solutions are singular:

```
              | Number of real solns  | Number of nonreal solns       | Total
--------------------------------------------------------------------------------
Nonsingular   |         0             |                0              |  0
Singular      |         2             |                8              |  10
--------------------------------------------------------------------------------
Total         |         2             |                8              |  10
```

The approximated condition numbers can be located for each point by perusing the file main_data, e.g.,

```
Estimated condition number: 3.385314942520330e+08
```

For such a small problem, one could simply rerun the homotopy using a different value of CondNumThreshold. However, by maintaining the raw_data file from this previous run, we can use the sharpening module by running Bertini again from the same folder with only minor modifications to the input code. In the following code, we instruct Bertini to run the sharpening module with the modified value of CondNumThreshold:

```
% HighCondNum_sharpen.input
CONFIG
 SharpenOnly:1; % use the sharpening module
 CondNumThreshold:1e12; % adjust the condition number threshold
END;
INPUT
 function f,g;
 variable_group x,y;

 f = y - x^10;
 g = y - 1e-10;
END;
```

When Bertini starts, rerunning the postprocessor is accomplished by selecting option 4 from the sharpening module menu:

```
Sharpening options:
  1. Sharpen all endpoints
  2. Sharpen endpoints listed in a file
  3. Manually input endpoints to sharpen
  4. Recreate output (i.e., run the postprocessor)
  5. Change the number of sharpening digits (currently 14 digits)
  6. Change the sharpening method (currently Newton's method)
  7. Quit
```

The summary printed to the screen is now

	Number of real solns	Number of nonreal solns	Total
Nonsingular	2	8	10
Singular	0	0	0
Total	2	8	10

7.2.2 ▪ Root Sharpening after a Homotopy

Since proper classification of endpoints can sometimes be accomplished only by recomputing the endpoints to higher accuracy, the sharpening module provides the option to sharpen the endpoints computed by a homotopy either using Newton's method or rerunning an endgame. For example, we can identify real endpoints by comparing the norm of the imaginary part of various approximations since the imaginary part must converge to zero as the endpoint is sharpened. We remark that for a nonsingular solution to a square polynomial system with real coefficients, α-theory [36, Chapter 8] can be used to *prove* the reality or nonreality of the solution [96], which is further discussed in §14.3.

In the sharpening module, the user can first change the number of sharpening digits and the sharpening approach. Then, the user can select whether to attempt to sharpen all endpoints, sharpen only the endpoints for which the path numbers are listed in a file, or manually enter the path numbers. These options are all clearly seen from the sharpening menu:

```
Sharpening options:
  1. Sharpen all endpoints
  2. Sharpen endpoints listed in a file
  3. Manually input endpoints to sharpen
  4. Recreate output (i.e., run the postprocessor)
  5. Change the number of sharpening digits (currently 14 digits)
  6. Change the sharpening method (currently Newton's method)
  7. Quit
```

When using Newton's method, which is the default for the sharpening module, Bertini uses the same two-stage approach discussed in §7.1. In summary, the first stage uses some sample Newton iterations to test for convergence, and upon success, the second stage then repeatedly performs Newton iterations.

To demonstrate, we consider the intersection of a hyperbola and parabola:

$$x^2 - y^2 = 1, \quad x = y^2 + 1. \tag{7.3}$$

The intersection consists of the nonsingular points $(0, \pm\sqrt{-1})$ and the multiplicity 2 point $(1,0)$. The following uses a total-degree homotopy to solve the system:

```
% HyperbolaParabola.input
 function hyperbola,parabola;
 variable_group x,y;

 hyperbola = x^2 - y^2 - 1;
 parabola = x - y^2 - 1;
END;
```

For comparison later, we note that the file `finite_solutions` contains the following approximation of $(0, \sqrt{-1})$:

```
-3.606140745660162e-16 2.610861849468747e-16
 2.021223900056102e-16 1.000000000000000e+00
```

Now, we want to use Newton's method to attempt to sharpen all the endpoints correct to 20 digits, which we expect to fail for the singular solution. Among the files generated is the file `raw_data`, which contains the data needed by the sharpening module for running Newton's method. The following instructs Bertini to start the sharpening module:

```
% HyperbolaParabola_sharpen.input
CONFIG
 SharpenOnly:1; % use the sharpening module
 SharpenDigits:20; % sharpen to 20 digits
END;
INPUT
 function hyperbola,parabola;
 variable_group x,y;

 hyperbola = x^2 - y^2 - 1;
 parabola = x - y^2 - 1;
END;
```

By selecting the first option of the sharpening menu, namely,

```
1. Sharpen all endpoints
```

Bertini will perform the two-stage sharpening approach using Newton's method. The onscreen summary, namely,

```
Endpoints Considered for Sharpening: 4
 Finite Sharpening Successes: 2
 Sharpening Failures: 2
     Singular Endpoints: 2
```

shows that four endpoints were considered, two were successfully sharpened, and two failed the first-stage singularity test. In the file `finite_solutions`, the point corresponding to $(0, \sqrt{-1})$ is now listed as

```
-0.432342662628787632173397376371e-77 0.335766689532499599836898630831e-77
 0.000000000000000000000000000000e0   0.100000000000000000000000000000e1
```

Since the sharpening approach based on Newton's method is designed to fail at singular endpoints, we now consider rerunning an endgame to sharpen singular endpoints. As stated above, the sharpening module needs the data stored in the file `raw_data`. When using this endgame sharpening approach, the file `midpath_data` is also needed since this contains the point on each path where endgame was started. With these files, together with the `input` file, Bertini is able to reconstruct the homotopy and rerun the endgame.

The endgame sharpening approach is utilized by selecting option 6 of the sharpening menu,

```
6. Change the sharpening method (currently Newton's method)
```

and then selecting option 2 from the sharpening method menu:

```
Sharpening method:
  1. Newton's method
  2. Endgame
```

The comment for option 6 of the sharpening menu will now be updated to

```
6. Change the sharpening method (currently an endgame approach)
```

Note that an endgame sharpening approach is not permitted when either a regeneration (see §5.4) or a diagonal equation-by-equation approach (see §E.11.1) was used since these are implemented in Bertini to compute only nonsingular solutions. Also, an endgame sharpening approach enforces an upper bound of 300 digits. For more accurate approximations beyond 300 digits, consider restoring the quadratic convergence of Newton's method using some deflation approach (see §11.6 and §13.2).

7.3 ▪ Evaluation and Newton Iterations

Independent of running a homotopy, Bertini can be used to evaluate a system and Jacobian, perform a Newton iteration, and approximate the condition number using arbitrary fixed precision. These operations are useful in many circumstances, e.g., removing extraneous solutions in §6.7.1. We discuss system and Jacobian evaluation in §7.3.1 and Newton iteration with condition number estimation in §7.3.2.

These operations need only a list of points with an input system and we will demonstrate them with the system in (7.1) with the points $(3/4,2)$ and $(10^{-15},1/4)$. The input file will be constructed as needed, but we use the following start file:

```
2

0.75 0
2 0

1e-15 0
0.25 0
```

Since evaluation and a Newton iteration are performed using computations done in fixed precision, this is the first time where we consider setting the precision. The default setting of adaptive multiprecision will use double precision for these computations. To demonstrate the use of fixed multiprecision, we will use 64-bit precision throughout this section. This is accomplished by adding the following lines to each input file:

```
 MPType:1;
 Precision:64;
```

The first indicates that a fixed multiprecision setting will be used, with the second selecting the number of bits of precision. See §E.2 for more details.

7.3.1 ▪ Evaluating a System and Its Jacobian

Since evaluation is independent of tracking a homotopy, using Bertini for evaluation requires modifying the TrackType configuration. In particular, a value of −4 corresponds to system evaluation while a value of −3 will evaluate both the system and its Jacobian. The following instructs Bertini to evaluate the system and its Jacobian for the circle and parabola system using 64-bit precision:

```
% CircleParabola_eval.input
CONFIG
 TrackType:-3; % evaluate system and Jacobian
 % use 64-bit precision
 MPType:1;
 Precision:64;
END;
INPUT
 function circle,parabola;
 variable_group x,y;

 circle = x^2 + (y-1)^2 - 1;
 parabola = y - 2*x^2;
END;
```

The onscreen summary shows that two files are created, one for the function evaluation and one for the Jacobian:

```
Evaluating point 0 of 2

------------------------------------------------------------------------
The following files may be of interest to you:

function:  The function values evaluated at the given points.
Jv:        The Jacobian w.r.t. variables evaluated at the given points.
------------------------------------------------------------------------
```

The files `function` and `Jv` both have the same structure as the `start` file. For each point, the values of each function are listed in the order in which the functions are declared. Similarly, the Jacobian matrix is printed based on the declared order of the functions and variables using a row major ordering. For example, the `Jv` file is in this case is

```
 2

 0.150000000000000000000e1  0.000000000000000000000e0
 0.200000000000000000000e1  0.000000000000000000000e0
-0.300000000000000000000e1 -0.000000000000000000000e0
 0.100000000000000000000e1  0.000000000000000000000e0

 0.199999999999999999999e-14  0.000000000000000000000e0
-0.150000000000000000000e1    0.000000000000000000000e0
-0.399999999999999999998e-14 -0.000000000000000000000e0
 0.100000000000000000000e1    0.000000000000000000000e0
```

which corresponds to the two matrices

$$\begin{bmatrix} 1.5 & 2 \\ -3 & 1 \end{bmatrix} \text{ and } \begin{bmatrix} 2\cdot 10^{-15} & -1.5 \\ -4\cdot 10^{-15} & 1 \end{bmatrix}.$$

7.3.2 ▪ Performing a Newton Iteration

Normally, Bertini executes Newton's method only near nonsingular solutions, as this is where good behavior can be expected. But for maximal flexibility, the user can also ask Bertini to perform a single Newton iteration on any given set of points. As with evaluation, this requires modifying the `TrackType` configuration. In particular, a value of −2 corresponds to performing a Newton iteration, while a value of −1 will perform a Newton iteration and simultaneously approximate the condition number. We remark that when using a group of variables that is defined on projective space, e.g., declared using `hom_variable_group`, the Newton iteration changes the patch based on the input point. In particular, for the given point y, the patch used is $y^H x - y^H y = 0$ where y^H is the Hermitian transpose (conjugate transpose) of y.

The following file performs a Newton iteration and approximates the condition number using 64-bit precision:

```
% CircleParabola_newton.input
CONFIG
 % Perform a Newton iteration & approximate the condition number
 TrackType:-1;
 % use 64-bit precision
 MPType:1;
 Precision:64;
END;
```

```
INPUT
 function circle,parabola;
 variable_group x,y;

 circle = x^2 + (y-1)^2 - 1;
 parabola = y - 2*x^2;
END;
```

The onscreen summary describes the four files that are created:

```
Evaluating point 0 of 2

------------------------------------------------------------------------
The following files may be of interest to you:

successFlag:     Success or failure of Newton iterations.
newPoints:       The point resulting from the Newton iteration.
newtonResidual: The Newton residual for each point.
CN:              Approximation of the condition number at each point.
------------------------------------------------------------------------
```

Since the condition number for the point $(10^{-15}, 1/4)$ is on the order of 10^{15}, we expect an error when performing a Newton iteration for this point using only 64 bits of precision (roughly 19 digits). In particular, the `successFlag` file is

```
2

0

-1
```

which indicates that two points were attempted (corresponding to 2); the first was successful (corresponding to 0), but the second failed (corresponding to -1). We can also see this failure by looking in the `CN` file, which is

```
2

4.725097896271715e+00

-1
```

This file also indicates that two points were attempted (corresponding to 2), the first has a condition number of approximately 4.7 and the failure of the second is indicated by a reported condition number of -1. We note that due to the random numbers used in approximating the condition number, the approximation of the condition numbers may change in subsequent runs.

To rectify this Newton iteration failure, we need to increase the precision, which is accomplished by changing the value of `Precision`. By performing the same computations using 128-bit precision, the `successFlag` file now indicates that both were successful with the approximate condition numbers in `CN` as

```
3.343744273523548e+00

1.627091328340154e+15
```

The following is the `newtonResidual` file which lists the Newton residuals for this computation:

```
2

4.902372793100671314134911519201955119926e-1

1.713709677419354838709677419753147762744e13
```

In particular, this shows that the second point has a large Newton step.

The `newPoints` file contains the points computed after the Newton iteration. To perform another Newton iteration, just rename `newPoints` to `start` and run Bertini again. However, to ensure proper convergence of Newton's method, the precision should be adjusted based on the Newton residual and condition number, which may vary from point to point.

7.4 ▪ Parameter Homotopies for Root Sharpening

In the sharpening module (see §7.2), solutions can be approximated to higher accuracy if they arise as the endpoint of a homotopy path. When approximations to solutions do not arise in this fashion, e.g., using some other numerical solving technique, one common option is to apply Newton iterations, as discussed in §7.3.2, to sharpen the approximation. Due to the behavior of Newton's method near singular solutions, another sharpening option is to use a fixed-point or Newton homotopy. Both of these define a homotopy with the given point as a startpoint of a path with the expectation that the endpoint of this homotopy path is the solution of interest. These homotopies generally work when the given point is close enough to a solution, either singular or nonsingular. As we will show, the parameter homotopy technique discussed in Chapter 6 can be adapted to implement these root sharpening homotopies.

One main difference between the fixed-point and Newton homotopies is the conditioning of the startpoint. For a fixed-point homotopy, the Jacobian at the startpoint is the identity matrix. The Jacobian at the startpoint of Newton homotopy is the Jacobian of the original system evaluated at the startpoint, which may be poorly conditioned, resulting in an immediate increase in the precision when using adaptive multiprecision tracking.

We will present these homotopies in the context of computing the intersection of the parabola $y = x^2$ and the intersection of two lines defined by $x^2 = y^2$, which has two nonsingular solutions at $(\pm 1, 1)$ and a multiplicity 2 solution at the origin.

7.4.1 ▪ Fixed-Point Homotopy

For a square polynomial system $f : \mathbb{C}^n \to \mathbb{C}^n$ and a point $z_0 \in \mathbb{C}^n$, a fixed-point homotopy for f starting at z_0 is of the form

$$H(z,t) = (1-t)f(z) + \gamma t(z - z_0), \tag{7.4}$$

where $\gamma \in \mathbb{C}$. Clearly, $H(z_0, 1) = 0$ and $H(z, 0) = f(z)$. Since this homotopy has exactly one startpoint and the system f can have arbitrarily many isolated solutions, H is not a *good homotopy* as defined in §2.1. However, it is still useful as a heuristic method of finding a root of f near z_0.

The following constructs the general framework for a fixed-point homotopy for the parabola and two-lines system using a parameter homotopy and the endgame to approximate the solution to 20 digits:

```
% FixedPoint.input
CONFIG
 % Use a fixed-point homotopy to sharpen
 ParameterHomotopy:2;
 FinalTol:1e-20;
END;
INPUT
 function parabola,twolines
 variable_group x,y;
 parameter t,x0,y0;
 random gamma;

 parabola = (1-t)*(y - x^2) + gamma*t*(x - x0);
 twolines = (1-t)*(x^2 - y^2) + gamma*t*(y - y0);
END;
```

The point-specific data for this parameter homotopy is contained in three additional files needed for a parameter homotopy, namely, `start`, `start_parameters`, and `final_parameters`. The `start` file contains the startpoint for our fixed-point homotopy. Here, we take the startpoint to be a point close to the origin:

```
1

9.714922892647063e-15 3.020012792979345e-14
2.423670784618578e-15 4.407625914310857e-15
```

For our homotopy, the parameter `t` starts at 1 and ends at 0, while the parameters `x0` and `y0` are actually constant, namely, equal to the values in `start`. We could have specified these in the input file using a `constant` declaration, but declaring them as parameters has the advantage that we can use the same `input` file for different starting values. Thus, the corresponding `start_parameters` file is as follows, with the leading 3 indicating that the system has three parameters:

```
3

1 0
9.714922892647063e-15 3.020012792979345e-14
2.423670784618578e-15 4.407625914310857e-15
```

The `final_parameters` file is identical to `start_parameters` except that the first parameter, t, is set to zero.

After running Bertini, the `finite_solutions` file lists the sharpened point:

```
0.135471534249197034448169167307520676 81e-23 -0.802065810704713260241316167201449886 01e-24
0.165433242516359130820999121672607761 53e-26  0.170742890560779447162728006331855380 03e-31
```

7.4.2 ▪ Newton Homotopy

For a square polynomial system $f : \mathbb{C}^n \to \mathbb{C}^n$ and a point $x_0 \in \mathbb{C}^n$, a Newton homotopy for f starting at x_0 is of the form

$$H(x,t) = (1-t)f(x) + \gamma t(f(x) - f(x_0)) = (1-t+\gamma t)f(x) - \gamma t f(x_0), \tag{7.5}$$

where $\gamma \in \mathbb{C}$. Clearly, $H(x_0, 1) = 0$ and $H(x, 0) = f(x)$.

For our running example, we utilize the same `start`, `start_parameters`, and `final_parameters` files as in §7.4.1. The following constructs the general framework for a Newton homotopy using a parameter homotopy and the endgame to approximate the solution to 20 digits:

```
% NewtonHomotopy.input
CONFIG
 % Use a Newton homotopy to sharpen
 ParameterHomotopy:2;
 FinalTol:1e-20;
END;
INPUT
 function parabola,twolines
 variable_group x,y;
 parameter t,x0,y0;
 random gamma;

 parabola = (1-t+gamma*t)*(y - x^2) - gamma*t*(y0 - x0^2);
 twolines = (1-t+gamma*t)*(x^2 - y^2) - gamma*t*(x0^2 - y0^2);
END;
```

When running this homotopy, due to the poorly conditioned startpoint, Bertini immediately increases the precision to 96 bits and uses this precision throughout the computation. After the computation, `finite_solutions` lists the sharpened point:

```
-0.444523288416970307123615965039e-22 -0.417777897572984509870687795826e-22
 0.341444083929845981448559930935e-22 -0.606698002721993667567718892852e-22
```

7.5 ▪ User-Defined Homotopies

In parameter homotopies in Bertini, as discussed in Chapter 6, the variables must appear in polynomial form, and the parameters evolve along straight lines in complex Euclidean parameter space. In contrast, Bertini's user-defined homotopies are more flexible in that they need not be polynomial in the variables and their parameters can be scheduled to move along nonlinear paths. In particular, user-defined homotopies allow for function evaluation, Newton iterations, and path tracking for systems which need not be polynomial or even differentiable everywhere; for example, they may involve sin(), cos(), and exp() and the arguments of these may be rational functions.

Even though user-defined homotopies are flexible, there are two requirements for their construction. First, each user-defined homotopy must declare exactly one `pathvariable`. As is standard in Bertini, the path variable will start at 1 and decrease to 0. Second, each user-defined homotopy must declare and define at least one `parameter`. It often happens that a user-defined homotopy depends upon one parameter which is set equal to the path variable and shown in the following:

```
pathvariable T;
parameter t;
t = T;
```

Another option is to use the parameter to consider the homotopy as moving from 0 to 1 by using the following:

```
pathvariable T;
parameter t;
t = 1-T;
```

The remainder of this section focuses on the two types of user-defined homotopies in Bertini. The first, which performs path tracking using affine coordinates, is discussed in §7.5.1. The other, which performs path tracking on a product of projective spaces, is discussed in §7.5.2.

7.5.1 ▪ User-Defined Homotopies in Affine Coordinates

A user-defined homotopy using affine coordinates is the most flexible option in Bertini in terms of the systems that can be used. Due to this flexibility, Bertini does not attempt to compute the adaptive precision configurations `CoeffBound` and `DegreeBound` as it does for polynomial systems (see §E.3 for more details). Moreover, to distinguish that affine coordinates are being used, we must include the configuration line

```
UserHomotopy:1;
```

and variables for the system must be declared by the keyword `variable`.

The capabilities of Bertini are demonstrated in the following example.

Example 7.1. As a demonstration of the power of affine coordinate user-defined homotopies in Bertini, consider computing a solution of

$$f(x) := \exp\left(\sin\left(\frac{x}{1+x}\right)\right) + \frac{1}{\cos(x-2)} - 1 = 0. \tag{7.6}$$

First, we will demonstrate function evaluation (see §7.3.1) with user-defined homotopies by computing $f(-1/2)$ using 64-bit precision. Function evaluation requires two files: one containing the system and one containing the points. The system is described in the following file:

```
% FunctionEval.input
CONFIG
  % Evaluate a user-defined homotopy using 64-bit precision
  UserHomotopy:1;
  TrackType:-4;
  MPType:1;
  Precision:64;
END;
INPUT
  function f;
  variable x;
  pathvariable T;
  parameter t;
  t = T;

  f = exp(sin(x/(1+x))) + 1/cos(x-2) - 1;
END;
```

We remark that even though the path variable T and parameter t are not used in the definition of f, they are needed by Bertini since user-defined homotopies require exactly one path variable and at least one parameter.

The `start` file containing the point is

```
1

-0.5 0
% FunctionEval.start
```

Running Bertini creates the file `function`, which includes the following

```
-0.181713970082322550628e1 0.000000000000000000000e0
```

which shows that $f(-1/2)$ is approximately -1.817.

Next, we will use a Newton homotopy (see §7.4.2) constructed with a user-defined homotopy starting at $x = -1/2$ to attempt to compute a solution of (7.6). We utilize the same `start` file from above with the following system, rounding $f(-1/2)$ computed above for brevity:

```
% UserDefined_NewtonHomotopy.input
CONFIG
  % Track a user-defined homotopy
  UserHomotopy:1;
END;
INPUT
  function f;
  variable x;
  pathvariable T;
  parameter t;
  t = T;

  f = exp(sin(x/(1+x))) + 1/cos(x-2) - 1 - t*(-1.81713970082);
END;
```

After running Bertini, the file `nonsingular_solutions` contains the following approximation of a solution to (7.6):

```
-7.980107744243672e-01 0.000000000000000e+00
```

Finally, we demonstrate performing a Newton iteration (see §7.3.2) for f using a user-defined homotopy with 64-bit precision starting with this approximate solution. To do this, we first change the name of the just-constructed file `nonsingular_solutions` to `start`. Second, we simply change the `TrackType` setting in the `input` file used for evaluation above from -4 to -2. Running Bertini generates the file `newPoints` which contains the following new approximation of a solution to (7.6):

```
-0.798010774424367164386e0 0.000000000000000000000e0
```

The file `newtonResidual` contains the Newton residual for the point, namely,

```
3.563604028029648844475e-17
```

∎

Remark 7.2. When evaluating or performing a Newton iteration on a user-defined homotopy, by default Bertini sets the path variable to 0 for each point. In the previous example, this is irrelevant since the function does not depend on the path variable. However, the user can change the value of the path variable for each point by creating a file called `start_time`. The format of this file is the same as a typical `start` file for a univariate system since there is only one path variable. The number of values in `start_time` must be no less than the number of startpoints in the `start` file. Extra values in `start_time` are simply ignored.

7.5.2 ▪ User-Defined Homotopies on a Product Space

As discussed in Chapter 4, paths in affine space which diverge to infinity are infinitely long and therefore must be truncated. To avoid infinitely long paths but still have some flexibility for constructing homotopies, Bertini permits user-defined homotopies on a product of spaces. Bertini will automatically homogenize each space as needed, creating a homotopy on a product of projective spaces. Hence, each path will have finite length, meaning that the endpoint could be computed accurately if needed. Similar to multihomogeneous homotopies (see §5.1), the variables in each affine space are declared using `variable_group` and in each projective space using `hom_variable_group`. In particular, since the system either must already be homogenized, i.e., when only using projective spaces declared by `hom_variable_group`, or will be homogenized by Bertini, it must be polynomial in the variables.

The following example demonstrates tracking paths for parameters defined along an analytic curve such that one endpoint is at infinity.

Example 7.3. For the parameterized system

$$F(x,y;a,b)=\begin{bmatrix} x^2-3ax-4 \\ xy+y-b \end{bmatrix}, \tag{7.7}$$

the 2-homogeneous Bézout number yields that $F(x,y;a,b)=0$ will have at most two isolated solutions for all $a,b\in\mathbb{C}$. For $a=b=0$, it is easy to compute the two solutions $(\pm 2,0)$. Consider tracking these two solutions as the parameters (a,b) change from $(0,0)$ to $(1,1)$ along the curve defined by $(\sin(t\pi/2),t)$ as t goes from 0 to 1. This requires constructing two files, one containing the system and one containing the startpoints. Since we want to investigate paths that appear to be heading to infinity, the following system file also changes the `SecurityLevel` configuration:

```
% UserDefined_ProductSpace.input
CONFIG
  % Track a user-defined homotopy on a product space
  UserHomotopy:2;
  SecurityLevel:1;
END;
INPUT
  function f,g;
  variable_group x;
  variable_group y;
  pathvariable T;
  t = 1-T; % t goes from 0 to 1
  parameter a,b;
  a = sin(t*Pi/2);
  b = t;

  f = x^2 - 3*a*x - 4;
  g = x*y + y - b;
END;
```

The `start` file contains the two startpoints corresponding to $(\pm 2, 0)$:

```
2

2 0
0 0

-2 0
0 0
% UserDefined_ProductSpace.start
```

Upon running Bertini, the file `nonsingular_solutions` contains two points:

```
4.000000000000000e+00 -9.036147481308876e-18
2.000000000000000e-01  1.066583887182615e-17

-1.000000000000000e+00 -2.572269654221859e-17
 9.294730337655424e+15  1.009658768658067e+16
```

The first corresponds to the finite point $(x,y) = (4,2)$. The second suggests that x is converging to -1 while y is diverging to infinity, which one can easily verify since

$$F(-1,y;1,1) = \begin{bmatrix} 0 \\ -1 \end{bmatrix}. \quad \blacksquare$$

7.6 ▪ Determining the Rank of a Matrix

The ability to accurately determine the rank of a matrix is essential in numerical algebraic geometry, e.g., with multiplicity and deflation computations in §11.6 and isosingular sets discussed in §13.2. The standard approach from numerical linear algebra is to use a singular value decomposition. One then peruses the singular values looking for large "gaps" as well as singular values larger than a prescribed tolerance, so as to make a judgment on the rank. Bertini uses such an approach when using fixed precision based on the configuration `SingValZeroTol` and the unit roundoff error for the working precision.

When working in adaptive precision, Bertini builds on this standard approach by comparing the behavior of the singular values as computed in different levels of precision. As a sharpening technique increases the accuracy of a solution point, the error between the numerical approximation of a matrix function evaluated at the computed point and the actual matrix at the true solution decreases. As this proceeds, the error in the differences of the singular values must also decrease, thereby revealing the correct rank of the matrix. For example, consider the matrix $A(0)$ where

$$A(x) = \begin{bmatrix} 1 & 0 \\ 1 & x \end{bmatrix}$$

using $x_1 = 10^{-10}$ and $x_2 = 10^{-20}$ as two approximations for 0. The largest singular value for both $A(x_1)$ and $A(x_2)$ is approximately 1.41, while the smallest singular values are $7.07 \cdot 10^{-11}$ and $7.07 \cdot 10^{-21}$, respectively. The relative change for the largest singular value is negligible, while the relative change for the smallest singular value is significant, indicating rank deficiency. The configuration used by Bertini to decide how much relative change is permitted and still be considered nonzero is `RatioTolerance`. Of course, we note that even though such a technique may fail on pathological examples with extremely small singular values, it is generally robust for well-scaled polynomial systems (see §7.7).

We conclude this section by returning to the example from §7.1, namely, the intersection of the circle of radius one centered at (0, 1) and the parabola $y = 2x^2$. The following is the `start` file we consider:

```
4

-8.660254037844386e-01 9.492190020110591e-17
 1.500000000000000e+00 6.120321263680673e-17

8.660254037844386e-01 2.186564731360141e-16
1.500000000000000e+00 -1.412715430748612e-16

-2.213743081408535e-13 -1.498562286411079e-13
 2.540827022107365e-15 -1.330798662382149e-15

2.878540067625655e-13  1.392857465195220e-13
3.125787117584775e-15 -1.703491954589669e-15
% CondNumCalc.start
```

We can use the Newton iteration code in Bertini (see §7.3.2) to compute an approximation of the condition number of the Jacobian for each of these points. Since the last two points are expected to have a relatively high condition number, we will perform the computation using 256-bit precision. Additionally, since we will recompute the condition numbers after sharpening, we will fix a seed for the random number generator. In particular, we run Bertini with the following file:

```
% CondNumCalc.input
CONFIG
 % Perform a Newton iteration & approximate the condition number
 TrackType:-1;
 % Utilize 256-bit fixed precision
 MPType:1;
 Precision:256;
 % fix a seed for the random number generator
 RandomSeed:1934835;
END;
INPUT
 function circle,parabola;
 variable_group x,y;

 circle = x^2 + (y-1)^2 - 1;
 parabola = y - 2*x^2;
END;
```

This yields the following four condition number approximations in the `CN` file:

```
3.312184663658138e+00

3.125333423874220e+00

5.074027090900738e+12

5.131874105824588e+12
```

Since these points did not arise from a homotopy in Bertini, we sharpen the points using a fixed-point homotopy (see §7.4.1) with `FinalTol` set to 10^{-30}, i.e., approximate to 30 digits. After updating `start`, the recomputed `CN` file now contains the following four condition number approximations:

```
3.312184663658138e+00

3.125333423874219e+00

4.268293437234362e+33

2.018845678007700e+33
```

The negligible change in the condition number for the first two indicates that these are nonsingular solutions, while the significant change in the condition number for the last two is indicative of singular solutions.

7.7 ▪ Scaling Polynomial Systems

In numerical linear algebra, preconditioning is used to improve the conditioning of the problem, which often leads to improved performance. In numerical algebraic geometry, function and variable scaling are the most commonly used preconditioners. Typically, the goal is to create a polynomial system in which all the coefficients are approximately of unit magnitude. For example, the algorithm SCLGEN [131] uses optimization to attempt to reduce the variability of the coefficients and cluster the coefficients around the unit circle.

Function scaling simply involves multiplying a polynomial f by a scalar α. This is particularly useful when all the coefficients are roughly of the same magnitude, namely, α^{-1}. For example, rescaling the polynomial

$$f(x) = 10^{-14}x^2 + 2\cdot 10^{-16}x - 10^{-15}$$

with the scaling factor $\alpha = 10^{15}$ produces the numerically amenable polynomial

$$\alpha f(x) = 10x^2 + \frac{1}{5}x - 1.$$

When the coefficients for a polynomial system differ by many orders of magnitude, rescaling each variable can be used to help reduce the variability of the coefficients. Variable scaling replaces each variable with a multiple of a new variable. This simply changes the unit length in each coordinate direction. After reducing the variability, function scaling is then used to rescale the coefficients to approximately unit magnitude.

For example, consider the polynomial system

$$f(x,y) = \begin{bmatrix} xy - 10^{-16} \\ x - 10^{-6}y - 10^{-10} \end{bmatrix}.$$

The coefficients of this system differ by 16 orders of magnitude, and the condition numbers of the Jacobian matrix at the solutions are approximately 10^{10}. These condition numbers may cause Bertini to use higher precision to accurately track the paths.

We first scale the variables by setting $x = 10^{-11}\hat{x}$ and $y = 10^{-5}\hat{y}$, where $\hat{x}$ and $\hat{y}$ are new variables. Then, we use a scaling factor of 10^{16} and 10^{11} for the two polynomials, respectively, yielding the polynomial system

$$\hat{f}(\hat{x},\hat{y}) = \begin{bmatrix} \hat{x}\hat{y} - 1 \\ \hat{x} - \hat{y} - 10 \end{bmatrix}.$$

Since we will use the solutions of $\hat{f}$ to return to the original coordinates, we sharpen the endpoints using the following file:

```
% ScaledSystem.input
CONFIG
 SharpenDigits:30; % sharpen endpoints to 30 digits
END;
INPUT
 function f1,f2;
 variable_group hatx,haty;

 f1 = hatx*haty - 1;
 f2 = hatx - haty - 10;
END;
```

The coefficients of the resulting polynomial system differ by only one order of magnitude, and the condition numbers of the Jacobian matrix at the solutions are approximately 10. Due to the improved conditioning, double precision suffices for tracking these paths. At the end of tracking, the sharpening module uses higher precision to deliver the requested 30 digits of accuracy.

To return to the original coordinates, we simply need to perform an evaluation using Bertini (see §7.3.1). We first rename `finite_solutions` as `start` and then run Bertini with the following `input`:

```
% Unscale.input
CONFIG
 TrackType:-4; % perform evaluation
 % Utilize 96-bit fixed precision
 MPType:1;
 Precision:96;
END;
INPUT
 function x,y;
 variable_group hatx,haty;

 x = 1e-11*hatx;
 y = 1e-5*haty;
END;
```

The following two points are printed in the resulting `function` file:

```
0.100990195135927848300282241091e-9 0.000000000000000000000000000000e0
0.990195135927848300282241090224e-6 0.000000000000000000000000000000e0

-0.990195135927848300282241090223e-12 0.000000000000000000000000000000e0
-0.100990195135927848300282241091e-3  0.293873587705571876992184134304e-43
```

7.8 ▪ Rounding Errors in Polynomial Construction

One underlying assumption made by the adaptive precision algorithms (see §3.2) in Bertini is that the system in working precision becomes closer to the true system as the precision is increased. If the coefficients are only provided to Bertini rounded to relatively few digits, the results may not be as expected. The underlying theoretical reason is that the effects of rounding may change which parameter subspace the coefficients lie in, causing the possibly unexpected behavior.

As a demonstration, consider the parameterized polynomial system

$$f(x,y;a,b) = \begin{bmatrix} xy^3 - ay^3 - 1 \\ x^2 - b \end{bmatrix}.$$

For general a and b, f has six roots, which is equal to the 2-homogeneous root count. However, for nonzero a and b with $b = a^2$, f has only three roots. Suppose that we want to solve when $a = \sqrt{2}$ and $b = 2$, but we only round $\sqrt{2}$ to 6 decimal places. Rather than using a parameter homotopy approach (see Chapter 6), we simply solve the system using a 2-homogeneous homotopy as described in the following file:

```
% RoundingError.input
 function f1,f2;
 variable_group x;
 variable_group y;
 constant a,b;

 a = 1.414214; % round sqrt(2) to 6 decimal places
 b = 2;

 f1 = x*y^3 - a*y^3 - 1;
 f2 = x^2 - b;
END;
```

For this system, the parameter point $(a,b) = (1.414214, 2)$ is still sufficiently far away from the exceptional set of parameters that Bertini computes six solutions to this system.

If we instead replace the definition of a with

```
 a = 2^(1/2);
```

the parameter point $(a,b) = (\sqrt{2}, 2)$ is on the exceptional set of parameters. In this case, Bertini now computes the expected three solutions. In summary, it is imperative that one ensures the given system is a member of the intended parameter space.

7.9 ▪ Bertini Guide: Newton Iterations and Sharpening

Newton Iteration Summary

Input files
- `input` = run configuration & system definition
- `start` = list of points

Output files
- `newPoints` = points obtained after one iteration
- `newtonResidual` = Newton residuals for each step
- `successFlag` = success or failure of the iteration

Key configurations:
- `TrackType:-2;` or `TrackType:-1;`

Notes:
- If using `TrackType:-1`, `CN` contains approximations of the condition numbers.

Execution:
- `>> bertini <input>`

Sharpening Module Summary

Input files
- `input` = run configuration & system definition
- `midpath_data` = points obtained at the start of the endgame
- `raw_data` = data for each path

Output files
- As directed interactively.

Key configurations:
- `SharpenOnly:1;`

Notes:
- Input file is exactly the same as the original run used to generate `midpath_data` and `raw_data` except for the configuration setting.

Execution:
```
>> bertini <input>
```

Part II

Positive-Dimensional Solution Sets

Chapter 8

Positive-Dimensional Components

In Part II, we return to the fundamental problem of "solving" a polynomial system,

$$f(z) := \begin{bmatrix} f_1(z_1,\ldots,z_N) \\ \vdots \\ f_n(z_1,\ldots,z_N) \end{bmatrix} = 0. \tag{8.1}$$

As opposed to the case of systems of linear equations, where a nonempty solution set will consist of exactly one linear space, the solution set of a polynomial system can have several separate components and these may have different dimensions[24] (points, curves, surfaces, etc). Thus, the problem of "solving" a polynomial system includes the detection of all such components no matter what dimension each has. Solution components of dimension greater than zero (i.e., everything except isolated points) are said to be positive-dimensional.

Consider the following system of three polynomials in the variables x, y, z:

$$f = \begin{bmatrix} (y-x^2)(x^2+y^2+z^2-1)(x-2) \\ (z-x^3)(x^2+y^2+z^2-1)(y-2) \\ (z-x^3)(y-x^2)(x^2+y^2+z^2-1)(z-2) \end{bmatrix}. \tag{8.2}$$

Whereas the solution set of three linear equations in three variables must be either a point, a line, *or* a plane, the solution set of (8.2) consists of a surface, four curves, *and* a point. If we were to apply one of the homotopies presented in Part I to solve this system, we would be assured only of finding the isolated solution, namely, $(2,2,2)$, although we may expect as well some points on some of the other components.

In brief, Part II presents how numerical algebraic geometry treats full solution sets, including their positive-dimensional components. This chapter introduces the basic concepts and shows results obtained with Bertini on some simple examples. The next two chapters, Chapters 9 and 10, which go into the details of the algorithms, are organized according to the two main stages of the computation: first, finding a sufficient set of points on the solution set, and then sorting these points into representations of the individual components. The final chapter of Part II, Chapter 11, covers several advanced topics concerning positive-dimensional solution sets, such as local dimension computations, which in a kinematics context means determining mechanism mobility.

[24] We gave a definition of the *dimension* of a component in §1.2.2.

This first chapter of Part II begins with a topic already discussed briefly in §1.2.1, the concept of an irreducible algebraic set. Also known as irreducible solution components or, when clear from the context, simply components, these are the individual points, curves, surfaces, etc., that together form an entire solution set. After defining irreducible components more carefully, we define two invariants they possess: dimension and degree. These concepts, paired with a few words about projective versus affine algebraic sets round out the first section of this chapter.

In the numerical domain, pure-dimensional algebraic sets—ones whose irreducible components all have the same dimension—are represented by a finite data structure called a *witness set*. Accordingly, the *numerical irreducible decomposition*, which is a list of witness sets, one for each irreducible component of a solution set, becomes the principal computational aim in numerical algebraic geometry. These are the topic of discussion in §8.2. Given the numerical representation of a solution set, component sampling (§8.3) and membership testing (§8.4) are two fundamental and useful tools available in the numerical algebraic geometry toolbox. The chapter concludes with a discussion in §8.5 of how inaccuracy in the coefficients of a system has an impact on the numerical irreducible decomposition.

Before getting into the decomposition of the solution set of a polynomial system, we should first comment on our choice of words. Such solution sets go by a wide range of other names, including algebraic set, variety, scheme, and manifold. We generally use the term *solution set* when we are dealing with an explicit polynomial system; otherwise we use *algebraic set*.

Furthermore, algebraic sets can be irreducible or not, multiplicity one or greater than one, and affine or projective.[25] We will discuss these distinctions further in a moment. In this book, unless otherwise specified at the beginning of a section or a chapter, we make no assumptions about which properties may be true for the algebraic set.

8.1 ▪ Irreducible Components of an Algebraic Set

For the moment, let us restrict our attention to affine algebraic sets. Such a set is the collection of points in a complex space that satisfy a system of polynomial equations. The projective setting is described briefly near the end of this section.

Thinking of solutions of polynomial systems over the real numbers, it is natural to decompose a solution set into connected components. However, over the complex numbers, the natural decomposition is into *irreducible components*, as briefly introduced in Chapter 1. These are the pieces of the solution set that, roughly speaking, correspond to algebraic factors. The definition can be made precise in either algebraic or geometric terms, and we will be going the geometric route in a bit. Before some technical details, let us turn to an illustrative example, adapted from [155], to build some intuition.

Example 8.1. Consider the equations of (8.2) given above. The solution set breaks into six irreducible components, as follows:

- *One surface.* The factor $x^2 + y^2 + z^2 - 1$ appears in each polynomial, so one obvious solution component arises when $x^2 + y^2 + z^2 - 1 = 0$. The solution set of a single nonconstant polynomial is always a hypersurface, that is, a set that is one dimension smaller than the ambient space. (In linear algebra, the solution set of

[25] Algebraic sets can even be quasi-projective. A quasi-projective algebraic set is a set of the form $X \setminus Y$ where X and Y are projective algebraic sets.

a single nonconstant linear equation is a hyperplane.) Since this example lives in three-space, the factor at hand is two-dimensional, and so it is a surface.[26]

- *Three lines*: All three polynomials are satisfied when $x = 2$ and $z = 8$, leaving y free. Similarly, setting $y = 2$ and $x = \pm\sqrt{2}$ leaves z free. Hence, we have a total of three lines.
- *One nonlinear curve*: The factors $y - x^2$ and $z - x^3$ together define a curve, the so-called twisted cubic curve.
- *One isolated point*: The point $(2,2,2)$ is clearly a solution. It is easy to show that it does not lie on any of the previous components. Thus, it is isolated.

One can verify that the surface intersects each of the four curves and that the twisted cubic also intersects each line, while the lines are skew. Thus, there are two connected components (one is just the isolated point) but six irreducible components. ∎

We see from this example that the subdivision into irreducible components is finer than the subdivision into connected components. Even so, connectedness plays a big role in the definition of irreducible components. The difference is that irreducible components are the closures of the sets that are connected after certain singular points are removed.

Let's be a bit more precise. Recall our notation for $\mathbf{V}(f)$. That is, if f is a system of polynomials defined on $\mathbb{C}^N$, then

$$\mathbf{V}(f) := \{z \in \mathbb{C}^N \mid f(z) = 0\}.$$

So, when we write $\mathbf{V}(f)$, we have thrown away any multiplicity information. This means that in $\mathbb{C}^2$, $\mathbf{V}(x-y)$ and $\mathbf{V}\big((x-y)^2\big)$ are both the same line, even though $(x-y)^2 = 0$ defines a "double line." (More on this in a moment.)

For a polynomial system f, the points making up the set $X = \mathbf{V}(f) \subset \mathbb{C}^N$ are called either *manifold points* (also known as *regular points* or *smooth points*) or *singular points*. We discussed this dichotomy in the first chapter in the case of isolated solutions. One way of recognizing singular points is that they are those solutions at which the tangent space to X at the point is not the same as other nearby points. The most common singularities are where X crosses itself. Removing all singularities of X leaves us with X_{reg}, the set of regular points of X. Said another way, a point $x \in X$ is regular if and only if the tangent space to X at x is a vector space with the same dimension as the tangent space at any other point in the immediate neighborhood of x.

With this setup, the *irreducible components* of X are the closures of the connected components of X_{reg}. Here, the closure just means that after seeing which pieces are connected absent the singularities, the singular points are put back.

An algebraic set consisting of just one irreducible component is called an *irreducible algebraic set*; otherwise, it is called *reducible*. Note that *reduced*, a term from algebraic geometry that we will not define here, is not the same as reducible.

Example 8.2. Consider the simple bivariate polynomial equation

$$xy = 0$$

[26]This is clearly a sphere in real space, but we instead use the term "surface" as we focus on the complex setting.

with $(x,y) \in \mathbb{C}^2$. The solution set consists of the x-axis and the y-axis, which intersect at the origin. Apart from the origin, the tangent space for any point of this solution set is the axis on which it lies. However, the tangent space at the origin consists of the whole plane, making the origin a singular point.

Excluding the origin, there are two connected components, namely, the two axes minus the origin. This might seem counterintuitive! After all, if we consider the situation in real space, $\mathbb{R}^2$, then after removing the origin, we are left with four connected pieces, these being the rays along the axes in the plus and minus directions. But here is the catch: the axes in $\mathbb{C}^2$ are each two-real-dimensional, having a real and an imaginary part. After removing the origin $x = 0$ from the complex plane[27] of $x = x_1 + x_2 i$, $(x_1, x_2) \in \mathbb{R}^2$, we can *go around* the origin. So the x-axis in $\mathbb{C}^2$ is still connected after removing $(x,y) = (0,0)$, and so is the y-axis. Furthermore, it is clear that the axes $x = 0$ and $y = 0$ only meet at $(x,y) = (0,0)$, so with that singular point removed, we have two punctured Argand planes, each of which is connected but which do not touch each other. The two irreducible components are these planes with the origin filled in.

For this simple example, it should have been obvious from the start that the set $\mathbf{V}(xy)$ naturally factors into two pieces: $\mathbf{V}(x)$ and $\mathbf{V}(y)$. ∎

Example 8.3. For a similar example with nonlinear irreducible components, consider the polynomial equation

$$\left(x^2 + y^2 + (z-1)^2 - 4\right)\left(x^2 + y^2 + (z+1)^2 - 4\right) = 0.$$

In $\mathbb{R}^3$, this equation defines the union of two spheres of radius 2 that intersect in a circle of radius 1 in the xy-plane, i.e., there is one connected component.

Over the complex numbers, the intersection of the two sets corresponding to the two factors is still a planar curve. Since this curve is singular, we remove it to see what connected pieces are left. In real space there are four spherical caps, but just as in the previous example, in complex space these connect into just two pieces. Accordingly, this reducible algebraic set has two irreducible components. ∎

In both previous examples, it is important to realize that our intuition over the real numbers is misleading when working with complex numbers. The complex space $\mathbb{C}^N$ has N complex dimensions, which means it is $2N$-real-dimensional, and this makes all the difference.

One might wonder why, when these factorable examples seem so obvious, we bother with this geometric definition of irreducibility depending on connectedness. This is because the algebraic means of breaking algebraic sets into irreducible components is quite difficult in more general examples, and we wish to pursue a geometric route of exploring connectedness via numerical homotopy. We will see in Chapter 10 just how this is done, but for now, we only want to convey the geometric picture of what these components mean.

For those with an algebraic bent, we note that the (geometric) irreducible decomposition of an algebraic set X corresponds to the (algebraic) prime decomposition of the radical of the ideal that vanishes on X. See [44, Chapter 4] for more details. We do not explicitly use this correspondence in this book.

Irreducible components of an algebraic set come with two basic invariants, dimension and degree, described in the next two sections.

[27] This is often called the Argand plane by engineers and the complex line by algebraic geometers.

8.1.1 ▪ Dimension and the Irreducible Decomposition

Every irreducible algebraic set Y has a *dimension*, denoted $\dim Y$. This dimension is defined to be the vector space dimension of the tangent space of Y at any smooth point of Y. We designate the dimension of a reducible algebraic set, i.e., an algebraic set that consists of multiple irreducible components, to be the maximum of the dimensions of the irreducible components of the set. If all irreducible components of a reducible algebraic set X have the same dimension, say, d, then we call X *pure d-dimensional* or *equidimensional*. These concepts were previously discussed in §1.2. Chapter 9 of [44] also gives a gentle introduction to both the algebraic and the geometric meanings of dimension.

Example 8.4. The algebraic set from Example 8.1 consists of six components in dimensions 0, 1, and 2. Thus, it is a two-dimensional algebraic set but not a pure two-dimensional algebraic set.

If we were to forget about the point and the surface, we would be left with four one-dimensional components, the union of which is a pure-one-dimensional reducible algebraic set. Although we might intuitively think of any one of these as curves, the union of all of these components is also called a curve. ∎

More generally, we use the standard convention of referring to (not necessarily irreducible) algebraic sets of pure-dimension 0 as *points*, dimension 1 as *curves*, dimension 2 as *surfaces*, and dimension $k > 2$ as *k-folds*. Threefolds are often called bodies in the classical literature.

It is convenient to organize the irreducible components of an algebraic set X into equidimensional algebraic subsets, X_i, and then to further subdivide each X_i into its irreducible components $X_{i,j}$. This yields the *irreducible decomposition* of X, which we may write as

$$X = \bigcup_{i=0}^{D} X_i = \bigcup_{i=0}^{D} \bigcup_{j \in \Lambda_i} X_{i,j}, \tag{8.3}$$

where $D = \dim X$, X_i is the pure-i-dimensional algebraic set consisting of the union of all i-dimensional irreducible components $X_{i,j}$, and Λ_i is simply a finite (possibly empty) indexing set for the i-dimensional irreducible components.

Example 8.5. With this notation, the solution set for Example 8.1 can be written as $X = X_2 \cup X_1 \cup X_0$, where

- $X_2 = X_{2,1}$ is just the surface;
- $X_1 = X_{1,1} \cup X_{1,2} \cup X_{1,3} \cup X_{1,4}$, where $X_{1,1}$ is the cubic curve (though there is no real reason for listing it first) and the other components are the three lines; and
- $X_0 = X_{0,1}$ is just the set containing the isolated point $(2,2,2)$.

We again highlight that the numbering of the irreducible curves $X_{1,j}$ is arbitrary. ∎

8.1.2 ▪ Slicing a Component and Degree

Another invariant attached to an irreducible algebraic set X comes from the intersection of X with a general linear space of appropriate dimension. In $\mathbb{C}^N$, a general linear space

L of dimension k will intersect an irreducible algebraic set X of dimension m in an algebraic set of dimension $k+m-N$. Here a negative number corresponds to an empty intersection. A version of this fact is proved in [161, Chapter 13].

Unfortunately, the following shows that our real intuition can be misleading.

Example 8.6. In $\mathbb{R}^2$, many lines will miss the unit circle $x^2+y^2-1=0$. However, in $\mathbb{C}^2$, almost every choice of a complex line will intersect the (complex) solution set of $x^2+y^2-1=0$ in two points.

Indeed, $y=ax+b$ may be substituted into x^2+y^2-1 to get

$$(1+a^2)x^2+2abx+(b^2-1)=0.$$

Accordingly, the line given by $y=ax+b$ intersects $\mathbf{V}(x^2+y^2-1)$ in two points unless $1+a^2=0$ or the discriminant of the quadratic is zero: $a^2b^2-(1+a^2)(b^2-1)=0$. In other words, the intersection will be two points unless we choose parameters a and b that lie on one of these two curves in $\mathbb{C}^2$, a measure zero set. ∎

A k-dimensional set in $\mathbb{C}^N$ is said to have *codimension* $N-k$. Setting $k+m-N=0$ and rearranging, we see that an m-dimensional irreducible algebraic set in $\mathbb{C}^N$ will intersect a random linear space of codimension m in a 0-dimensional set, i.e., a set of points, with probability one. The *degree* of an m-dimensional irreducible algebraic set X, written $\deg X$, is the number of points in the intersection of X with a general linear space of codimension m.

Example 8.7. With probability one, the intersection of two random linear spaces whose dimensions sum to the ambient dimension will be a single point. Thus, $\deg X=1$ for any linear algebraic set X. ∎

Example 8.8. Based on the analysis in Example 8.6, the algebraic set given by $x^2+y^2-1=0$ has degree 2. ∎

Example 8.9. In Example 8.5, $\deg X_{2,1}=2$, $\deg X_{1,1}=3$, and all other components have degree 1. ∎

As we shall see throughout the remainder of Part II, the idea of slicing an algebraic set with a linear space of some dimension plays a central role in numerical algebraic geometry.

8.1.3 ▪ Irreducible Algebraic Sets of Multiplicity > 1

So far, we have considered only affine algebraic sets of multiplicity one; the possibility of encountering algebraic sets with higher multiplicities was swept under the rug, along with the idea of working with projective algebraic sets (treated next). Let's start by considering a simple example of an algebraic set having multiplicity 2.

Example 8.10. The solution set of the equation $f(x,y)=(x+y-1)^2=0$ is simply the line $x+y=1$. However, at each point on the line $x+y=1$, the Jacobian matrix of partial derivatives

$$Jf(x,y)=2(x+y-1)\cdot\begin{bmatrix}1 & 1\end{bmatrix}$$

also vanishes. Since the tangent space (the right nullspace of the Jacobian) is two-dimensional while the set itself is only one-dimensional, every point of the set is singular and there are no smooth points. This may cause one to worry how exactly dimension is defined for this sort of "double line."

It is easy to tell (without relying on the knowledge that the solution set is a line!) that this irreducible set has dimension 1. Indeed, we know that it is not two-dimensional, as $f(x,y)$ will not vanish at a generic point chosen from $\mathbb{C}^2$. However, a generic line in $\mathbb{C}^2$ will clearly intersect the solution set, meaning that the dimension must be at least one. ■

This example makes an important point. If we encounter an algebraic set of multiplicity greater than one, it is useful to consider the associated multiplicity one algebraic set (which is the same set of points, but without the extra multiplicity information). To distinguish between the two, we write $f^{-1}(0)$ to mean the solution set of f with multiplicity included, and we write $\mathbf{V}(f)$ for the set of points with multiplicities dropped. The numerical algebraic geometry technique used to reduce $f^{-1}(0)$ to $\mathbf{V}(f)$ is called *deflation*.[28] Chapter 11 goes into further detail about how we handle algebraic sets of higher multiplicity.

One should not conclude from simple examples that higher multiplicity always arises from polynomials that have factors raised to some power greater than one. While Chapter 11 gives a technical definition of multiplicity, a simpler notion suffices for many cases. Working in $\mathbb{C}^N$, consider cutting a k-dimensional irreducible component $X \subset f^{-1}(0)$ with a generic linear space of codimension k, which we do by appending a system of k generic linear equations, say, L. As we know, this intersection will consist of isolated points. The multiplicity of X is the multiplicity of these points as isolated solutions of the augmented system $\{f, L\}$. If f consists of $N-k$ polynomials, this augmented system is square, and the interpretation of multiplicity is straightforward. The nonsquare case is subtle, so that is where the technical definition comes into play.

As an aside, we note that multiplicity is closely related to, but different from, the dichotomy between reduced and nonreduced algebraic sets. The precise definition of reduced algebraic sets, which relates to radical ideals, is not needed in this book. However, *multiplicity one* and *generically reduced* are synonymous, and so we may occasionally use the latter term.

8.1.4 ▪ Projective Irreducible Algebraic Sets

Little changes from the previous definitions if we move to the projective setting. The dimension and degree of an irreducible algebraic set are defined in projective space as in affine space with minor modifications. The dimension of a projective irreducible algebraic set X is the dimension of the projective tangent space at a general point of X. If $\dim X = m$, the degree of X is simply the number of points in the intersection of X with a general projective linear space of codimension m.

Projective irreducible algebraic sets often arise by taking the closure of affine irreducible algebraic sets in projective space. In this case, the dimension and degree of both sets are equal, which one can understand via homogenization (see §4.3). In particular, let $Y \subset \mathbb{C}^N$ be an irreducible affine algebraic set and X be the projective closure of Y. On $\mathbb{C}^N$, homogenization introduces a new coordinate, thus forming $\mathbb{C}^{N+1}$. In doing so, each point of Y now corresponds to a line, so the dimension increases by 1. All points on the same line through the origin are taken to be equivalent in projective space, thereby reducing the apparent dimension back to its original value. The statement regarding

[28] As different irreducible components of $f^{-1}(0)$ can have different multiplicity structures, deflation must be carried out component by component.

degree follows from the fact that the intersection of X with the hyperplane that was added at "infinity" is a proper algebraic subset of X. Hence, if $\dim Y = \dim X = m$, a general projective linear space of codimension m will intersect X outside of this subset.

As fair warning, we note that if a polynomial system is homogeneous, then the dimension of each irreducible component of its solution set when cast into complex space by using `variable_group` will be one larger than if the problem is cast into projective space by using `hom_variable_group`. We refer the interested reader to [44, Chapter 9] for more information about projective dimension. Nonetheless, from a practical standpoint, most Bertini users will consider polynomial systems as having solution sets living in affine space, in which case there is no need to think about the projective setting. Those who choose to use homogeneous polynomial systems with solution sets living in projective space will be well aware of this distinction, so we say no more of it here.

8.2 ▪ Witness Sets and the Numerical Irreducible Decomposition

One of the fundamental goals of numerical algebraic geometry is the detection of all positive-dimensional irreducible components of an algebraic set. The methods for doing so are described in the next two chapters, and the goal here is to describe the data structure for storing such information.

Let Z be an algebraic set with irreducible decomposition

$$Z = \bigcup_{i=0}^{D} \bigcup_{j \in \Lambda_i} Z_{i,j},$$

where the irreducible components $Z_{i,j}$ of Z are indexed by dimension (up to $D = \dim Z$) and within each dimension i by some indexing set Λ_i. This decomposition was described in some detail in §1.2.1. Though we want to define a numerical data type for all of Z, we will begin with one for a single irreducible component.

A *witness set* for an irreducible component $Z_{i,j}$ is a 3-tuple,

$$\mathcal{W}_{i,j} = \{f, L_{i,j}, W_{i,j}\},$$

where f is a polynomial system such that $Z_{i,j}$ is an irreducible component of $\mathbf{V}(f)$, $L_{i,j}$ is a system of i general linear polynomials (here, $i = \dim Z_{i,j}$), and $W_{i,j}$, called a *witness point set*, is the set of points constituting the intersection of $Z_{i,j}$ with the linear space $\mathbf{V}(L_{i,j})$. By virtue of the manner in which they are constructed, each point in $W_{i,j}$, called a *witness point*, is a *generic* point of $Z_{i,j}$, i.e., a point that satisfies no special properties. We keep track of the polynomials f in the witness set data type, since different components in the same algebraic set Z might be defined by variations of the original defining equations for Z, depending on the context. It's worth noting that we often have $L_{i,j} = L_{i,j'}$ for $j \neq j'$, though this may not *always* be the case.

The *numerical irreducible decomposition* $\mathcal{W}$ of an algebraic set Z mirrors the structure of the irreducible decomposition of Z:

$$\mathcal{W} = \bigcup_{i=0}^{D} \bigcup_{j \in \Lambda_i} \mathcal{W}_{i,j}. \tag{8.4}$$

Note that this simply means a formal union of witness sets and not the union of witness *point* sets. Indeed, the union of witness point sets would technically be a large set of witness points with no particular order or decomposition into components!

With the configuration setting `TrackType:1`, Bertini will compute a numerical irreducible decomposition in a single run. Full details regarding input and output may be found in Appendix B and §D.2.1, respectively, but here is how Bertini handles our running example.

Example 8.11. An input file for computing the numerical irreducible decomposition of the algebraic set from Example 8.1 is as follows:

```
% Illustrative.input
CONFIG
 TrackType:1;
END;
INPUT
 variable_group x,y,z;
 function f1,f2,f3;
 S = x^2+y^2+z^2-1;
 T = y-x^2;
 U = z-x^3;
 f1 = T*S*(x-2);
 f2 = U*S*(y-2);
 f3 = T*U*S*(z-2);
END;
```

The use of subfunctions S, T, and U is not necessary but will slightly improve the efficiency of the run and reduces typing. The final lines of the output to the screen are

```
************** Decomposition by Degree **************

Dimension 2: 1 classified component
-----------------------------------------------------
degree 2: 1 component

Dimension 1: 4 classified components
-----------------------------------------------------
degree 1: 3 components
degree 3: 1 component

Dimension 0: 1 classified component
-----------------------------------------------------
degree 1: 1 component

*****************************************************
```

which matches the expected decomposition of the algebraic set.

This computation produces a machine-readable format of the witness sets in the file `witness_data`. This file is essential for any subsequent runs that make use of the numerical irreducible decomposition, e.g., sampling or membership, described below. Bertini can also use this file to print the linear system and witness points to files, a topic further discussed in §11.1. A human-readable format of the witness point sets may be found atop the file `main_data`.

```
Number of variables: 3
Variables:  x y z
Rank: 3
----------DIMENSION 2----------
NONSINGULAR SOLUTIONS
---------------
Path number: 0
Component number: 0
Estimated condition number: 1.000000e+00
 2.143717399441825e+00 -6.389004635682655e-01
 6.460359426234658e-01  1.663885109739101e+00
-3.056476217806470e-01 -9.641575264301632e-01
Multiplicity: 1
Deflations needed: 0
---------------
Path number: 1
Component number: 0
Estimated condition number: 1.000000e+00
 5.924673442090049e-02  2.484588169711845e-01
-4.580900495876303e-01  1.494426800100002e-01
-9.348862229651018e-01 -5.748061083008665e-02
Multiplicity: 1
Deflations needed: 0
----------DIMENSION 1----------
NONSINGULAR SOLUTIONS
---------------
Path number: 1
Component number: 0
Estimated condition number: 3.232563e+00
 7.071067811865470e-01  1.387778780781446e-16
 4.999999999999997e-01 -1.110223024625157e-16
-1.280399813644957e+00  8.698150074681951e-01
Multiplicity: 1
Deflations needed: 0
---------------
Path number: 2
Component number: 1
Estimated condition number: 1.172638e+00
-7.071067811865481e-01  4.718447854656915e-16
 5.000000000000001e-01 -1.526556658859590e-16
-2.828480701302487e+00 -1.952031067829314e-01
Multiplicity: 1
Deflations needed: 0
 .... (more output in file)
```

The points are organized by dimension. Here, we show only two points on the surface $X_{2,1}$ and one point on each of two lines. The remainder of this portion of the `main_data` file (omitted to save space) includes one point on the other line, three points on the cubic curve, and the isolated solution. Further details regarding the format of this file may be found in Appendix D. ■

The following is a simpler example.

Example 8.12. Consider the polynomial system

$$g(x,y,z) = \begin{bmatrix} xz+y \\ yz+x \end{bmatrix}.$$

What is $\mathbf{V}(g)$? Clearly, if $x = y = 0$, z is free, so we get a line. Also, if $x = y \neq 0$, then we must have $z = -1$, so we get another line. Similarly, if $x = -y \neq 0$, then $z = 1$, yielding a third line.

Here is a Bertini input file for this system:

```
% ThreeLines.input
CONFIG
  TrackType:1;
END;
INPUT
  variable_group x,y,z;
  function f,g;

  f = x*z+y;
  g = y*z+x;
END;
```

Running this yields the following screen output.

```
************** Decomposition by Degree **************

Dimension 1: 3 classified components
-----------------------------------------------------
   degree 1: 3 components

*****************************************************
```

Finally, here is the main part of main_data.

```
Number of variables: 3
Variables:  x y z
Rank: 2

----------DIMENSION 1----------
NONSINGULAR SOLUTIONS
---------------
Path number: 0
Component number: 0
Estimated condition number: 7.501121e+01
-1.650608413443591e+00  3.815469193355912e+00
 1.650608413443591e+00 -3.815469193355912e+00
 1.000000000000000e+00  0.000000000000000e+00
Multiplicity: 1
Deflations needed: 0
---------------
Path number: 1
Component number: 1
```

```
Estimated condition number: 2.158128e+01
-1.357699715292874e-01  5.967195365831728e-01
-1.357699715292877e-01  5.967195365831723e-01
-1.000000000000000e+00 -2.220446049250313e-16
Multiplicity: 1
Deflations needed: 0
---------------
Path number: 2
Component number: 2
Estimated condition number: 4.848235e+00
 1.469054550401450e-71 -6.757407106908999e-72
-2.052450101206459e-72  3.299711931912004e-72
-1.913461079499319e-01 -1.443858569282686e-01
Multiplicity: 1
Deflations needed: 0
```

∎

Next, we consider a series of examples drawn from kinematics. The associated Bertini input files are available on the website for this book, `www.siam.org/books/se25`.

Example 8.13 (Bricard six-revolute mechanism). Suppose a 6R serial-link robot arm were to grasp its own base. Then, we expect the resulting closed-loop structure to be rigid. Indeed, we know from Example 5.2 that 6R robots have at most 16 isolated poses that put the end-effector in a given location.

For general 6R cases, isolated solutions are all that exist, but for special cases, a solution curve can exist. One such special case is the Bricard 6R mechanism. It can be specified from the formulation of Example 5.2 by specializing the given data as follows:

$$\mathbf{z}_1 = \begin{bmatrix} 1 \\ 0 \\ 0 \end{bmatrix}, \quad \mathbf{z}_6 = \begin{bmatrix} 0 \\ 0 \\ 1 \end{bmatrix}, \quad \mathbf{p} = \begin{bmatrix} 0 \\ 1 \\ 0 \end{bmatrix},$$
$$a_i = 1, c_i = 0, \qquad i = 1,2,3,4,5,$$
$$d_i = 0, \qquad i = 2,3,4,5.$$

Placing all variables in one `variable_group` and setting `TrackType` to 1, we compute the numerical irreducible decomposition of the system (5.2)–(5.4). The data printed to the screen includes the following:

```
************** Decomposition by Degree **************

Dimension 1: 1 classified component
-----------------------------------------------------
   degree 8: 1 component

*****************************************************
```

This informs us that the motion of the Bricard 6R mechanism is a single irreducible curve of degree 8. ∎

Example 8.14 (Griffis–Duffy Type I). In Example 6.3, we found the isolated solutions of the forward kinematics of general Stewart–Gough platforms. Members of a subfamily of these devices are called Griffis–Duffy platforms. These have their base joints at the vertices and edges of a triangle and similarly for the moving platform. Legs connect the vertices of one to the edges of the other in a cyclical fashion. The forward kinematics problem of general members of this subfamily is particularly easy to solve, having only 16 nonsingular isolated solutions, which was the motivation for noticing them [70]. Something surprising occurs if the triangles are equilateral with joints at the vertices and midpoints: the device is no longer rigid when the leg lengths are locked. Instead, this Griffis–Duffy Type I device moves with one degree of freedom, meaning that the problem has a solution curve. Griffis–Duffy Type I and Type II (see below) platforms were first discussed in [104].

Using the notation of Example 6.3, Griffis–Duffy Type I platforms are given as

$$\begin{bmatrix} \vec{a}_1 & \cdots & \vec{a}_6 \end{bmatrix} = \begin{bmatrix} -\sqrt{3} & 0 & \sqrt{3} & \sqrt{3}/2 & 0 & -\sqrt{3}/2 \\ -1 & -1 & -1 & 1/2 & 2 & 1/2 \\ 0 & 0 & 0 & 0 & 0 & 0 \end{bmatrix}$$

with $\vec{b}_i = \alpha \vec{a}_{i+3}$ (cyclic) for some arbitrary scaling factor α. After picking a random platform location, e.g., random Study coordinates $[e, g]$, (6.8) can be used to compute compatible leg lengths, ℓ_i, $i = 1, \ldots, 6$.

The numerical irreducible decomposition of a general problem of this class gives the following:

```
************** Decomposition by Degree **************

Dimension 2: 1 classified component
-----------------------------------------------------
   degree 2: 1 component

Dimension 1: 13 classified components
-----------------------------------------------------
   degree 1: 12 components
   degree 28: 1 component

*****************************************************
Witness Points Deflated: 2
```

The two-dimensional quadric is a double surface which Bertini deflated (see §11.6). All points on this surface and the 12 lines satisfy $e * e' = 0$, which means they do not correspond to meaningful rigid-body transforms. (For an explanation, see the discussion in Example 4.3.) To check this, one may obtain a witness point on these components in `main_data` and note that $e * e'$ evaluates to a tiny value. Setting these nonphysical components aside, the device has a one-dimensional motion curve that has degree 28. ∎

Example 8.15 (Griffis–Duffy Type II). Griffis–Duffy Type II platforms are a subfamily of the Type I platforms. For these, the two triangles are congruent (scaling factor $\alpha = 1$

above) and the leg lengths are equal. For these specializations, the Bertini screen output reports the following:

```
************* Decomposition by Degree *************

Dimension 2: 1 classified component
-----------------------------------------------------
   degree 2: 1 component

Dimension 1: 17 classified components
-----------------------------------------------------
   degree 1: 12 components
   degree 4: 1 component
   degree 6: 4 components

*****************************************************
Witness Points Deflated: 2
```

We have the same nonphysical surface and 12 lines, but the degree 28 curve of the Type I family now factors into five irreducible components. The device has a three-way symmetry that the solutions must obey. The quartic and one of the sextics are self-symmetric, while the remaining three sextics are symmetric to each other. Using parenthetical grouping to indicate symmetries, we have the decomposition:

$$28 = 4 + 6 + (6 + 6 + 6). \quad \blacksquare$$

Example 8.16 (foldable Griffis–Duffy). The nonlinear curves of the Type II platforms factor into even more components if the common leg length is equal to the altitude of the triangles, i.e., $\ell_i = 3$ for $i = 1, \ldots, 6$. The numerical irreducible decomposition is as follows:

```
************* Decomposition by Degree *************

Dimension 2: 1 classified component
-----------------------------------------------------
   degree 2: 1 component

Dimension 1: 22 classified components
-----------------------------------------------------
   degree 1: 15 components
   degree 2: 3 components
   degree 4: 4 components

*****************************************************
Witness Points Deflated: 5
```

In this case, the degree 28 curve of the Type I family factors into three double lines, three quadrics, and four quartics, which, counting multiplicities and indicating symmetric groups with parentheses, can be summarized as

$$28 = (1 \cdot 2 + 1 \cdot 2 + 1 \cdot 2) + (2 + 2 + 2) + (4 + 4 + 4) + 4. \quad \blacksquare$$

8.3 ▪ Sampling a Component

It is often useful to obtain points on an irreducible component X not contained in the same linear space as is the case with a witness point set. We refer to this as *sampling* X, and the concept is quite simple. Since the witness set W corresponding to component X includes points in the intersection of X with a linear space $\mathbf{V}(L)$, we can find more points on X by moving $\mathbf{V}(L)$ to some other linear space $\mathbf{V}(L')$ of the same dimension as $\mathbf{V}(L)$.

This shifting of the linear space from L to L' is accomplished via the following simple homotopy:

$$H_{sample}(z,t) := \begin{bmatrix} f(z) \\ tL(z)+(1-t)L'(z) \end{bmatrix}. \tag{8.5}$$

In this way, the solutions of the start system are exactly the previously computed witness points of W. If L' is also generic, this homotopy will yield $\deg X$ new points on X. One final note is that this homotopy may not be square since f may consist of more polynomials than the codimension of X. In this case, one simply replaces f with a randomization of f, a topic that will be addressed in more detail in §9.2.

Sampling is implemented in Bertini, using `TrackType:2`. The user (interactively or using a script) specifies a component to sample and the number of points to be computed. The user can determine whether she wants the points printed to the screen or placed in a file where the name of this file is provided by the user. Algorithmically, Bertini randomly picks a new L' for each sample point requested. Further details may be found in §B.1 and §C.3.

8.4 ▪ Membership Testing

Given a point $x^* \in \mathbb{C}^N$ and the numerical irreducible decomposition of a possibly reducible algebraic set X, one may wish to determine whether $x^* \in X$ and, if so, which irreducible component(s) contain x^*.

If Y is an irreducible component of X, the algorithm for determining whether x^* lies on Y is as follows. The witness point set W for Y consists of points in the intersection of Y and a linear space $\mathbf{V}(L)$ defined by linear polynomials L. We may easily choose linear polynomials L' corresponding to a linear space $\mathbf{V}(L')$ passing through x^* and of the same dimension as $\mathbf{V}(L)$. Indeed, such a linear system is $L'(x) = A\cdot(x - x^*)$, where A is a random matrix of the appropriate size. Then, $x^* \in Y \cap \mathbf{V}(L')$ if and only if $x^* \in Y$.

By following exactly the same homotopy (8.5) above, starting with the witness points of W, we may determine whether $x^* \in Y \in \mathbf{V}(L')$ by simply comparing the coordinates of x^* with the endpoints of the homotopy paths.

As described in the next chapter, Bertini computes all witness sets for components of the same dimension using the same linear space. As a result, to determine whether a point x^* lies on any irreducible component of an algebraic set X, one may simply use a homotopy of the type above for each dimension in which irreducible components exist. All this is automated in Bertini using the `TrackType:3`. Details regarding the three input files needed for membership testing in Bertini are provided in §B.1 and §B.3, while the output is described in §D.2.3.

8.5 ▪ A Brief Note on the Expected Accuracy of Coefficients

We regard the system we are solving as exact. This is more true than is often realized. Typical problems arising in practice lead to systems of polynomials $f(z;q)$ with $(z;q) \in \mathbb{C}^N \times \mathbb{C}^M$ with the functions that give the coefficients in terms of the parameters being exact expressions involving small integers and elementary analytic functions. Inexactness enters when these expressions are evaluated numerically. When this happens, points that appear to be in a small cluster are really a single multiple point. Indeed, as we go to a higher precision the cluster will get tighter. For this to happen, the parameter values need to stay on the parameter space, e.g., the IPP problem mentioned in §3.2.4 is a nontrivial example of this. Therefore, when solving a system, coefficients need to be given to as many digits as will occur in the computations. Accordingly, it is recommended that the input file for Bertini give the exact expressions for the coefficients, but if they are precomputed instead, this should be done in high enough precision.

It is important to realize that from a numerical standpoint the irreducibility of a set depends on numerical decisions about smallness. As a simple example consider

$$f(x,y) = xy - \epsilon = 0. \tag{8.6}$$

Algebraically, if $\epsilon = 0$, then $\mathbf{V}(f) = \mathbf{V}(x) \cup \mathbf{V}(y)$, but for $\epsilon \neq 0$, $\mathbf{V}(f)$ is irreducible. Of course, for ϵ close to zero, $\mathbf{V}(f)$ lies very close to $\mathbf{V}(x) \cup \mathbf{V}(y)$. This is similar to how the numerical determination of the rank of a matrix depends on a cutoff value below which singular values are regarded as zero. We regard this as a strength of numerical methods.

To see this explicitly consider the following system:

```
% SmallConstant.input
CONFIG
  TrackType:1;
END;
INPUT
  variable_group x,y;
  function f;
  f = x*y-1e-11;
END;
```

The screen output ends with the following information:

```
************** Decomposition by Degree **************

Dimension 1: 1 classified component
-----------------------------------------------------
   degree 2: 1 component
*****************************************************
```

In our experiment, if we change `1e-11` to `1e-12`, the decomposition becomes

```
************** Decomposition by Degree **************

Dimension 1: 2 classified components
-----------------------------------------------------
   degree 1: 2 components
*****************************************************
```

The cutoff here for deciding irreducibility is related to the default configurations and may slightly change based on the random numbers used in the computation.

In short, the key issue in this example is whether the small constant term is mathematically meaningful or an artifact of numerical errors. If it is mathematically meaningful, then one needs to take measures to ensure this. One option is to adjust the tolerances used by Bertini to be smaller than the mathematically meaningful values but larger than the numerical error in generating the system (see Appendix F for more details). However, the best option for revealing the mathematically meaningful values is to employ scaling (see §7.7). For example, if $\epsilon = 10^{-12}$, one should instead use Bertini to numerically compute $\mathbf{V}(g)$, where

$$g(x,y) = 10^{12} \cdot f(10^{-6}x, 10^{-6}y) = 10^{12}((10^{-6}x)(10^{-6}y) - 10^{-12}) = xy - 1.$$

Any additional computations one may wish to perform on $\mathbf{V}(f)$ can be performed on $\mathbf{V}(g)$ using the appropriate scaling.

8.6 ▪ Bertini Guide: Positive-Dimensional Solving

Numerical Irreducible Decomposition Summary

Input files
- `input` = run configuration & system definition

Output files
- `main_data` = summary of witness points and components
- `witness_data` = machine-readable file containing witness sets
- `witness_superset` = witness point superset

Key configurations:
- `TrackType:1;`

Key declarations:
- `variable_group` or `hom_variable_group`

Notes:
- Only one variable group is allowed.

Execution:
```
>> bertini <input>
```

Sampling Summary

Input files
- `input` = run configuration & system definition
- `witness_data` = contains the witness set information

Output files
- As directed interactively.

Key configurations:
- `TrackType:2;`

Notes:
- Use redirection (<) to automate the process through the menus.

Execution:
```
>> bertini <input>
```
Then follow the interactive menus.

Membership Testing Summary

Input files

`input`	= run configuration & system definition
`member_points`	= contains the points to test
`witness_data`	= contains the witness set information

Output files

Output goes to screen and to `incidence_matrix`

Key configurations:

`TrackType:3;`

Notes:

Use redirection (>) to save screen output to file.

Execution:

```
>> bertini <input>
```

Witness Set Printing Summary

Input files

`input`	= run configuration & system definition
`witness_data`	= contains the witness set information

Output files

As directed interactively.

Key configurations:

`TrackType:4;`

Execution:

```
>> bertini <input>
```

Then follow the interactive menus.

Chapter 9

Computing Witness Supersets

The computation of the numerical irreducible decomposition of the solution set of a polynomial system f requires the computation of a witness set $\mathscr{W}_{i,j} = \{f, L_{i,j}, W_{i,j}\}$ for each irreducible component, as defined in the previous chapter. This computation has two main phases for each possible dimension. In particular, for a single dimension i these are as follows:

1. **Compute a witness point superset in dimension i:** Find a finite set of points $\widehat{W}_i \subset \mathbf{V}(f)$ that contains a witness point set $W_{i,j}$ for each i-dimensional component in $\mathbf{V}(f)$.

2. **Reduce the witness point superset $\widehat{W}_i$ to witness point sets $W_{i,j}$:** Remove unnecessary points from $\widehat{W}_i$ and decompose the resulting set W_i into a union of witness point sets, one for each i-dimensional irreducible component.

From witness point sets, it is easy to construct a witness set for each irreducible component and thus the full numerical irreducible decomposition.

This chapter focuses on the first step, and the next chapter focuses on the second. Fortunately, Bertini will carry out all of these operations automatically, so neither this chapter nor the next is essential for the casual user.

In §9.1, we take up the problem of bounding the possible dimensions of the irreducible components of $\mathbf{V}(f)$. In linear algebra, the dimension of the span of a set S of vectors depends on the number of linearly independent vectors in S, i.e., the rank of the matrix having the vectors of S as columns. Similarly, the rank of the Jacobian matrix of f, evaluated at a randomly chosen point, will provide a bound on the dimensions of the irreducible components. Section 9.2 returns to the topic of randomization, first touched on in §1.1.4 as a step in finding isolated roots to overdetermined systems, but this time discussed in terms of its effect on positive-dimensional solution sets.

Section 9.3 covers the simplest approach, the dimension-by-dimension method, for generating witness point supersets. This method (and all others for computing witness point supersets) relies on the idea of taking linear slices to reduce the problem of computing $\widehat{W}_i$ to a sequence of homotopies aimed at computing isolated points, such as those in Part I. (Linear slicing is described in the previous chapter.) Unfortunately, the simple approach can be very inefficient.

Cascade methods, described in §9.4, offer a significant improvement in computation time over the dimension-by-dimension method by using a sequence of homotopies to "cascade" down the possible dimensions. We describe the two forms of cascades: classical and regenerative.

9.1 ▪ Rank of a System

The methods used to compute the numerical irreducible decomposition require some computation in each dimension for which there *could* be an irreducible component. Accordingly, in this section we concern ourselves with the following question:

> How can we bound, a priori, the dimensions in which the solution set of a polynomial system may have irreducible components?

While some applications come with special knowledge about the feasible dimensions, e.g., a roboticist might be able to guarantee that there will only be solutions in a particular dimension for some mechanism, the goal here is to produce bounds not specific to any particular application.

As a starting point, we know that the solution set of a single nonconstant polynomial in N variables will be a *hypersurface* in $\mathbb{C}^N$, i.e., a component of dimension $N-1$ (equivalently, codimension 1). For example, a single nonconstant linear equation yields a hyperplane. Similarly, the solution set of $x^2+y^2+z^2-1=0$ is a two-dimensional set in a three-dimensional space.

We next consider the intersection of an irreducible algebraic set X with an irreducible hypersurface $\mathscr{H}$. There are three possible outcomes for $X \cap \mathscr{H}$. The first is that X is contained in $\mathscr{H}$ so that $X \cap \mathscr{H} = X$. The second is that X and $\mathscr{H}$ do not intersect (as in the case of two parallel planes) so that $X \cap \mathscr{H} = \emptyset$. If X is not contained in $\mathscr{H}$ and the intersection is nonempty, then the last possibility is that $0 \le \dim(X \cap \mathscr{H}) = \dim X - 1$.

By always considering just the irreducible components, the previous discussion generalizes to the case of a polynomial system f consisting of $n \ge 1$ polynomials in $N \ge n$ variables.[29] In particular, for each irreducible component $X \subset \mathbf{V}(f)$, we have

$$N - n \le \dim X \le N - 1.$$

Note that here and throughout this chapter, we are assuming that at least one polynomial in f is nonzero. We can test, with probability one, whether any polynomial is identically zero by evaluating it at a random point. (See §1.2.3 for a discussion.)

It can happen that the lower bound, $N-n$, for the dimension of each irreducible component is not achieved, such as, for example, the polynomials of f are linearly dependent. To sharpen this lower bound, consider the *rank of the system* $f : \mathbb{C}^N \to \mathbb{C}^n$, denoted $\operatorname{rank} f$, defined as the rank of the $n \times N$ Jacobian matrix Jf evaluated at a general point of $\mathbb{C}^N$. Since $\operatorname{rank} f \le \max\{N, n\}$, we can remove the restrictions on N and n, i.e., the system f can be overdetermined, square, or underdetermined. Considering f as a mapping from $\mathbb{C}^N$ to $\mathbb{C}^n$, Lemma A.4.9 of [161] shows that $\operatorname{rank} f = \dim \overline{f(\mathbb{C}^N)}$. It follows from the upper semicontinuity of dimension [161, Theorem A.4.5] that for each irreducible component $X \subset \mathbf{V}(f)$,

$$N - \operatorname{rank} f \le \dim X \le N - 1.$$

The number $N - \operatorname{rank} f$ is called the *corank* of f, denoted $\operatorname{corank} f$.

Example 9.1. For a linear system $f(z) = Az + b$ where $A \in \mathbb{C}^{n \times N}$ and $b \in \mathbb{C}^n$, $\operatorname{rank} f = \operatorname{rank} A$ so that either $\mathbf{V}(f) = \emptyset$ or $\mathbf{V}(f)$ is a linear space of dimension $N - \operatorname{rank} f$. ∎

[29] Note that we are allowing underdetermined polynomial systems.

The following example shows that the lower bound obtained need not be sharp.

Example 9.2. For the polynomial system

$$f(x,y,z) = \begin{bmatrix} y - x^2 \\ xy - z \\ xz - y^2 \end{bmatrix},$$

one can verify that $N - \operatorname{rank} f = 3 - 3 = 0$, whereas $\mathbf{V}(f)$ is an irreducible curve. ∎

The computation of the rank of a polynomial system in Bertini is based on the probability-one statement that the rank of $Jf(z^*)$ at a randomly selected point $z^* \in \mathbb{C}^N$ is equal to $\operatorname{rank} f$. Since widely scaled coefficients can lead to incorrectly determining the true rank, one should consider rescaling such systems (see §7.7).

9.2 ▪ Randomization and Bertini's Theorem*

In §1.2.3, we considered the use of randomization to find the isolated solutions of an overdetermined polynomial system. Bertini's theorem yields that the set of isolated solutions upon randomizing down to a square system is the set of isolated solutions of the original system together with possibly some additional nonsingular isolated solutions that do not satisfy the original system. We again turn to Bertini's theorem for the positive-dimensional case.

There are many variants of Bertini's theorem, including Theorem 1.15, with the following being a special case of [161, Theorem A.8.7].

Theorem 9.3 (Bertini's theorem for systems). *Given a polynomial system $f : \mathbb{C}^N \to \mathbb{C}^n$, there is a Zariski open dense set $U \subset \mathbb{C}^{k\times n}$ of matrices A such that*

1. $\mathbf{V}(A\cdot f)\setminus\mathbf{V}(f)$ *is either empty or smooth of pure-dimension $N-k$, and*
2. *the number of irreducible components of $\mathbf{V}(A\cdot f)\setminus\mathbf{V}(f)$ is independent of A and, since they are smooth, they must also be disjoint.*

Notice that any extraneous irreducible components (ones that are solutions to the randomized system $A\cdot f$ but are not solutions to f) can exist only at the lowest possible dimension, $N-k$.

Since there is a Zariski open dense set of $k\times k$ complex matrices that are invertible, it may be assumed that the first k columns of A are invertible. It follows that if $k < n$, Theorem 9.3 is true for matrices

$$A = [I_k \ \ C]$$

with I_k the $k\times k$ identity matrix and with C in a Zariski open dense set of $k\times(n-k)$ complex matrices. Additionally, one may reorder the functions of f in any convenient way. For efficiency, we often choose to list the highest-degree equations first. Otherwise, these higher-degree equations will be a summand in each of the randomized equations, which raises the total degree of the system and usually results in many more paths to track.

Bertini's theorem provides that the k-codimensional components of $\mathbf{V}(f)$ are k-codimensional components of $\mathbf{V}([I_k \ \ C]f)$ for generic $C\in\mathbb{C}^{k\times(n-k)}$. As a very simple example, consider the system

$$\begin{bmatrix} x \\ x(x-1) \end{bmatrix} = 0. \tag{9.1}$$

Clearly, the system has only the single solution, namely, $x = 0$, but for general $c \in \mathbb{C}$, the randomization $x(1 + c(x - 1))$ is a quadratic polynomial with two roots. One is $x = 0$, while the other is $x = (c-1)/c$, which is nonsingular, as it must be by Bertini's theorem. This example is zero-dimensional, but simply replace x by $x-y$ and consider the solution set in $\mathbb{C}^2$ to see that randomization to one equation gives a solution set consisting of two lines, one satisfying the original system and the other being a nonsingular extraneous solution set.

The multiplicities with respect to f of the k-codimensional irreducible components of $\mathbf{V}(f)$ are not less than their multiplicities with respect to $A \cdot f$. Moreover, if the multiplicity of a k-codimensional irreducible component of $\mathbf{V}(f)$ with respect to f is one, then the multiplicity of the component with respect to $A \cdot f$ is also one.

As an example of a multiplicity that increases, consider the system

$$f(x,y) = \begin{bmatrix} x^2 \\ xy \\ y^2 \end{bmatrix} = 0. \tag{9.2}$$

The multiplicity of the origin, which (see §11.3) is the same as $\dim \mathbb{C}[x,y]/(x^2, xy, y^2)$, is three, but its multiplicity as a solution of $A \cdot f$ is four.

The key advantage of randomization is numerical stability. In particular, changing the coefficients of the system $A \cdot f$ will change the union of the k-codimensional components of $A \cdot f$ only slightly when considering the solution set in a bounded domain of $\mathbb{C}^N$ or the closure in projective space. In practice, it is also numerically advantageous to choose A or C with columns close to orthonormal matrices. Additionally, note that even if we start out with an exact system, we may (for example, because of rounding) end up with a system that is slightly different. Since algebraic properties often change radically with small changes of coefficients, this could at first sight be a showstopper. As a toy example, suppose that the original system of equations is

$$\begin{bmatrix} x-1 \\ x-1 \end{bmatrix} = 0,$$

which obviously has a root at $x = 1$. Suppose further that instead of the original system, some numerical process produces the perturbed system

$$\begin{bmatrix} x-1+\epsilon \\ x-1 \end{bmatrix} = 0, \tag{9.3}$$

where ϵ is some small number. Of course, this perturbed system has no exact solution, but if we randomize it to a single equation using $A = [1 \;\; c]$, the resulting system has a solution, namely,

$$1 - \frac{\epsilon}{1+c}.$$

Of course, there is trouble if we happen to choose $c = -1$, but this is a zero-probability event. For any other c, as we go back to the original system and use more digits, we will obtain numerically perturbed systems with ever smaller $|\epsilon|$, and thus the computed value of the randomized root approaches the true root at $x = 1$. Moreover, to check whether the solution of the randomized system is an actual solution of the original system, we substitute it back into (9.3) and find that the residuals also approach zero in proportion to $|\epsilon|$.

For a multivariate example, the condition number of the Jacobian matrix at the root comes into play, but the basic phenomenon remains the same. Under perturbations that destroy exact solutions of an overdetermined system, randomization produces a square system with solutions near those of the exact system, and as higher-precision arithmetic is employed, these solutions satisfy the original system to ever greater accuracy. By checking results at several levels of precision, the decision of whether a computed point is an approximate solution or a nonsolution of the original system can be made robustly.

Since $\mathbb{R}^N$ is Zariski dense in $\mathbb{C}^N$, the matrices A and C can be real. This fact is useful when the original system has real coefficients; e.g., see [124, 34].

The main advantage of all this is that we can always reduce down to solving systems where the number of polynomials is equal to the rank of the system.

9.3 ▪ Working Dimension by Dimension

With randomization and rank in hand, we may now describe the computation needed in each feasible dimension i in order to compute a requisite witness point superset $\widehat{W}_i$ for a nonzero polynomial system $f : \mathbb{C}^N \to \mathbb{C}^n$. Recall from §8.1.2 that the intersection of a general hyperplane with a pure k-dimensional algebraic set is a pure $(k-1)$-dimensional algebraic set.[30] Since the intersection is pure $(k-1)$-dimensional, we can again slice with another general hyperplane to obtain a pure $(k-2)$-dimensional algebraic set. In particular, by intersecting with a total of k general hyperplanes, the intersection contains finitely many points, namely, the degree of the original algebraic set. If we intersect with fewer than k hyperplanes, the intersection will be positive-dimensional, but it will be empty if we slice with more than k general hyperplanes.

To summarize, if X is an algebraic set of pure-dimension k and L is a linear space defined by m general linear equations, then

1. if $m < k$, $X \cap L$ is dimension $k - m > 0$;
2. if $m = k$, $X \cap L$ is zero-dimensional, i.e., a set of $\deg X$ isolated points; and
3. if $m > k$, $X \cap L$ is empty.

Recall that the process of intersecting X with a linear space is called *slicing*, and $X \cap L$ is called a *slice* of X.

Thus, by item 2, when we slice with the proper number of hyperplanes, namely, $m = k$, there are finitely many isolated points that we need to compute. Recall that a standard homotopy such as those in Part I is guaranteed to find all isolated solutions of a polynomial system, perhaps along with points that lie on positive-dimensional irreducible components.

By §9.1, the dimension of each irreducible component of $\mathbf{V}(f)$ is between $N - \operatorname{rank} f$ and $N - 1$, so we need to consider intersecting $\mathbf{V}(f)$ with $N - \operatorname{rank} f \le k \le N - 1$ hyperplanes and computing the isolated points.[31] Let $\ell_1, \dots, \ell_{N-1}$ be general linear polynomials and, for $i = 1, \dots, \operatorname{rank} f$, consider

$$g_i(z) := \begin{bmatrix} f(z) \\ \ell_1(z) \\ \vdots \\ \ell_{N-i}(z) \end{bmatrix}.$$

[30] A negative dimension indicates the set is empty.

[31] It is natural to think of irreducible sets and witness sets in terms of dimension, rather than codimension. However, in the remainder of this chapter, we work with algebraic sets of dimensions $N-1$ down to $N - \operatorname{rank} f$, in that order. It is thus more natural to let i now represent the codimension, letting i run from 1 up to $\operatorname{rank} f$.

Clearly, $\mathbf{V}(g_i)$ is indeed the intersection of $\mathbf{V}(f)$ with a general linear space of dimension i resulting from the intersection of $N-i$ general hyperplanes. This intersection will consist of isolated points on each $(N-i)$-dimensional component of $\mathbf{V}(f)$, along with positive-dimensional subsets on each component of $V(f)$ of dimension larger than $N-i$. In fact, by solving $g_i(z)=0$ after randomization using a standard homotopy as in Part I, we will obtain the isolated points in $\mathbf{V}(g_i)$ along with at most finitely many points on any higher-dimensional components. By solving for each i from 1 up to rank$f$, we obtain the witness point supersets, $\widehat{W}_{N-i}$, $i=1,\ldots,\mathrm{rank} f$, for which we have been searching.

At the conclusion of this process, for each dimension k, the witness superset $\widehat{W}_k$ contains points that lie on components of exactly dimension k (i.e., actual *witness points*), along with points that lie on higher-dimensional components. We do not know which points are true witness points nor do we know which witness points lie on which components of $\mathbf{V}(f)$; sorting this out is the topic of the next chapter.

Example 9.4 (adjacent minors). To demonstrate this process, let's run the following polynomial system through Bertini. Let $x_{i,j}$, $i=1,2,3$, $j=1,2,3,4$, be 12 variables arranged in a 3×4 matrix. Then, the system

$$f(x)=\begin{bmatrix} x_{1,1}x_{2,2}-x_{1,2}x_{2,1} \\ x_{1,2}x_{2,3}-x_{1,3}x_{2,2} \\ x_{1,3}x_{2,4}-x_{1,4}x_{2,3} \\ x_{2,1}x_{3,2}-x_{2,2}x_{3,1} \\ x_{2,2}x_{3,3}-x_{2,3}x_{3,2} \\ x_{2,3}x_{3,4}-x_{2,4}x_{3,3} \end{bmatrix}$$

consists of the 2×2 adjacent minors of that matrix. Although our current focus in on computing a witness point superset, we mention that this polynomial system has six irreducible components: a codimension 4 linear space, two codimension 5 linear spaces, two components of codimension 5 and degree 3, and one of codimension 6 and degree 10. (Remember, a set of codimension k in $\mathbb{C}^N$ has dimension $N-k$.)

To run Bertini using the dimension-by-dimension solving approach, stopping after computing the witness point supersets,[32] include the configurations

```
TrackType:1;
WitnessGenType:1; % dimension-by-dimension
WitnessSupersetOnly:1; % only compute a witness point superset
```

With these settings, Bertini writes the following table to the screen:

codim	paths	witness superset	nonsingular	singular	nonsolutions	infinite endpoints	other bad endpoints
1	2	0	0	0	2	0	0
2	4	0	0	0	4	0	0
3	8	0	0	0	8	0	0
4	16	1	1	0	15	0	0
5	32	14	8	6	18	0	0
6	64	64	10	54	0	0	0
total	126						

The rank of this system is 6, so we need to consider from codimension 1 down to codimension 6. Since each polynomial in f has degree 2, the number of paths tracked for

[32]The default is to proceed on to compute the numerical irreducible decomposition.

codimension i (listed in the second column) is 2^i. This is the total degree of a randomization of g_i formed by randomizing the six polynomials in f down to i polynomials and retaining the $12-i$ linear equations yielding a square system.

The third column shows the number of points in the witness point superset for each codimension, and the fourth and fifth columns partition these as nonsingular and singular, respectively. Each nonsingular point must be a witness point, as there aren't enough linear equations to slice out an isolated point on any higher-dimensional set. In general, the singular points might or might not be witness points, but in this particular example, none of them are because each irreducible component has multiplicity one with respect to f. ∎

It's worth noting that the system of Example 8.1 from the previous chapter could be run dimension by dimension as well. However, this generates many paths going off to infinity, so the regenerative cascade method described later in this chapter is much more effective.

In the dimension-by-dimension approach, solving for each dimension is independent of the others. When searching for components that lie in a specific dimension (see §11.2), this fact can be exploited. However, these systems g_i are clearly related to each other. In particular, one can *cascade* through the various systems using a sequence of homotopies starting from the largest dimension down to the smallest. This is described in the next section.

9.4 ▪ Cascade Methods

While dimension-by-dimension solving is a straightforward application of slicing, it is typically not the most efficient method for computing the numerical irreducible decomposition. In this section, we outline the basic steps of two cascade methods, with technical details left to the references. In particular, the first version of a cascading approach, hereafter referred to as the classical cascade, appeared in [154] with [161] containing many details about the method up to 2005, while the Bertini default—the regenerative cascade method—was first presented in [94].

Classical cascade For simplicity, we shall assume that $f : \mathbb{C}^N \to \mathbb{C}^n$, $n \leq N$, is a polynomial system of rank n. The general case where n might be larger than rank$f$ can be reduced down to this case by using randomization (see §9.2). Of course, this maneuver may introduce extraneous components of codimension equal to rank$f$, but we will check for these and discard them at the end of the procedure. Thus, we know $N \geq n$ and each irreducible component of $\mathbf{V}(f)$ has dimension at least $N-n \geq 0$. Let $\ell_1,\ldots,\ell_{N-1}$ be general linear polynomials on $\mathbb{C}^N$ and, for a general $A \in \mathbb{C}^{N\times(n-1)}$, consider the polynomial system

$$h(z,t_1,\ldots,t_{n-1}) = \begin{bmatrix} f(z) \\ \ell_1(z) \\ \vdots \\ \ell_{N-n}(z) \end{bmatrix} + A \cdot \begin{bmatrix} t_1\ell_{N-n+1}(z) \\ \vdots \\ t_{n-1}\ell_{N-1}(z) \end{bmatrix}.$$

For $i = 1,\ldots,n$, define

$$g_i(z) := h(z,\underbrace{1,\ldots,1}_{n-i},\underbrace{0,\ldots,0}_{i-1}).$$

From a witness point superset standpoint, there are two key facts. First, if S_i is any finite superset of the isolated points in $\mathbf{V}(g_i)$, then $\widehat{W}_{N-i} := S_i \cap \mathbf{V}(f)$ is a witness point superset that we seek for dimension $(N-i)$. The second is that Bertini's theorem ensures that each point in the set of nonsolutions $\mathscr{N}_i := S_i \setminus \widehat{W}_{N-i}$ is nonsingular with respect to g_i.

From a computational standpoint, the "cascade" arises from the fact that given $\mathscr{N}_i$ for $i < n$, a finite superset S_{i+1} of the isolated points in $\mathbf{V}(g_{i+1})$ is the set of finite endpoints obtained using the homotopy

$$H_i(z,t) := h(z, \underbrace{1,\ldots,1}_{n-i-1}, t, \underbrace{0,\ldots,0}_{i-1})$$

with startpoints $\mathscr{N}_i$ at $t = 1$. To start the cascade, the set S_1, and hence $\mathscr{N}_1$, is obtained using a standard homotopy discussed in Part I.

Regenerative cascade In the regenerative cascade, randomization on $f : \mathbb{C}^N \to \mathbb{C}^n$ is directly employed so the simplification that $\operatorname{rank} f = n$ can be dropped. Let $\ell_1, \ldots, \ell_{N-1}$ be general linear polynomials as above. Let $A \in \mathbb{C}^{\operatorname{rank} f \times n}$ be a general matrix and, for $i = 1, \ldots, \operatorname{rank} f$, let A_i be the matrix consisting of the first i rows of A. For $i = 1, \ldots, \operatorname{rank} f$, consider the polynomial systems

$$g_i(z) := \begin{bmatrix} A_i \cdot f(z) \\ \ell_1(z) \\ \vdots \\ \ell_{N-i}(z) \end{bmatrix}.$$

As with the classical cascade, if S_i is any finite superset of the isolated points in $\mathbf{V}(g_i)$, then $\widehat{W}_{N-i} := S_i \cap \mathbf{V}(f)$ is a witness point superset for dimension $(N-i)$. Each point in the set of nonsolutions $\mathscr{N}_i := S_i \setminus \widehat{W}_{N-i}$ is also nonsingular with respect to g_i.

To start the regenerative cascade, one needs to compute $S_1 := \mathbf{V}(g_1)$, which can be accomplished using a standard homotopy from Part I. In fact, by working intrinsically on the linear space $\mathbf{V}(\ell_1, \ldots, \ell_{N-1})$, this can be reduced to solving a univariate polynomial. From S_1, one can then compute $\mathscr{N}_1$. To cascade from $\mathscr{N}_i$ to S_{i+1}, one simply regenerates (see §5.4) from the linear polynomial ℓ_{N-i} to $a_{i+1} \cdot f$, where a_{i+1} is the $(i+1)$st row of A.

Comparison The cascade methods benefit when the set of nonsolutions $\mathscr{N}_i$ is smaller than one would expect, e.g., based on the relevant total degree. One advantage of both the regenerative cascade and the dimension-by-dimension approach from §9.3 is the ease of working intrinsically on the linear space $\mathbf{V}(\ell_1, \ldots, \ell_{N-i})$. A disadvantage of the classical cascade is the number of paths tracked in the first step, i.e., to compute S_1. In Bertini, the number of paths for this step is equal to the total degree, but one could possibly reduce this by taking advantage of additional structure in the particular system of interest. Nonetheless, the main advantage of the cascade methods over the dimension-by-dimension approach is that by cascading down the dimensions, the witness point supersets usually contain fewer points that lie on higher-dimensional components. This will have an impact on the efficiency of computing witness point sets from the witness point supersets discussed in the next chapter.

Example 9.5 (adjacent minors revisited). The configuration to use the classical cascade mode in Bertini is `WitnessGenType:0`. The following is the screen output for the adjacent minors problem of Example 9.4:

```
|codim|paths  |witness  |nonsingular|singular|nonsolutions|infinite
|     |       |superset|           |        |            |endpoints
-----------------------------------------------------------------
| 1   | 64    | 0       | 0          | 0       | 49          | 15
| 2   | 49    | 0       | 0          | 0       | 49          | 0
| 3   | 49    | 0       | 0          | 0       | 49          | 0
| 4   | 49    | 1       | 1          | 0       | 48          | 0
| 5   | 48    | 12      | 8          | 4       | 36          | 0
| 6   | 36    | 36      | 10         | 26      | 0           | 0
-----------------------------------------------------------------
|total| 295
```

Despite a larger number of paths in this case, this run takes approximately the same amount of time as the dimension-by-dimension run.

The regenerative cascade is the default setting in Bertini but can be explicitly called by using `WitnessGenType:2`. The following is the screen output in this case:

```
|codim|paths  |witness  |nonsingular|singular|nonsolutions|infinite
|     |       |superset|           |        |            |endpoints
-----------------------------------------------------------------
| 1   | 2     | 0       | 0          | 0       | 2           | 0
| 2   | 4     | 0       | 0          | 0       | 4           | 0
| 3   | 8     | 0       | 0          | 0       | 8           | 0
| 4   | 16    | 1       | 1          | 0       | 15          | 0
| 5   | 30    | 12      | 8          | 4       | 18          | 0
| 6   | 36    | 36      | 10         | 26      | 0           | 0
-----------------------------------------------------------------
|total| 96
```

This run takes about 30% less time than the other two approaches.

In both cascade runs, we see that a total of 30 singular points were computed in the witness point supersets compared with a total of 60 using the dimension-by-dimension approach summarized in Example 9.4. ∎

Chapter 10
The Numerical Irreducible Decomposition

As noted at the beginning of the previous chapter, the numerical irreducible decomposition of the solution set of a polynomial system $f(z)$ consists of a witness set $\mathscr{W}_{i,j} = \{f, L_{i,j}, W_{i,j}\}$ for each component. For each dimension i of the solution set, the computation of the witness sets has two main phases:

1. **Compute a witness point superset in dimension i:** Find a finite set of points $\widehat{W}_i \subset \mathbf{V}(f)$ that contains a witness point set $W_{i,j}$ for each i-dimensional component in $\mathbf{V}(f)$.

2. **Reduce the witness point superset $\widehat{W}_i$ to witness point sets $W_{i,j}$:** Remove unnecessary points from $\widehat{W}_i$ and decompose the resulting set W_i into a union of witness point sets, one for each i-dimensional irreducible component.

From witness point sets, it is easy to construct a witness set for each irreducible component and thus the full numerical irreducible decomposition.

The previous chapter focused on step 1, while this chapter focuses on step 2. One key maneuver used in various parts of this chapter is moving a linear slice, a topic that was presented in §8.3 and used in §8.4 for developing a membership test. As discussed in §10.1.1, the membership test forms the basis for one approach to extracting the set W_i from $\widehat{W}_i$, a process called *junk removal*. The other approach for junk removal uses a local dimension test, as summarized in §10.1.2 and expanded upon in §11.5.

Once W_i has been computed, it is then partitioned into witness point sets, one for each i-dimensional irreducible component. This is presented in §10.2 using traces (§10.2.1) and monodromy loops (§10.2.2).

The topics of this chapter, although fundamental to the computation of the numerical irreducible decomposition, are in no way necessary for running Bertini.

10.1 ▪ Junk Removal

The witness point superset $\widehat{W}_i$ consists of points that lie on components of $\mathbf{V}(f)$ of dimension at least i. In this section, we present two ways to identify the points in $\widehat{W}_i$ that lie on some component of dimension strictly larger than i, called *junk points*. Once these are removed from $\widehat{W}_i$, the resulting set W_i is a witness point set for the pure i-dimensional component of $\mathbf{V}(f)$. The two methods for junk removal are membership testing and local dimension testing. By default, Bertini first uses the local dimension test to find points in $\widehat{W}_i$ that are easy to classify and then uses membership testing on the remaining points.

10.1.1 ▪ Membership Test and Junk Removal

Since $\widehat{W}_i$ consists of points that lie on components of $\mathbf{V}(f)$ of dimension at least i, it follows that $\widehat{W}_i = \emptyset$ for $i > \dim \mathbf{V}(f)$. So, after generating a superset $\widehat{W}_i$ at every possible dimension using a method from Chapter 9, $\dim \mathbf{V}(f)$ is revealed as the largest i for which $\widehat{W}_i \neq \emptyset$. Moreover, each point in $\widehat{W}_{\dim \mathbf{V}(f)}$ lies on a component of dimension equal to $\dim \mathbf{V}(f)$, as there are no components of higher dimension to accommodate junk points. This implies that $W_{\dim \mathbf{V}(f)} := \widehat{W}_{\dim \mathbf{V}(f)}$. This initializes a procedure that recursively descends through the dimensions starting at $i = \dim \mathbf{V}(f) - 1$ to extract W_i from $\widehat{W}_i$ assuming that W_k for $k > i$ have already been computed. For each $w \in \widehat{W}_i$, the membership test of §8.4 determines whether w lies on a component of dimension $k > i$ using the known sets W_k for $k > i$. A point $w \in \widehat{W}_i$ is a junk point if this test determines that it does indeed lie on some component of dimension $k > i$. Removing these junk points from $\widehat{W}_i$ produces W_i.

This approach requires knowledge of the union of all higher-dimensional components. This so-called global information is available in Bertini during the process of computing a full numerical irreducible decomposition. However, if the user is interested in only a specific dimension (see §11.2), such information may not have been computed. Moreover, if a higher-dimensional component has multiplicity greater than one, Bertini must use deflation (see §11.6) to facilitate the path tracking in the membership test.

If $w \in \widehat{W}_i$ is indeed a witness point, i.e., $w \in W_i$, then the number of paths the membership test will track is the sum of the degrees of all components of dimension strictly larger than i. All the components will have to be tested because that is how it is determined that w is not a member of any of them. In contrast, if $w \in \widehat{W}_i \setminus W_i$, i.e., w is a junk point, then the membership test will track paths until it encounters one that ends at w. In short, the number of paths tracked by using the membership test for junk removal could be quite large.

Remark 10.1. The first junk removal algorithm (presented in [155]) was based on interpolation. For some larger problems, this interpolation-based approach can be many orders of magnitude slower than the homotopy-based membership test. This occurs because the number of monomials in N variables of degree at most d grows rapidly as N and d increase.

10.1.2 ▪ Local Dimension and Junk Removal*

The *local dimension*[33] of a point x^* with respect to an algebraic set X is the maximum of the dimensions of the irreducible components of X that contain x^*. In the context of junk removal, each point in $\widehat{W}_i$ has local dimension at least i. A point $w \in \widehat{W}_i$ is a witness point, i.e., $w \in W_i$, if and only if the local dimension of w with respect to $\mathbf{V}(f)$ is exactly i. The membership testing approach of §10.1.1 determines local dimension using information about every higher-dimensional component of $\mathbf{V}(f)$. This leads us to the following fundamental question, which is explored in more detail in §11.5:

> Given $x^* \in \mathbf{V}(f)$, can one determine the local dimension of $\mathbf{V}(f)$ at x^* using only local information?

[33]The alternative phrase "the local dimension of algebraic set X at point x^*" means the same thing.

In the context of junk removal, suppose that L_i is a system of i general linear polynomials so that $\widehat{W}_i \subset \mathbf{V}(f, L_i)$. For a given $w \in \widehat{W}_i$, we know that $w \in W_i$ if and only if w is isolated in $\mathbf{V}(f, L_i)$. Therefore, junk removal from $\widehat{W}_i$ simply means removing the points in $\widehat{W}_i$ that are not isolated in $\mathbf{V}(f, L_i)$.

Fix $w \in \widehat{W}_i$. The homotopy used to compute w incidentally produces information about w as a point in $\{f, L_i\}^{-1}(0)$. In particular, if B_w is the number of paths that ended at w, then B_w is a bound on the multiplicity of w with respect to $\{f, L_i\}$ provided that w is isolated in $\mathbf{V}(f, L_i)$. Therefore, as described in [19], this bound B_w provides a stopping criterion for an algorithm to determine if w is isolated in $\{f, L_i\}$. By using Macaulay dual spaces, described in §11.4, one can compute the actual multiplicity of w with respect to $\{f, L_i\}$ if w is isolated. If w is not isolated, the dimension of these dual spaces grows without bound. Thus, one simply needs to demonstrate that the dual space dimension is larger than B_w to conclude that w is not isolated.

The current implementation in Bertini performs the local dimension test using the Macaulay dual space approach presented in [47], but more efficient methods are presented in [85, 187]. Nonetheless, Bertini uses a bounded form (based on the configuration `LDTDepth`) of the local dimension test (see §11.5 for more details) to quickly partition the set $\widehat{W}_i$ into three sets: known witness points, known junk points, and points which will be classified using the membership test.

Remark 10.2. An approach for determining the local dimension was described in [160, §3.3] by attempting to deform from w along a family of linear spaces. Although this article did not provide a numerical procedure to perform this computation, a useful heuristic approach was presented in [109]. For testing at smooth points of an irreducible component, such an approach can be very effective. However, multiple solution branches passing through the point in question could lead to incorrectly identifying the local dimension as shown by the example in [19, §3.2] consisting of two cubic equations in two variables. Due to this deficiency, this approach cannot be trusted, so Bertini uses only the homotopy membership test and the Macaulay dual space approach for local dimension testing.

10.2 ▪ Pure-Dimensional Decomposition

Upon removing the junk points from $\widehat{W}_i$, the remaining set W_i is a witness point set for the union of the irreducible i-dimensional components. The final task is to partition W_i into sets $W_{i,j}$, each of which is a witness point set for an irreducible component. This equidimensional decomposition is carried out using *traces* and *monodromy loops*, as described, respectively, in the next two subsections. The final subsection puts these two pieces together to complete the pure-dimensional decomposition.

10.2.1 ▪ Trace Test

The trace test is a consequence of a generalized Lefschetz theorem. It builds from results in [151], and an efficient numerical version of this method first appeared in [158]. The following provides a brief overview of the trace test, with more details provided in [161, §15.5].

Suppose that $X \subset \mathbb{C}^N$ is a pure-dimensional algebraic set of dimension k and $\mathcal{L}$ is a general linear space of codimension k. Thus, $X \cap \mathcal{L}$ will consists of $\deg X$ points. The centroid of these intersection points will move linearly as the linear space $\mathcal{L}$ is moved parallel to itself. In particular, X is reducible if and only if the centroid of some nonempty proper subset of these intersection points also moves linearly.

Example 10.3. To illustrate, consider intersecting the curve $X = \mathbf{V}(x^2 + y^2 - 1) \subset \mathbb{C}^2$ with the line defined by $y = ax + b$, where $a, b \in \mathbb{C}$ are general. The two points of intersection are

$$(x_1, y_1) = \left(\frac{-ab + \sqrt{1+a^2-b^2}}{1+a^2}, \frac{b + a\sqrt{1+a^2-b^2}}{1+a^2} \right),$$

$$(x_2, y_2) = \left(\frac{-ab - \sqrt{1+a^2-b^2}}{1+a^2}, \frac{b - a\sqrt{1+a^2-b^2}}{1+a^2} \right).$$

Since neither of (x_j, y_j), $j = 1, 2$, is a linear function of b, X is irreducible. Additionally, the centroid of these two points, namely,

$$\left(\frac{x_1 + x_2}{2}, \frac{y_1 + y_2}{2} \right) = \left(\frac{-ab}{1+a^2}, \frac{b}{1+a^2} \right),$$

is linear in b. In fact, the centroid lies on the line $x = -ay$, which is perpendicular to every line parallel to the original line $y = ax + b$. ∎

Since each coordinate of the centroid moves linearly, every linear combination of the coordinates moves linearly. This will form the foundation of the trace test implemented in Bertini for pure-dimensional decomposition. To that end, suppose that L_i is a system of general linear polynomials such that W_i is the set of isolated points of $\mathbf{V}(f, L_i) \subset \mathbb{C}^N$. Bertini first selects $\gamma_1, \gamma_2 \in \mathbb{C}$, $\alpha \in \mathbb{C}^N$, and $v \in \mathbb{C}^i$ at random. Then, for $j = 1, 2$, Bertini uses the homotopy defined in (8.5) to linearly interpolate between $L = L_i$ and $L' = L_i + \gamma_j \cdot v$ and, for each $w \in W_i$, follows the solution path from w to its endpoint w_j. Define the trace test for w as

$$\sigma_w := \alpha \cdot \left(\frac{w_1 - w}{\gamma_1} - \frac{w_2 - w}{\gamma_2} \right) \in \mathbb{C}.$$

Note that if the solution path from w is a linear function of γ, then the trace test σ_w must be zero. The randomness of the constants in the test implies, with probability one, that the converse implication also holds: a zero trace test implies that the path is linear. Moreover, since the test is linear in the sample points, the test for a sum of points is the sum of their trace tests. Accordingly, with probability one, a set $T \subset W_i$ is a witness point set for an irreducible component if and only if

$$\sum_{w \in T} \sigma_w = 0 \text{ and } \sum_{w \in S} \sigma_w \neq 0 \text{ for all nonempty } S \subsetneq T.$$

In particular, the first statement implies that T forms a witness point set for some pure-dimensional algebraic set, whereas the second statement implies irreducibility. From trace tests, the linear and quadric irreducible components are easy to identify. A linear i-dimensional component exists if and only if there is $w \in W_i$ with $\sigma_w = 0$. A quadric i-dimensional component exists if and only if there is a pair $w, w' \in W_i$ with $\sigma_w = -\sigma_{w'}$ and $\sigma_w, \sigma_{w'} \neq 0$. Of course, since each σ_w is numerically approximated in Bertini, the decision regarding zero and nonzero for traces is made based on a tolerance, which in this case is `FinalTol*TargetTolMultiplier`.

If $k = \#W_i$, the total number of subsets of W_i is 2^k, so this is the maximum number of sums, $\sum_{w \in T} \sigma_w$, $T \subset W_i$, that have to be tested in order to find all the irreducible witness point sets in W_i. Actually, this can be cut in half, because the sum of the trace tests for T and the sum for its complement, $W_i \setminus T$, must add to zero, so one is zero if and only if the other one is too. As we check these sums, if any is found to be zero, that group of points is removed from W_i and stored as an irreducible witness point set $W_{i,j}$, which greatly reduces the number of further combinations to be checked. The worst case of checking 2^{k-1} sums occurs only when the whole set W_i represents a single irreducible component. When k is small, checking at most 2^{k-1} sums is practical and efficient. However, for larger k, the exponential growth eventually makes such a combinatorial search impractical. Fortunately, as the next subsection shows, monodromy loops help to simplify this search.

We conclude by illustrating a combinatorial trace test.

Example 10.4. Although trace tests computed by Bertini are floating-point complex numbers, let's consider a toy example with integer trace tests to illustrate the main points. Suppose we know W_{12} consists of eight points. Thus, a priori, there could be anywhere from 1 to 8 irreducible components, with degrees ranging from 8 down to 1. If there is only one component, no proper subset of the eight trace tests would sum to zero. At the other end of the spectrum, if there are eight linear components, each trace test would be zero (numerically, of course).

Suppose the trace tests are as follows:

$$10, 0, 5, 10, -20, 0, -5, 0,$$

corresponding, respectively, to points $w_1, \ldots, w_8$. We can easily spot the ones which correspond to linear components—the trace is zero. Thus, we record these: $W_{12,1} := \{w_2\}$, $W_{12,2} := \{w_6\}$, and $W_{12,3} := \{w_8\}$. It is of course easy to see how the remaining trace tests would be grouped in a minimal way to force zero sums, but let's belabor the point a bit to illustrate how we would proceed algorithmically.

Since all of the remaining trace tests are nonzero, we next need to consider all pairs: $\sigma_{w_1} + \sigma_{w_3} = 10 + 5 \neq 0$, etc. After several quick sums, we see that $\sigma_{w_3} + \sigma_{w_7} = 0$, producing the witness point set $W_{12,4} := \{w_3, w_7\}$. Since no other pair works, the remaining triple represents an irreducible component, $W_{12,5} := \{w_1, w_4, w_5\}$.

In the end, this test shows that there are five components of dimension 12: three of degree 1 and one each of degree 2 and 3. ∎

10.2.2 ▪ Monodromy

Recall from §8.1 that the set of smooth points of an irreducible component is path connected. This means that if $w, w' \in W_i$ lie on the same irreducible component X, then there is a path in X_{reg} between w and w'. Also, recall from the discussion of the Cauchy endgame in §3.3.3 that walking around a branch point in the parameter space can make points in the fiber permute. (This phenomenon is also seen in §2.3.) Since the branch points are unknown, one can use randomly generated loops to attempt to find ones where the points permute. Any such permutation establishes connectivity between witness points, so one obtains information about which points in W_i must lie on the same irreducible component.

As above, suppose that L_i is a system of general linear polynomials such that W_i is the set of isolated points of $\mathbf{V}(f,L_i) \subset \mathbb{C}^N$. Then, one simply constructs a loop in the space of systems of i general linear polynomials that starts (and ends) at L_i. Starting from each point in W_i, this loop defines a path whose endpoint is also in W_i. Since the startpoint and endpoint must lie on the same irreducible component, new information is gathered when a startpoint and endpoint are different.

In Bertini, the loops are constructed by a succession of two homotopies of the form of (8.5), which linearly interpolates between two linear slicing systems, L and L'. For this purpose, Bertini picks random $\gamma_j \in \mathbb{C}$ for $j = 1,2$ and $v \in \mathbb{C}^i$. The first homotopy moves from $L = \gamma_1 \cdot L_i$ to $L' = L_i + v$, while the second moves back using $L = \gamma_2(L_i + v)$ and $L' = L_i$. For each $w \in W_i$, Bertini tracks the path starting at w through the two homotopies to arrive at some endpoint w'. If $w' \neq w$, this loop has indeed encircled at least one branch point producing a nontrivial monodromy action.

In these loops used in Bertini, the linear space $\mathbf{V}(L_i)$ is always moved parallel to itself, so one might be concerned that these maneuvers are not general enough to find sufficient connections to complete the irreducible decomposition of W_i. However, the use of the gamma trick via γ_j (see §6.1 for more details) ensures that if any nontrivial monodromy action exists, then there exist choices of γ_1, γ_2, and v that give one. Moreover, there exist finite sets of loops of this type that generate the corresponding *monodromy group*, meaning that all connections can be deduced to complete the decomposition. (See Chapter 15 for additional uses of this.) Unfortunately, we do not know which loops to take to get this result; we can only guess at them randomly.

The insufficiency of depending on monodromy alone can be seen by considering a path through the parameter space of a family of systems such that two branch points approach each other. It might happen that to generate the monodromy group, one must find a loop that encircles just one of the branch points. In the limit as the branch points merge, the expected number of random loops needed to find such a loop could be infinite. Moreover, there is no stopping criterion, except in the special case that W_i represents a single irreducible component and the loops have established connections between all the witness points, i.e., the full symmetric group has been confirmed.

This shortcoming of the monodromy method does not mean that we should abandon it. Quite the contrary, because any nontrivial monodromy loops we can find can be used to simplify the combinatorial trace test, as described in the next subsection.

10.2.3 ▪ Connecting Traces and Monodromy

The previous two subsections described ways to facilitate the partitioning of W_i into a collection of witness points sets, one for each irreducible i-dimensional component. Bertini uses a combination of these approaches to produce this partition.

First, using the process described in §10.2.1, the trace test σ_w is computed for each $w \in W_i$. The ones which are zero, i.e., corresponding to a linear space, are classified accordingly. After that, if there are more than two points left, we must identify which subsets of them lie on the same irreducible component. If there are few enough of them, it is best to perform an exhaustive combinatorial search based on the trace tests; otherwise, it is advisable to try finding connections via monodromy loops. In Bertini, if the number of points is at most `MaxNumPtsForTrace`, then the combinatorial search is employed; otherwise a monodromy loop is performed.

After a monodromy loop, points that are identified to lie on the same component are merged together, forming collections of points. Trace tests are used to identify collections that form a witness point set for an irreducible component, and these are removed from further consideration. Additional loops are performed until either

- the number of remaining collections of points is at most `MaxNumPtsForTrace`,
- the number of loops performed since the last time new information was produced via a monodromy loop is larger than `MaxNumBadMonoLoops`, or
- the total number of loops performed is larger than `MaxNumMonoLoops`.

Afterward, the combinatorial search is performed on any remaining collections.

From a computational standpoint, there are two items to consider. First, when the degrees of the components are large, the numerical error in the sum of the traces could accumulate. This can be mitigated by decreasing `FinalTol`, increasing `TargetTolMultiplier`, or using sharpening via `SharpenDigits`.

The second issue is how to handle the case when one or more points are missing in W_i due to a failure in path tracking. If the set of missing points contains the corresponding witness point set for some irreducible component, this component will be missed entirely. If only a subset of the witness points for a component is missing, the trace test will confirm that this is indeed the case. In Bertini, each point in the remaining subset of the witness points will be listed as *unclassified*. The missing points could be recovered using monodromy loops, but such recovery is not currently employed in Bertini. When unclassified points occur, it may be best to rerun the problem, perhaps first using scaling (see §7.7) to reformulate the system. Adjusting the configuration settings and rerunning Bertini several times using different random numbers can help. Using a priori knowledge about the expected solution set can also help, such as by recovering a missing point using a known symmetry relation. An a posteriori certification that each irreducible component has indeed been found is an open topic for further inquiry with some progress made in [17] by recovering exact generators for components and symbolically testing the computed decomposition.

Chapter 11

Advanced Topics about Positive-Dimensional Solution Sets

Throughout Chapters 8–10, the focus has been on computing a numerical irreducible decomposition. Since many computations on an irreducible component can be accomplished using a user-defined homotopy given the witness points and linear slices, in §11.1 we discuss printing this information, a topic that was briefly mentioned in §8.2. This data is extracted from the `witness_data` file.

When extra knowledge about the polynomial system is known, such as the dimension of the solution components of interest, one might be satisfied with only partial information rather than the full numerical irreducible decomposition. In §11.2, we describe how to compute the numerical irreducible decomposition of the union of the irreducible components of dimension either at least k or equal to k. In the latter case, this depends upon using local information computed at the points in the witness superset to determine if they are true witness points or lie on a higher-dimensional component. (See §11.5 for more information on the local dimension test.) To fully describe this local information, we explore the formal definition of multiplicity in §11.3 and present Macaulay dual spaces in §11.4.

To conclusively determine the true local dimension at a solution z^* (of a polynomial system $f(z) = 0$) using the local dimension test, we need to know a positive integer μ_{z^*} with the property that if z^* is an isolated solution of $f(z) = 0$, then μ_{z^*} bounds the multiplicity of z^*. During the numerical irreducible decomposition computation, such a bound μ_{z^*} is given by the number of paths ending at the point z^*. However, in applications, one may know a solution without such a bound and want to investigate its local dimension. This can be done using global information by performing the membership test (see §8.4) after a numerical irreducible decomposition has been computed. But since computing a numerical irreducible decomposition may be impractical for many large problems arising in applications, we explore what information can be obtained using only local information in §11.5. To keep the discussion concrete, this is presented in the context of the mobility of mechanisms.

Since components of multiplicity greater than one arise in applications, we explore in §11.6 how Bertini performs computations on these components via deflation.

11.1 ▪ Printing a Witness Set

Since many algorithms depend on the manipulation of witness sets, Bertini has the functionality to print the witness sets in a form that can be easily reused by Bertini, particularly for a user-defined homotopy. In particular, Bertini can print the linear slicing equations and witness points for each irreducible or pure-dimensional component resulting from a previous numerical irreducible decomposition computation. This simply requires modifying the TrackType configuration in the input file and making available the witness_data file created during a previous run of Bertini.

We will demonstrate by considering the union of the line $x = y$ and the unit circle. Bertini will generate a numerical irreducible decomposition for the line and unit circle by using the following file:

```
% LineCircle.input
CONFIG
 TrackType:1;
END;
INPUT
 function f;
 variable_group x,y;

 f = (x - y)*(x^2 + y^2 - 1);
END;
```

By maintaining the witness_data file, we are able to print the linear slicing equations and witness points by rerunning Bertini with the TrackType configuration changed to 4. It is essential to keep witness_data files at the end of any run for which post-processing is planned. One needs to be careful since at the beginning of each run, Bertini will overwrite all output files except those that are to be used as data input. For presentation purposes, we use MPType:0 in the following file to print the data using double precision:

```
% LineCircle.print
CONFIG
 TrackType:4; % print witness set
 MPType:0;    % print using double precision
END;
INPUT
 function f;
 variable_group x,y;

 f = (x - y)*(x^2 + y^2 - 1);
END;
```

In particular, if a fixed precision setting is used, i.e., setting MPType to either 0 or 1, then each floating-point number in witness_data is simply converted to the precision specified by the user (see §E.2). If adaptive precision is used, MPType:2, then the witness points are printed in the precision used to store them in witness_data and the coefficients for the linear slicing equations are printed in the precision corresponding to the AMPMaxPrec configuration.

Upon running Bertini to print the witness set from above, the following summarizes the interactive menu used for printing the linear slicing equations to linear_system and witness point set to witness_points for the union of the two curves in $\mathbf{V}(f)$:

```
*************** Components to Print ****************

Dimension 1: 2 classified components
-------------------------------------------------
   degree 1: 1 component
   degree 2: 1 component

Please select a dimension to print (-1 to quit): 1

Dimension 1: 2 classified components
-------------------------------------------------
   component 0 has degree 1
   component 1 has degree 2

Please select a component to print (-1 to quit, -2 to print all): -2

Enter the name of the file where to write the witness points: witness_points

Enter the name of the file where to write the linear system: linear_system
```

For our computation, the resulting `witness_points` file is as follows:

```
3

 9.784356699813977e-01 7.816295973707499e-02
-3.235088278717050e-01 2.363998175295627e-01

-7.000703142380933e-01 5.415691377883562e-01
 9.766736330540036e-01 3.881915756111992e-01

2.468893515298665e-01 2.933483564687335e-01
2.468893515298665e-01 2.933483564687335e-01
```

In this case, the first two correspond to the unit circle, while the third corresponds to the line $x = y$. The resulting `linear_system` file is as follows:

```
variable_group x,y;
function linear1;
constant const1_1,const1_2,const1_3;

const1_1 = 2.924295155952777e-01-4.555587159230302e-01*I;
const1_2 = 4.935145472375286e-02+6.151231406177551e-01*I;
const1_3 = 3.685838280706287e-01+7.334894485514644e-01*I;

linear1 = const1_1 + const1_2*x + const1_3*y;

END;
```

This file consists of one linear polynomial in two variables, using the names of the variables from the `input` file. Note that the three constants are the random numbers chosen by Bertini when it selected the linear slicing equation that was used to cut out the particular points printed in `witness_points`.

11.2 ▪ Computing Specific Codimensions

A numerical irreducible decomposition describes all the irreducible components for a given polynomial system. A common occurrence in many applications is that a full numerical irreducible decomposition is not needed since it is known a priori that the dimensions of the components of interest either are exactly k or are bounded below by k. This section describes how Bertini can be used in both these situations by using the configurations `SpecificCodimension` and `MaxCodimension`, respectively, to compute the components of the relevant dimension(s) without needing to compute the full numerical irreducible decomposition.

For an illustrative example, consider the following polynomial system on $\mathbb{C}^3$:

$$f(x,y,z) = \begin{bmatrix} x(x^2-y-z^2) \\ x(x+y+z^2-2)(y-5) \\ x(x^2+x-2)(z-2) \end{bmatrix}. \tag{11.1}$$

It is easy to verify that $\mathbf{V}(f)$ consists of the following irreducible components:

- the plane $\{(0,y,z) \mid y,z \in \mathbb{C}\}$,
- two quadratic curves $\{(1,y,z) \mid y+z^2=1\}$ and $\{(-2,y,z) \mid y+z^2=4\}$, and
- two points $(3,5,2)$ and $(-3,5,2)$.

Suppose we are interested only in computing the numerical irreducible decomposition for the union of the irreducible components of dimension at least 1. Since the ambient space, namely, $\mathbb{C}^3$, is three-dimensional, the maximum codimension under consideration is 2. This is accomplished using the following code:

```
% MaxCodimension.input
CONFIG
 TrackType:1;       % compute a numerical irred. decomp.
 MaxCodimension:2; % for components of codim <= 2
END;
INPUT
 variable_group x,y,z;
 function f1,f2,f3;

 f1 = x*(x^2-y-z^2);
 f2 = x*(x+y+z^2-2)*(y-5);
 f3 = x*(x^2+x-2)*(z-2);
END;
```

After running Bertini, the summary of the decomposition is as follows:

```
************* Decomposition by Degree *************

Dimension 2: 1 classified component
----------------------------------------------------
   degree 1: 1 component

Dimension 1: 2 classified components
----------------------------------------------------
   degree 2: 2 components

****************************************************
```

In this computation, which used the `MaxCodimension` option, the witness superset generation algorithm is simply truncated once that codimension is reached. The junk removal and decomposition steps utilize the computed witness supersets as if the lower-dimensional components simply did not exist.

We now turn to using Bertini to compute the numerical irreducible decomposition of the union of the irreducible components of $\mathbf{V}(f)$ of a specific dimension. Before presenting the input file for this case, we need to first consider how Bertini performs this computation to understand which configurations to use. The first step in this computation is to compute a witness superset for the specific dimension. There are three options for this: the dimension-by-dimension approach (see §9.3) and the classical and regenerative cascades (see §9.4). The dimension-by-dimension approach (`WitnessGenType:1`) treats each dimension independently, meaning that it can directly compute a witness superset for a given dimension using one homotopy. The classical cascade (`WitnessGenType:0`) and the regenerative cascade (`WitnessGenType:2`), as implemented in Bertini, will compute a witness superset for the given dimension by cascading down to that dimension.

The advantage of using a cascading approach instead of the dimension-by-dimension approach is in the next step, namely, turning a witness superset into a witness set by removing the so-called junk points. The number of junk points in a witness superset is typically much smaller when using a cascade. Second, since Bertini has cascaded down to the given dimension, Bertini will have computed whether the given dimension is equal to the dimension of $\mathbf{V}(f)$. If it is, the witness superset is actually a witness set, meaning that the junk removal step can be bypassed. This will be of significant advantage if components of high multiplicity exist in this top dimension.

Suppose that either the dimension-by-dimension approach was used, meaning that Bertini has no knowledge about the dimension of $\mathbf{V}(f)$, or the requested dimension is not the dimension of $\mathbf{V}(f)$. Since witness sets for the higher-dimensional components are not computed, Bertini will always use the local dimension test (see §10.1.2) for junk removal. Since the local dimension test in Bertini is *depth bounded* (see §11.4), this test may fail to yield a conclusive decision based on the bound described by the configuration `MaxLDTDepth`. In this case, the point is considered to be junk and removed. Since this action may yield invalid results, a message similar to the following is printed to the screen, alerting the user to this action:

```
NOTE: The local dimension test was terminated before it could
      determine if 1 point was isolated.  Consider either
      increasing MaxLDTDepth or using MaxCodimension instead
      of SpecificCodimension.
```

Since the decomposition of the pure-dimensional witness set into irreducible components is always performed independently in each dimension, this third step proceeds as usual.

We now return to the illustrative example (11.1) and consider computing just the curves (codimension 2) first and then the isolated points (codimension 3). For comparison, we will use both the dimension-by-dimension and the regenerative cascade witness superset generation approaches for these computations. The following code uses the dimension-by-dimension approach to compute the one-dimensional components:

```
% SpecificCodimension.input
%    for components of codim = 2
CONFIG
 TrackType:1;              % compute a numerical irred. decomp.
```

```
 SpecificCodimension:2; % for components of codim = 2
 WitnessGenType:1;      % using dimension-by-dimension
END;
INPUT
 variable_group x,y,z;
 function f1,f2,f3;

 f1 = x*(x^2-y-z^2);
 f2 = x*(x+y+z^2-2)*(y-5);
 f3 = x*(x^2+x-2)*(z-2);
END;
```

The summary printed to the screen in this case is as follows:

```
************** Decomposition by Degree **************

Dimension 1: 2 classified components
-----------------------------------------------------
   degree 2: 2 components

*****************************************************
```

If we change `SpecificCodimension` to 3, Bertini will use the dimension-by-dimension approach to compute the isolated points. If we opt not to specify `WitnessGenType`, Bertini will use the regenerative cascade.

Table 11.1 provides a summary of the computations for specifically computing the one- and zero-dimensional components using the regenerative cascade and dimension-by-dimension approaches. In both cases, the regenerative cascade produced a witness superset with fewer junk points at the expense of cascading down through the dimensions and moving additional linear slices. For more complicated examples, the computational cost of the different methods will be directly affected by the difference between the Bézout number and the degree of the system randomized down to the specific dimension. The dimension-by-dimension approach is often faster when these two values are equal. When one expects components of higher dimension, including components at infinity in the affine case, the regenerative cascade approach is often faster.

Table 11.1. *Summary of computations associated to solving* (11.1).

	One-dimensional solving		Zero-dimensional solving	
	Regen cascade	Dim-by-dim	Regen cascade	Dim-by-dim
Paths tracked	16	16	36	64
Slices moved	9	0	24	0
Junk points	3	5	10	22

11.3 ▪ Multiplicity*

Multiplicity has arisen in various situations throughout previous chapters. For example, in §3.1, we state that the multiplicity of an isolated solution of a square system of polynomial equations is equal to the number of homotopy paths for which it is the endpoint. In this section, we provide a formal definition of the multiplicity of irreducible components in terms of the convergent power-series ring by considering first isolated points and then positive-dimensional components.

Let f be a polynomial system of n polynomials on $\mathbb{C}^N$. Suppose that $x^* \in \mathbb{C}^N$ is an isolated zero of f, meaning that $f(x^*) = 0$, and there is an open neighborhood $U \subset \mathbb{C}^N$ of x^* such that f is nonzero on $U \setminus \{x^*\}$. We note that this immediately implies that f is not an underdetermined system, i.e., we have $n \geq N$. The *multiplicity* of x^* with respect to f is defined as

$$\text{mult}(f, x^*) := \dim \mathcal{O}_{x^*}/(f), \tag{11.2}$$

where $\mathcal{O}_{x^*}$ is the ring of convergent power series centered at x^* and (f) is the ideal in $\mathcal{O}_{x^*}$ generated by the polynomial system f. In particular, each element in $\mathcal{O}_{x^*}$ is a power series centered at x^* with a positive radius of convergence.

Example 11.1. In the univariate case, this definition agrees with the usual notion of multiplicity based on the exponent of the corresponding linear factor. To illustrate this on a concrete example, consider the polynomial $f(x) = x^2(x-1)$ for which one has $\text{mult}(f, 0) = 2$ and $\text{mult}(f, 1) = 1$.

First, consider $x^* = 0$. Since $x - 1$ is nonzero at x^*, it has a convergent power-series expansion centered at x^*. This implies that $(f) = (x^2)$ so that $\mathcal{O}_0/(f)$ is two-dimensional with basis $\{1 + (f), x + (f)\}$.

Similarly, for $x^* = 1$, we have $(f) = (x - 1)$ since x^2 is nonzero at x^*. Therefore, $\mathcal{O}_1/(f)$ is one-dimensional and spanned by $1 + (f)$. ■

As noted in §3.1, if f is a square polynomial system, i.e., $n = N$, then for each isolated zero x^*, a total-degree homotopy will have exactly $\text{mult}(f, x^*)$ paths ending at x^*.

Example 11.2. If $f(x,y) = [x^a, y^b]^T$ for $a, b \geq 1$, then the total degree of f is ab and all ab paths of a total degree homotopy will end at the origin. It is also not difficult to use (11.2) to confirm that $\text{mult}(f, 0) = ab$. ■

As discussed in §1.1.4, for an overdetermined polynomial system f, i.e., $n > N$, Bertini will compute the isolated zeros of f by randomizing down to a square system. For the randomized square system, the number of homotopy paths ending at each isolated zero is an upper bound on the multiplicity with respect to original system f. However, the geometric meaning of the multiplicity is not so clear in this case.

Example 11.3. Consider $f(x,y) = [x^2, xy, y^2]^T$ from (9.2) and the sufficiently randomized square system $g(x,y) = [x^2 - y^2, xy + 2y^2]^T$. One can easily verify that $\text{mult}(g, 0) = 4$, but $\text{mult}(f, 0) = 3$. ■

We conclude this section by defining the multiplicity of an irreducible component. Let f be a polynomial system defined on $\mathbb{C}^N$ and $V \subset f^{-1}(0)$ be an irreducible component of dimension k. Then, the multiplicity of V with respect to f is

$$\text{mult}(f, V) := \text{mult}(\{f, L_1, \ldots, L_k\}, x^*), \tag{11.3}$$

where $L_1, \ldots, L_k$ are general linear polynomials on $\mathbb{C}^N$ and x^* is any point contained in the finite set $V \cap \mathbf{V}(L_1, \ldots, L_k)$.

Example 11.4. Consider $f(x,y) = x - y$ and $g(x,y) = (x-y)^2$. As sets, we clearly have $\mathbf{V}(f) = \mathbf{V}(g) = \{(t,t) \mid t \in \mathbb{C}\}$. However, the multiplicity with respect to f is 1 and that with respect to g is 2. ■

In §11.6, we consider the process of deflating, i.e., regularizing, irreducible components of multiplicity greater than 1. We also consider regularizing other irreducible algebraic sets which may not be irreducible components in §13.2.

11.4 ▪ Macaulay Dual Spaces*

For univariate polynomials, the multiplicity uniquely determines the local structure of the polynomial. As an example, every univariate polynomial f which has a multiplicity 4 zero at the origin can be written as $f(x) = x^4 p(x)$, where $p(x)$ is some polynomial with $p(0) \neq 0$. In particular, following the notation in §11.3, $p(x)^{-1} \in \mathcal{O}_0$ so that $\{1+(f), x+(f), x^2+(f), x^3+(f)\}$ is a basis for $\mathcal{O}_0/(f)$. However, this is false in higher dimensions, as can be seen by considering the local structure of the multiplicity 4 zero at the origin of $f(x,y) = [x^2, y^2]^T$ and $g(x,y) = [x, y^4]^T$. An easy way to see this is by simply comparing the dimension of the nullspace of the Jacobian matrix of f and g at the origin, namely, 2 and 1. So clearly the multiplicity by itself does not uniquely determine the local structure.

In order to completely understand the local structure, we generalize the Jacobian matrix to matrices arising from higher-order derivatives. These matrices correspond to the higher-order derivatives that appear in the classical treatment of univariate polynomials. In particular, a univariate polynomial f has a multiplicity μ zero at x^* if and only if

$$f(x^*) = f'(x^*) = \cdots = f^{(\mu-1)}(x^*) = 0 \text{ and } f^{(\mu)}(x^*) \neq 0.$$

This implies that the Taylor series of $f(x)$ centered at x^* is

$$f(x) = f^{(\mu)}(x^*)(x - x^*)^\mu/\mu! + \text{higher-order terms}.$$

We note that the use of Taylor series expansions is quite natural since the definition of multiplicity presented in (11.2) is in terms of convergent power series.

Let f be a polynomial system in N variables and $x^* \in \mathbb{C}^N$. For each $d \geq 0$, the *dth-order Macaulay matrix* of f at x^*, denoted $\mathcal{M}_d(f, x^*)$, is the matrix whose rows consist of the coefficients of the Taylor series centered at x^* up to degree d for each of the polynomials in the system f multiplied by a monomial of degree up to $\max\{0, d-1\}$. The zeroth- and first-order Macaulay matrices are $\mathcal{M}_0(f, x^*) = f(x^*)$ and $\mathcal{M}_1(f, x^*) = [f(x^*)\, Jf(x^*)]$, respectively, where $Jf(x^*)$ is the Jacobian matrix of f evaluated at x^*.

The nullspace of $\mathcal{M}_d(f, x^*)$ is the *dth-order Macaulay dual space*. For $d \geq 1$, it follows from the relationship between $\mathcal{M}_{d-1}(f, x^*)$ and $\mathcal{M}_d(f, x^*)$ that

$$\dim \operatorname{null} \mathcal{M}_{d-1}(f, x^*) \leq \dim \operatorname{null} \mathcal{M}_d(f, x^*).$$

By commutative algebra, $\dim \operatorname{null} \mathcal{M}_{d-1}(f, x^*) = \dim \operatorname{null} \mathcal{M}_d(f, x^*) = m$ implies

$$\lim_{d\to\infty} \dim \operatorname{null} \mathcal{M}_d(f, x^*) = m.$$

The following is a famous theorem derived from Macaulay [125] relating multiplicity to the dimension of the Macaulay dual spaces.

Theorem 11.5. *Let f be a polynomial system and $x^* \in \mathbf{V}(f)$. Then, x^* is isolated in $\mathbf{V}(f)$ if and only if* $\lim_{d\to\infty} \dim \operatorname{null} \mathcal{M}_d(f, x^*) < \infty$. *Moreover,* $\operatorname{mult}(f, x^*) = m$ *if and only* $\lim_{d\to\infty} \dim \operatorname{null} \mathcal{M}_d(f, x^*) = m$.

Example 11.6. Consider the polynomial system

$$f(x,y) = \begin{bmatrix} y - x^2 \\ y - 2x + 1 \end{bmatrix},$$

which is the intersection of a parabola with a line that is tangent to the parabola at the point $(x^*, y^*) = (1,1)$. Rewriting f in terms of $(x - x^*) = (x-1)$ and $(y - y^*) = (y-1)$ gives

$$f(x,y) = \begin{bmatrix} -2(x-1)+(y-1)-(x-1)^2 \\ -2(x-1)+(y-1) \end{bmatrix}.$$

Accordingly, the second-order Macaulay matrix $\mathscr{M}_2(f,(1,1))$ is

$$\left(\begin{array}{r|c|cc|ccc} & 1 & (x-1) & (y-1) & (x-1)^2 & (x-1)(y-1) & (y-1)^2 \\ \hline f_1 & 0 & -2 & 1 & -1 & 0 & 0 \\ f_2 & 0 & -2 & 1 & 0 & 0 & 0 \\ \cline{2-4} (x-1)f_1 & 0 & 0 & 0 & -2 & 1 & 0 \\ (x-1)f_2 & 0 & 0 & 0 & -2 & 1 & 0 \\ (y-1)f_1 & 0 & 0 & 0 & 0 & -2 & 1 \\ (y-1)f_2 & 0 & 0 & 0 & 0 & -2 & 1 \end{array}\right).$$

The lines show how $\mathscr{M}_0(f,(1,1))$ and $\mathscr{M}_1(f,(1,1))$ appear as submatrices. In particular, one can easily verify that $\dim\operatorname{null} \mathscr{M}_0(f,(1,1)) = 1$ and

$$\dim\operatorname{null} \mathscr{M}_1(f,(1,1)) = \dim\operatorname{null} \mathscr{M}_2(f,(1,1)) = 2,$$

showing that $\operatorname{mult}(f,(1,1)) = 2$. ∎

Although the approach described above, which follows [47], is very instructive and is currently utilized within Bertini, more efficient methods exist, e.g., [85, 187], based on closedness subspace approaches.

11.5 ▪ Local Dimension and Mechanism Mobility

Given a polynomial system f and a point $x^* \in \mathbf{V}(f)$, a fundamental question is to determine the *local dimension*, $\dim_{x^*} \mathbf{V}(f)$, of x^* with respect to f. This is the maximum dimension of the set of irreducible components of $\mathbf{V}(f)$ passing through x^*. In engineering applications, local dimension is typically called *degrees of freedom*, whereas multiplicity corresponds to *infinitesimal degrees of freedom*. If a numerical irreducible decomposition of $\mathbf{V}(f)$ has been computed, the local dimension can be computed via membership tests (see §8.4). This approach for computing the local dimension is called a global approach since it requires a witness set for each irreducible component of $\mathbf{V}(f)$. This section describes a local approach which uses the Macaulay dual space together with a multiplicity bound. We note that one always has access to a multiplicity bound using root counting methods such as the total-degree bound. However, since such bounds are typically impractical, we also consider the *depth-bounded local dimension* following [180]. The Bertini module `LocalDimFinder` (available at `bertini.nd.edu/LocalDimFinder`) can perform this computation.

The local dimension $\dim_{x^*} \mathbf{V}(f)$ is equal to the minimum $k \geq 0$ such that x^* is an isolated zero of $\mathbf{V}(f, \mathscr{L}_1, \ldots, \mathscr{L}_k)$, where $\mathscr{L}_1, \ldots, \mathscr{L}_k$ are general linear polynomials passing through x^*. Hence, the key step in computing the local dimension is testing whether a root is isolated. Given a bound B on the multiplicity if it is isolated, e.g., the number of paths ending at x^*, Theorem 11.5 yields an algorithm for deciding if a point is an isolated solution [19]. Namely, one computes finitely many linearly independent elements in the Macaulay dual space until either

- the elements form a basis, meaning that the point is isolated of multiplicity at most B, or

- $B+1$ elements have been computed, meaning that the point is not isolated since it cannot have multiplicity larger than the bound B.

We now turn our attention to the concept of *depth-bounded local dimension*, which is focused on the case when an effective multiplicity bound is not known a priori. We motivate the idea in the following example.

Example 11.7. Compare the behavior of the polynomial systems

$$f(x,y)=\begin{bmatrix} y-x^2 \\ y^{100} \end{bmatrix} \text{ and } g(x,y)=y-x^2$$

locally near the origin. It is easy to verify that $\dim(f,0)=0$, $\text{mult}(f,0)=200$, and $\dim(g,0)=1$. This is the exact local behavior for these systems.

From a local structure standpoint, each dth-order Macaulay dual space for f and g at the origin has dimension 1 for $0 \leq d \leq 199$. Therefore, up to $d=199$, the local structure suggests that both f and g have a positive-dimensional component passing through the origin. It is only after considering $d=200$ that we find the correct dimension for f at the origin.

From a numerical standpoint, consider the set

$$C=\{(x,x^2) \mid |x|<1/2\}.$$

Clearly, g restricted to C is zero. Moreover, the absolute value of f restricted to C is bounded above by 10^{-60}. Therefore, from a local numerical perspective, it appears that f has a positive-dimensional solution set passing through the origin. ∎

This example was made extreme to show that it may be difficult from both a local structure and a numerical standpoint to distinguish between high-multiplicity solutions and true positive-dimensional components. In fact, the apparent local dimension for small d may given a more realistic measure of a device's freedom to move than the exact local dimension. Additionally, the computation of the dth-order Macaulay dual space becomes unfeasible for large d.

The depth-bounded local dimension has a direct manifestation in applications in the presence of construction tolerances and elasticity, as demonstrated by the unexpected motion of a parallel-link robot built at Seoul National University circa 2001 by F. Park and colleagues [73]. As discussed in [177, 180], in the algebraic model of this device, the central pose of this robot turns out to be an isolated root of multiplicity 4. Since the root is isolated, the exact algebra predicts that the device should not move, but the high multiplicity of the root is an indicator that the physical device is likely to allow some wobble. Indeed, the physical device exhibited a substantial motion.

If $\text{mult}(f,x^*)=m$, then $\lim_{d\to\infty} \dim \text{null}\, \mathcal{M}_d(f,x^*)=m$. Since this limit must be obtained for some $d \geq 0$, we define the *depth* to be the smallest such $d \geq 0$ with $\dim\text{null}\, \mathcal{M}_d(f,x^*)=m$. Since the dimension of the Macaulay dual space either must strictly increase or be equal to the multiplicity, the depth is at most $m-1$.

Example 11.8. One can easily check that the origin has depth 2 with respect to $f(x,y)=[x^2,y^2]^T$ and depth 3 with respect to $g(x,y)=[x,y^4]^T$. ∎

For each $d \geq 1$, the d-bounded local dimension of f at x^*, denoted $\dim_d(f,x^*)$, is the minimum k such that the depth of $\{f,\mathcal{L}_1,\ldots,\mathcal{L}_k\}$ at x^* is strictly less than d, where each $\mathscr{L}_i$ is a general linear polynomial through x^*. In other words, $\dim_d(f,x^*)$ is the minimum k such that

$$\dim\mathrm{null}\,\mathcal{M}_{d-1}(\{f,\mathscr{L}_1,\ldots,\mathscr{L}_k\},x^*)=\dim\mathrm{null}\,\mathcal{M}_d(\{f,\mathscr{L}_1,\ldots,\mathscr{L}_k\},x^*),$$

which is a certificate that x^* has dimension at most k. Clearly,

- $\dim_{x^*}\mathbf{V}(f)\leq\dim_{d+1}(f,x^*)\leq\dim_d(f,x^*)$ and
- $\lim_{d\to\infty}\dim_d(f,x^*)=\dim_{x^*}\mathbf{V}(f)$.

In particular, the 1-bounded local dimension is simply the dimension of the nullspace of the Jacobian matrix of f at x^*.

Example 11.9. For f as in Example 11.7, we have

$$\dim_d(f,0)=\begin{cases}1, & 0\leq d\leq 199,\\ 0, & d\geq 200.\end{cases}$$ ∎

Example 11.10 (foldable Griffis–Duffy platform). Let us return to the example of the foldable Griffis–Duffy platform, treated earlier in Example 8.16. The assembly configurations of this device include a pose where the base and platform triangles coincide, with the six legs crisscrossing as altitudes. The Study coordinates for this "folded-in" pose are $[e,g]=[1,0]$. Letting f be the system of (6.6), (6.9), we have $\mathrm{rank} f=6$ while $[e,g]\in\mathbb{P}^7$, so it is clear the mechanism has at least one degree of freedom $(7-6=1)$. Using `LocalDimFinder` to test dimensions $k=1$, 2, and 3 up to depth $d\geq 4$, we obtain the following three stabilized sequences for

$$\dim\mathrm{null}\,\mathcal{M}_d(\{f,\mathscr{L}_1,\ldots,\mathscr{L}_k\},[1,0]):$$

k	Sequence
1	1,4,7,8,8
2	1,3,3
3	1,2,2

In particular, this shows that the d-bounded local dimension is 2, with multiplicity 3, for $d=2$ and $d=3$, but is 1, with multiplicity 8, for $d=4$. The high multiplicity is due to the fact that several of the solution components meet at this pose. ∎

11.6 ▪ Deflation of Irreducible Components

Throughout this book and in Bertini, Newton's method is used extensively, e.g., as a correcting method for path tracking (see §2.2) and to sharpen nonsingular solutions (see Chapter 7). Due to the local quadratic convergence of Newton's method for nonsingular solutions (see §1.1.2), one can readily obtain accurate numerical approximations of them to thousands (or even millions) of digits. Moreover, due to the numerical stability of performing linear algebra operations on full-rank matrices, we aim to reduce computations to the nonsingular case. This regularization process is called *deflation* and was already used for computing numerical irreducible decompositions of various Griffis–Duffy platforms in Examples 8.14, 8.15, and 8.16. In this section, we focus on deflating at a given general point of an irreducible component, but we return to deflation in a more general context in §13.2.

To demonstrate deflation, consider the polynomial system

$$f(x,y,z) = \begin{bmatrix} x^2+y^2-1 \\ x^2+y^2+z^2-1 \end{bmatrix}. \tag{11.4}$$

The following Bertini code computes its numerical irreducible decomposition:

```
% Deflation.input
CONFIG
 TrackType:1;
END;
INPUT
 variable_group x,y;
 function cylinder,sphere;

 cylinder = x^2 + y^2 - 1;
 sphere = x^2 + y^2 + z^2 - 1;
END;
```

It happens that the cylinder and sphere are tangent along a great circle of the sphere, so their intersection is a circle of multiplicity 2. This is reflected in the output of Bertini, wherein the last line printed to the screen, shown below, informs the user that two points had to be deflated.

```
Witness Points Deflated: 2
```

The remainder of this section describes how Bertini deflates such components.

We start with the isolated case, that is, let $f : \mathbb{C}^N \to \mathbb{C}^n$ be a polynomial system and x^* be an isolated zero of multiplicity $m > 1$. By *deflating* f at x^*, we mean to construct a new polynomial system $g : \mathbb{C}^{N+N'} \to \mathbb{C}^{n+n'}$ and point (x^*, λ^*) with $\text{mult}(g,(x^*,\lambda^*)) = 1$. That is, Newton's method applied to g near (x^*,λ^*) will be locally quadratically convergent. The modern study of restoring quadratic convergence of Newton's method in this context arises from work presented in [142, 141], which was formalized into algorithms in [113] and [117] following a symbolic and numerical approach, respectively, with variants of deflation presented in [47, 93, 118]. We describe the extrinsic deflation approach of [47] since this is the approach currently used by Bertini.

Let d be the dimension of the nullspace of the Jacobian of f at x^*, namely, $d = \dim \text{null}\, Jf(x^*)$, and let $\mathcal{K} : \mathbb{C}^N \to \mathbb{C}^d$ be a system of general linear polynomials. Consider the polynomial system $g : \mathbb{C}^{2N} \to \mathbb{C}^{2n+d}$ defined by

$$g(x,\lambda) = \begin{bmatrix} f(x) \\ Jf(x)\cdot\lambda \\ \mathcal{K}(\lambda) \end{bmatrix}.$$

Based on the construction, $g(x^*,\lambda) = 0$ is a system of linear equations that has a unique solution, say, λ^*. In particular, since x^* is an isolated zero of f, it follows that (x^*,λ^*) is an isolated zero of g. If (x^*,λ^*) is nonsingular with respect to g, the process is terminated. Otherwise, we simply repeat the process upon renaming g as f and (x^*,λ^*) as x^*. The following theorem shows this process must terminate by showing that the multiplicity (shown in [117]) and depth (shown in [47]) are reduced after each iteration.

Theorem 11.11. *Assume that x^* is isolated with multiplicity $m > 1$ and depth $\delta \geq 1$ with respect to f. Let g be the system and (x^*,λ^*) be the point constructed from one deflation* *iteration as above. Then,* (x^*, λ^*) *is isolated with multiplicity at most* $m-1$ *and depth at most* $\delta - 1$ *with respect to* g.

We next turn to the positive-dimensional case which, as first shown in [161, §13.3.2, §15.2.2], reduces to the isolated case. Let V be an irreducible component of $f^{-1}(0)$ of multiplicity $m > 1$ with witness set $\{f, \mathcal{L}, W\}$. For each $w \in W$, we know that w is an isolated solution of $\{f, \mathcal{L}\}^{-1}(0)$ of multiplicity m. We deflate V by first fixing $w^* \in W$ and applying deflation to $\{f, \mathcal{L}\}$ at w^*. It immediately follows that for each $w \in W$, there is a point (w, λ_w) which is a nonsingular solution for the resulting system g. Due to the choice of generic linear polynomials $\mathcal{K}$ in the deflation process, this deflated system g will work generically on V. That is, consider deforming g starting at these nonsingular solutions (w, λ_w) by changing the witness slices $\mathcal{L}$. If this deformation is sufficiently generic, then the resulting paths will be nonsingular at all points except possibly at the endpoint, which could have higher multiplicity or lie on a positive-dimensional irreducible component.

Example 11.12. Reconsider the polynomial system $f(x,y,z)$ from (11.4). Since $V = f^{-1}(0)$ is an irreducible curve, for simplicity we take the linear slicing polynomial $\mathcal{L}(x,y,z) = x + 2y + 3z - 1$. The system $f(x,y,z) = \mathcal{L}(x,y,z) = 0$ has two solutions, each of multiplicity 2 and depth 1. Therefore, one iteration of the deflation process will suffice to regularize these two points, which produces a system of the form

$$g(x,y,z,\lambda_1,\lambda_2,\lambda_3) = \begin{bmatrix} x^2+y^2-1 \\ x^2+y^2+z^2-1 \\ x+2y+3z-1 \\ 2x\lambda_1+2y\lambda_2 \\ 2x\lambda_1+2y\lambda_2+2z\lambda_3 \\ \lambda_1+2\lambda_2+3\lambda_3 \\ \square\lambda_1+\square\lambda_2+\square\lambda_3+\square \end{bmatrix}. \quad \blacksquare$$

Example 11.13. As a demonstration that deflation works only generically, consider the polynomial $f(x,y) = (y^2 - x^3)^2$ with linear slicing polynomial $\mathcal{L}(x,y) = 2x + 3y - 1$. Since $V = f^{-1}(0)$ is an irreducible curve of multiplicity 2 and depth 1, one iteration will regularize the three points satisfying $f(x,y) = \mathcal{L}(x,y) = 0$. The resulting system is of the form

$$g(x,y,\lambda_1,\lambda_2) = \begin{bmatrix} (y^2-x^3)^2 \\ 2x+3y-1 \\ (y^2-x^3)(4y\lambda_2-6x^2\lambda_1) \\ 2\lambda_1+3\lambda_2 \\ \square\lambda_1+\square\lambda_2+\square \end{bmatrix}.$$

For concreteness, suppose that we take the last polynomial to be $3\lambda_1 - 5\lambda_2 - 2$. We can track along V by deforming $\mathcal{L} = 2x + 3y - 1$ in a parallel fashion by simply changing the constant term. That is, we will consider homotopies of the form

$$H(x,y,\lambda_1,\lambda_2,t) = \begin{bmatrix} (y^2-x^3)^2 \\ 2x+3y-s(t) \\ (y^2-x^3)(4y\lambda_2-6x^2\lambda_1) \\ 2\lambda_1+3\lambda_2 \\ 3\lambda_1-5\lambda_2-2 \end{bmatrix},$$

where $s : [0,1] \to \mathbb{C}$ is smooth with $s(1) = 1$.

Suppose we utilize the gamma trick (see §6.1) to create a path $s(t)$ so that $s(1) = 1$ and $s(0) = 0$, namely, $s(t) = \frac{\gamma t}{1-t+\gamma t}$ for random $\gamma \in \mathbb{C}$. Then, two of the three paths will converge to the point corresponding to $(0,0)$, namely, $\left(0, 0, \frac{6}{19}, -\frac{4}{19}\right)$. We note that there is an additional singularity at the point $\left(\frac{16}{81}, -\frac{64}{729}, \frac{6}{19}, -\frac{4}{19}\right)$ if $s(t^*) = \frac{32}{243}$ for some $t^* \in (0,1)$. This demonstrates both the need for the gamma trick and the fact that deflation works only generically on V. ∎

Part III

Further Algorithms and Applications

Chapter 12

Intersection

Up to this point, we have concentrated on finding the solution set of a given polynomial system, that is, given a polynomial system $f(z)$, we find a numerical representation of $\mathbf{V}(f)$. In Part I, the objective is to find only the isolated solutions, while in Part II, a numerical irreducible decomposition of the whole set is computed. Once a witness set for an irreducible component, say, A, is in hand, one can sample points on it (see §8.3) or test if a given point is a member of A (see §8.4).

In this chapter, we explore additional operations that can be invoked once witness sets are in hand. The first of these is to compute intersections (§12.1): given witness sets for irreducible sets $A \subset \mathbb{C}^N$ and $B \subset \mathbb{C}^N$, find a numerical irreducible decomposition of $A \cap B$. Notice that if $x \in A \cap B$, then there is a corresponding point (x, x) in the set

$$\{(x,y) \mid x \in A, y \in B, x - y = 0\}.$$

Though this may seem like a trivial observation, it is key to reframing the intersection problem in a way that facilitates its solution by homotopy methods [159]. A variant on this capability, for a polynomial system $g(x, y)$, is to compute

$$\{(x,y) \mid x \in A, y \in \mathbb{C}^k, g(x,y) = 0\}$$

without first decomposing $\mathbf{V}(g)$, as discussed in §12.2. Finally in §12.3, we discuss how to compute

$$\{(x,y) \mid x \in A, y \in B, g(x,y) = 0\},$$

which includes the former two as special cases. A mild reformulation of the regeneration technique discussed in §5.4 and §9.4 is effective for solving these problems [99].

These capabilities allow a user to approach a problem in stages. Rather than queueing up all the polynomials into one big system to be solved, the analyst can first solve a subsystem and examine its solution components. Perhaps only one or two out of many are of interest. Then the user can proceed to operate on only the interesting components and ignore the rest. This can bring substantial gains in efficiency and also more clarity to the results, as a plethora of possible intersections between uninteresting components can be left out of the computations and kept out of the solution files.

Moreover, a general intersection capability facilitates some calculations that are not possible if the only tool available is computing a numerical irreducible decomposition of a single system. For example, suppose A and B are two distinct irreducible components of $\mathbf{V}(f)$. Then, $A \cap B$ is not an irreducible component of $\mathbf{V}(f)$, yet one might very well be interested in finding where A and B meet.

Fiber products, discussed in §12.4, are another kind of construction that generalizes intersection. In this case, we wish to find pairs of points in sets A and B that map to the same point in a third set C. This construction is useful in discovering exceptional loci where the dimensions of components of fibers of an algebraic map are greater than the generic fiber dimension of the map. For example, we have seen that when the leg lengths of a general Stewart–Gough platform are locked, the device has only isolated solutions—at most 40 of them—so the device cannot move. Yet, the foldable Griffis–Duffy platform of Example 11.10 has several one-dimensional motions. As discussed in §12.4.1, the problem of finding such special cases can be reduced to computing irreducible sets in an iterated fiber product.

12.1 ▪ Intersecting Irreducible Components

Suppose that we wish to intersect two irreducible multiplicity one algebraic sets, A and B, for which witness sets have already been found. Let's say $A \subset \mathbb{C}^N$ has witness set $W_A = \{F_A, L_A, S_A\}$, where A is an irreducible component of $\mathbf{V}(F_A)$, $L_A : \mathbb{C}^N \to \mathbb{C}^{\dim A}$ is a set of $\dim A$ generic affine linear functions, and $S_A = \mathbf{V}(L_A) \cap A$ is the set of witness points. Similarly, let the witness set for B be $W_B = \{F_B, L_B, S_B\}$. We assume that $\dim A > 0$ and $\dim B > 0$, because if either A or B is zero-dimensional, the intersection can be completed by testing membership (see §8.4) in the other set. Moreover, we also assume $A \not\subset B$ and $B \not\subset A$ (which can also be checked by testing membership), so that $\dim A \cap B < \min(\dim A, \dim B)$.

It is tempting to try to find $A \cap B$ by somehow analyzing the combined system

$$F(z) := \begin{bmatrix} F_A(z) \\ F_B(z) \end{bmatrix},$$

but this does not suffice. For example, as mentioned in the introduction to this chapter, it might be that A and B are both irreducible components of $\mathbf{V}(F_A) = \mathbf{V}(F_B) = \mathbf{V}(F)$. In that case, $A \cap B$ is not an irreducible component of $\mathbf{V}(F)$ (except in the trivial case $A = B$).

The key to performing the intersection operation is to use a maneuver called "reduction to the diagonal." Our objective is to find

$$Z := A \cap B = \{x \in \mathbb{C}^N \mid x \in A, x \in B\},$$

but instead we may aim at finding

$$Z' := \{(x,y) \in \mathbb{C}^{2N} \mid x \in A, y \in B, x = y\}.$$

A more abbreviated way of writing this is

$$Z' := \{(x,y) \in A \times B \mid x = y\}.$$

It is clear that under the maps $\phi : (x) \mapsto (x,x)$ and $\psi : (x,y) \mapsto (x)$, Z and Z' are isomorphic to each other. The set $\Delta := \mathbf{V}(x-y)$ is known as the "diagonal" of $\mathbb{C}^N \times \mathbb{C}^N$, hence the name of the maneuver. With that notation, we may write

$$Z' := (A \times B) \cap \Delta.$$

12.1.1 ▪ Witness Supersets for Intersection

Following the main idea of [159], let's abstract away from analyzing Z or Z' as defined above and assume that we desire to find the irreducible decomposition of

$$\mathscr{Z} := X \cap \mathbf{V}(f) = \{z \in X \mid f(z) = 0\},$$

where $X \subset \mathbb{C}^M$ is an irreducible algebraic set. Recall that in Chapter 9, the objective was to generate witness sets for the irreducible components of

$$\mathbf{V}(f) = \{z \in \mathbb{C}^M \mid f(z) = 0\},$$

so our current objective is similar, except $\mathbb{C}^M$ has been replaced by X. (We have changed the usual N to M so as not to cause confusion with the notation just above.)

To find the irreducible components of $\mathscr{Z}$, it is useful to form the usual sort of randomization of f. Now, f is a map from $\mathbb{C}^M$ to, say, $\mathbb{C}^m$, but since $X \subset \mathbb{C}^M$, we can consider f as a map $f : X \to \mathbb{C}^m$. Recall that in §9.1, we defined rank$f$ as the rank of the Jacobian of f evaluated at a generic point in $\mathbb{C}^M$. When we are restricting f to the domain X, it makes more sense to consider $\mathrm{rank}_X f$ as the rank of the Jacobian of f evaluated at a generic point of X. This rank may be smaller, and it is an upper bound on the codimension of $\mathbf{V}(f)$ as a subset of X. Of course, the codimension is also at most $\dim X$. Accordingly, we form a new system

$$g(z) = \begin{bmatrix} g_1(z) \\ \vdots \\ g_\eta(z) \end{bmatrix} = \Lambda \cdot f(z),$$

where Λ is chosen generically from $\mathbb{C}^{\eta \times m}$ and $\eta = \min(\mathrm{rank}_X f, \dim X)$. In what follows, by the usual argument, it is allowable for Λ to be chosen with all zeros below the main diagonal.

Since we aim to slice out witness points on $\mathscr{Z}$, we introduce a generic linear system $L(z) = \{L_1(z), \ldots, L_{\dim X}(z)\} : \mathbb{C}^M \to \mathbb{C}^{\dim X}$.

With this setup, for $i = 0, \ldots, \eta$, the following sets are important:

$$\widetilde{W}_i := \{z \in X \mid g_1(z) = \cdots = g_i(z) = L_{i+1}(z) = \cdots = L_{\dim X}(z) = 0\}. \tag{12.1}$$

Let $\mathscr{Z}_k$ be the pure k-dimensional component of $\mathscr{Z}$. It follows from the theory presented in §9.2 that the genericity of Λ and L implies that with probability one $\widetilde{W}_i$ consists of the following:

1. a set of $\deg \mathscr{Z}_{\dim X - i}$ points, namely, $W_i := \mathscr{Z}_{\dim X - i} \cap \mathbf{V}(L_{i+1}, \ldots, L_{\dim X})$;
2. a finite set of nonsingular nonsolutions, N_i, i.e., points that are not in $\mathscr{Z}$; and
3. positive-dimensional components $\mathscr{Z}_k \cap \mathbf{V}(L_{i+1}, \ldots, L_{\dim X})$ for $k > \dim X - i$.

Our objective will be to generate witness point supersets by constructing homotopies to return finite sets that include W_i and N_i plus possibly some points, say, J_i, on the positive-dimensional components $\mathscr{Z}_k \cap \mathbf{V}(L_{i+1}, \ldots, L_{\dim X})$, $k > \dim X - i$. That is, we wish to generate finite sets

$$E_i = J_i \cup W_i \cup N_i, \qquad i = 0, \ldots, \eta. \tag{12.2}$$

We do not need to consider E_i for $i > \eta$, as there can be no solution components with codimension greater than η. By definition $\widehat{W}_i = J_i \cup W_i$ is a witness point superset for $\mathscr{Z}_{\dim X - i}$, and W_i is a witness point set for the same.

12.1.2 ▪ Witness Generation for Diagonal Intersection

We will shortly return to the general setup of $X \cap \mathbf{V}(f)$, but first let us consider specifically witness superset generation for the diagonal intersection, where $X = A \times B$, $M = 2N$, and $\dim X = \dim A + \dim B$. Furthermore, let $f(z) : \mathbb{C}^{2N} \to \mathbb{C}^N$ be $f(z) = f(x,y) = x - y$. Our procedure will take advantage of the fact that f is linear. With these choices, we have $\mathscr{X} = Z'$.

We begin with witness sets for A and B, namely, $\{F_A, L_A, S_A\}$ and $\{F_B, L_B, S_B\}$ as described above. Since A and B are both irreducible sets, so is $X = A \times B$. Without loss of generality, we may assume that F_A and F_B consist of $N - \dim A$ and $N - \dim B$ polynomials, respectively. If not, they may be reduced to this number by randomization. Moreover, $\deg X = \deg A \cdot \deg B$.

Letting $z = (x,y)$, consider the homotopy

$$h_0(z,t) = \begin{bmatrix} F_A(x) \\ F_B(y) \\ t\begin{pmatrix} L_A(x) \\ L_B(y) \end{pmatrix} + (1-t)L(z) \end{bmatrix} = 0, \tag{12.3}$$

where $L(z)$ is the generic affine linear system from §12.1.1. The $\deg X$ points $S := S_A \times S_B = \{(x,y) \mid x \in S_A, y \in S_B\}$ are all nonsingular solutions of $h(z,1) = 0$.

While we will not argue it closely, the fact is that since $\mathbf{V}(L_A, L_B)$ and $\mathbf{V}(L)$ cut X in $\deg X$ points and since $L(z)$ has generic coefficients, the paths of h_0 emanating from S are all nonsingular for $t \in [0,1]$ and their endpoints form the witness point set $W_X = X \cap \mathbf{V}(L)$ for X.

By the definitions of the previous subsection, the set W_X is synonymous to E_0. We must sort E_0 into $J_0 \cup W_0 \cup N_0$. Since Z' is isomorphic to $A \cap B$, we have

$$\dim Z' = \dim A \cap B < \min(\dim A, \dim B)$$

so that $\dim Z' < \dim A + \dim B = \dim X$. Therefore, J_0 and W_0 must be empty, and $E_0 = N_0$, a set of $\deg X$ nonsolutions.

We are now ready to descend through the dimensions. For each i, E_i is the set of endpoints of the homotopy

$$h_i(z,t) = \begin{bmatrix} F_A(x) \\ F_B(y) \\ g_1(x,y) \\ \vdots \\ g_{i-1}(x,y) \\ tL_i(x,y) + (1-t)g_i(x,y) \\ L_{i+1}(x,y) \\ \vdots \\ L_{\dim A + \dim B}(x,y) \end{bmatrix} = 0, \tag{12.4}$$

starting at $t = 1$ with the points N_{i-1} and proceeding to $t = 0$. Note that $g(x,y) = \Lambda \cdot [x - y]$ is a randomization of the system $[x - y]$ and is linear. Accordingly, the number of variables in the homotopy can be reduced by working intrinsically on the linear space defined by the linear equations that do not change during the homotopy. That is, one may find a basis for $\mathbf{V}(g_1, \ldots, g_{i-1}, L_{i+1}, \ldots, L_{\dim A + \dim B})$ and work with coordinates associated to that basis.

After finding E_i, the nonsolutions N_i are those for which $x \neq y$. These become the startpoints for the next homotopy, h_{i+1}.

12.1.3 ▪ Skipping Dimensions

In the intersection $Z = A \cap B$, we have an upper bound on the dimension as $\min(\dim A, \dim B) - 1$. Since $Z' = (A \times B) \cap \Delta$ is isomorphic to Z, it has the same upper bound. The codimension in $A \times B$ of this largest possible solution set is

$$c := \dim A + \dim B - (\min(\dim A, \dim B) - 1) = \max(\dim A, \dim B) + 1.$$

Accordingly, there is no need to compute E_i for $i < c$. Instead, it is valid to follow a homotopy directly from the initial slice $(A \times B) \cap \mathbf{V}(L_A, L_B)$ to E_c. Since the slices at either end of this homotopy are not in general position, a linear interpolation might not be general enough to guarantee that singularities will be missed. But since the initial slice has the maximum root count, the gamma trick of (6.5) applies, and we can go directly to E_c using the homotopy

$$\hat{h}_c(x,y,t) = \begin{bmatrix} F_A(x) \\ F_B(y) \\ \gamma t \begin{pmatrix} L_A(x) \\ L_B(y) \end{pmatrix} + (1-t) \begin{pmatrix} g_1(x,y) \\ \vdots \\ g_c(x,y) \\ L_{c+1}(x,y) \\ \vdots \\ L_{\dim A + \dim B}(x,y) \end{pmatrix} \end{bmatrix} = 0, \tag{12.5}$$

where γ is chosen at random on the unit circle in $\mathbb{C}$. This homotopy starts with the set $S := S_A \times S_B$ at $t = 1$ and goes to $t = 0$.

Often in practice, only the sets of lowest possible dimension are of interest, in which case one may skip all higher dimensions using the homotopy above with $c = \eta$.

12.1.4 ▪ Witness Sets for $A \cap B$

Once one has the witness point supersets $\widehat{W_i} := E_i \setminus N_i = J_i \cup W_i$, one may proceed in the usual way to remove the junk points J_i and then decompose the witness points W_i according to the irreducible components of the $(\dim A + \dim B - i)$-dimensional part of Z'. Suppose $W_{i,j}$ is one of these point sets. Then,

$$\{\{f_A(x), f_B(y), g_1(x,y), \ldots, g_i(x,y)\}, \{L_{i+1}(x,y), \ldots, L_{\dim A + \dim B}(x,y)\}, W_{i,j}\}$$

is a witness set for an irreducible component, say, $Z'_{i,j}$, of $Z' = (A \times B) \cap \Delta$. Let $\pi : (x,y) \to (x)$, and remember that any point in Z' is of the form (x,x). By the isomorphism between Z' and $Z = A \cap B$, we know that there is an irreducible component of Z given by $Z_{i,j} := \pi(Z'_{i,j})$.

This is well and good, and it suffices for many purposes, but the original problem was to decompose Z, not Z'. Define $P(x)$ as

$$P(x) := \{f_A(x), f_B(x), L_{i+1}(x,x), \ldots, L_{\dim A + \dim B}(x,x)\}.$$

In "nice" cases, not only does each witness point $w \in W_{i,j}$ satisfy $P(\pi(w)) = 0$ but also $\pi(w)$ is a nonsingular root of $P(x)$. (Since the points are sliced out generically, if one satisfies this condition, all of them do.) In such cases, a witness set for $Z_{i,j}$ is

$$\{\{f_A(x), f_B(x)\}, \{L_{i+1}(x,x), \ldots, L_{\dim A + \dim B}(x,x)\}, \pi(W_{i,j})\}.$$

Not every case is so cooperative. A common case is when $f_A \equiv f_B$, that is, when A and B are two irreducible components of the same polynomial system. In that case, points of $A \cap B$ are not regular points of $\mathbf{V}(f_A, f_B) = \mathbf{V}(f_A)$, so any irreducible component $Z_{i,j}$ is not an irreducible component of $\mathbf{V}(f_A)$. Such singular sets can arise in other situations as well (see [98]).

When the witness points are singular solutions of $P(x) = 0$, a deflation operation is needed to move from a witness set for $Z'_{i,j}$ to one for $Z_{i,j} = \pi(Z'_{i,j})$. However, the standard deflation approach of §11.6 does not necessarily suffice in this situation. Standard deflation reduces the multiplicity of an isolated point to one. For a witness set, say, $\{f_A, L_A, W_A\}$, each witness point in W_A is an isolated point of $\mathbf{V}(f_A, L_A)$, so these can be deflated. Since the witness points are generic, the deflation holds on a Zariski-open subset of A and therefore suffices as a deflation of A. But in the case of $Z'_{i,j}$, we may have that, although its witness points $W_{i,j}$ are isolated in Z' by the slicing system $\{L_{i+1}(x,y), \ldots, L_{\dim A + \dim B}(x,y)\}$, the corresponding witness points $\pi(W_{i,j})$ for $Z_{i,j}$ are not isolated points of $\mathbf{V}(P(x))$. In particular, this happens if A and B are two irreducible components of the same $\mathbf{V}(f)$.

The way around this trouble is isosingular theory (see [98] and §13.2). In essence, a Zariski open subset of points in $Z_{i,j}$ are all singular points of $\mathbf{V}(P)$ in the same way. The points $\pi(W_{i,j})$ are generic in $Z_{i,j}$, and so an isosingular deflation at one of these suffices to produce a deflated version of $\{f_A, f_B\}$ that can be used as the first element of a witness set for $Z_{i,j}$. Details of applying isosingular theory to intersection can be found in [97].

12.1.5 ▪ Intersection Examples

To demonstrate computing intersections by tracking homotopies using Bertini, we consider two examples. The first is an intersection of two lines. The second is an intersection of two irreducible components of the same polynomial system.

Example 12.1. As an illustrative example, consider the algebraic sets $A = \mathbf{V}(x_1) \subset \mathbb{C}^2$ and $B = \mathbf{V}(x_2) \subset \mathbb{C}^2$ with witness sets $W_A = \{F_A, L_A, S_A\}$ and $W_B = \{F_B, L_B, S_B\}$, respectively, where

$$\begin{aligned} F_A(x_1,x_2) &= x_1, & F_B(x,y) &= x_1 x_2, \\ L_A(x_1,x_2) &= 2x_1 + x_2 - 2, & L_B(x_1,x_2) &= x_1 + 3x_2 + 4, \\ S_A &= \{(0,2)\}, & S_B &= \{(-4,0)\}. \end{aligned}$$

Of course, we don't need homotopy to solve this toy example, but let's see how it would apply. Since we know that $A \cap B$ is either zero-dimensional or empty, we will simply track using the homotopy defined in (12.5) with $c = 2$. This uses a user-defined homotopy (see §7.5) as shown in the input file

```
% DiagonalIntersection.input
CONFIG
 UserHomotopy:2;
END;
INPUT
 function FA,FB,L1,L2;
 variable_group x1,x2,y1,y2;
 pathvariable T;
 parameter t;
 random gamma;
```

```
  t = T;
  FA = x1;
  FB = y1*y2;
  L1 = gamma*t*(2*x1 + x2 - 2) + (1-t)*(x1 - y1);
  L2 = gamma*t*(x1 + 3*y2 + 4) + (1-t)*(x2 - y2);
END;
```

with the `start` file containing $S_A \times S_B$:

```
1

 0 0
 2 0
-4 0
 0 0
```

Upon running Bertini, the file `finite_solutions` contains a point:

```
4.477895110271487e-20 -5.954210472282593e-21
5.748124494856867e-18 -1.413059645633656e-17
5.599526916534281e-18 -6.071638098278914e-18
0.000000000000000e+00 -0.000000000000000e+00
```

That is, this computation yields the origin in $\mathbb{C}^2 \times \mathbb{C}^2$ so $A \cap B = \{(0,0)\}$. We remark that for this system, the origin is actually a smooth point on the one-dimensional set $\mathbf{V}(F_A(x), F_B(x)) = \mathbf{V}(x_1)$. ∎

Example 12.2. As an illustrative example of intersecting two irreducible components of the same polynomial system, consider the sets $A = \mathbf{V}(x_1x_2 - x_3) \subset \mathbb{C}^3$ and $B = \mathbf{V}(x_1^2 - x_2) \subset \mathbb{C}^3$ with witness sets $W_A = \{F, L, S_A\}$ and $W_B = \{F, L, S_B\}$, respectively, where

$$F(x_1,x_2,x_3) = (x_1x_2 - x_3)(x_1^2 - x_2), \quad L(x_1,x_2,x_3) = \begin{bmatrix} x_1 + 2x_3 - 2 \\ x_2 + x_3 - 1 \end{bmatrix},$$
$$S_A = \{(-2,-1,2),(1,0.5,0.5)\}, \qquad S_B = \{(0.5,0.25,0.75),(0,0,1)\}.$$

Consider tracking the homotopy defined in (12.5) with $c = 3$ and

$$L_4(x_1,x_2,x_3,y_1,y_2,y_3) = x_1 - 2x_2 + 5x_3 - 3y_1 - 4y_2 + y_3 - 3.$$

We use a user-defined homotopy (see §7.5) as shown in the `input` file

```
% IntersectComponents.input
CONFIG
 UserHomotopy:2;
END;
INPUT
 function FA,FB,L1,L2,L3,L4;
 variable_group x1,x2,x3,y1,y2,y3;
 pathvariable T;
 parameter t;
 random gamma;

 t = T;
 FA = (x1*x2 - x3)*(x1^2 - x2);
```

```
  FB = (y1*y2 - y3)*(y1^2 - y2);
  L1 = gamma*t*(x1+2*x3-2) + (1-t)*(x1-y1);
  L2 = gamma*t*(x2+x3-1) + (1-t)*(x2-y2);
  L3 = gamma*t*(y1+2*y3-2) + (1-t)*(x3-y3);
  L4 = gamma*t*(y2+y3-1) + (1-t)*(x1-2*x2+5*x3-3*y1-4*y2+y3-3);
END;
```

with the `start` file consisting of the four points in $S_A \times S_B$. Running Bertini yields that one path diverges to infinity, whereas the other three converge to distinct points which, under $\Phi(x,y)=x$, yield the intersection of the twisted cubic curve with a plane. ∎

12.2 ▪ Regeneration Extension

Let us now return to the setup of §12.1.1, where we considered how to approach $\mathscr{Z} = X \cap \mathbf{V}(f)$, assuming a witness set $\{f_X, L, W_X\}$ for X (irreducible and multiplicity one) is available. Since the objective is to extend the analysis of X to $X \cap \mathbf{V}(f)$, we call this the *extension problem*. We wish to compute set E_i as in (12.2), from which we can extract witness sets for the irreducible components of $\mathscr{Z}$. Our approach uses a generalization of the regeneration technique of §5.4.

The key is to notice that $\widetilde{W}_i$ as defined in (12.1) differs from $\widetilde{W}_{i-1}$ only in that L_i is swapped for g_i. In the intersection case of §12.1.2, L_i and g_i are both linear, with L_i being generic, so the single homotopy (12.4) suffices. In the more general setting, g_i may have degree 2 or greater. Then, as in §5.4, we choose $\deg g_i$ generic linear equations $L_{i,1},\dots,L_{i,\deg g_i}$ and move L_i to each $L_{i,j}$ in a separate homotopy, i.e., for $j=1,\dots,\deg g_i$,

$$h_{i,j}(z,t)=\begin{bmatrix} F_X(z)\\ g_1(z)\\ \vdots\\ g_{i-1}(z)\\ tL_i(z)+(1-t)L_{i,j}(z)\\ L_{i+1}(z)\\ \vdots\\ L_{\dim X}(z)\end{bmatrix}=0. \tag{12.6}$$

(We can always choose $L_{i,1}\equiv L_i$ to make the first of these homotopies trivial.) Then, a final homotopy for level i deforms the product $\prod_{j=1}^{\deg g_i} L_{i,j}$ into g_i,

$$h_i(z,t)=\begin{bmatrix} F_X(z)\\ g_1(z)\\ \vdots\\ g_{i-1}(z)\\ t\prod_{j=1}^{\deg g_i} L_{i,j}+(1-t)g_i(z)\\ L_{i+1}(z)\\ \vdots\\ L_{\dim X}(z)\end{bmatrix}=0, \tag{12.7}$$

using for startpoints the union over j of all the finite endpoints of the $h_{i,j}$ homotopies. The finite endpoints of homotopy h_i form the set E_i, and we proceed in the same fashion to stage $i+1$, always initiating the homotopy paths for $h_{i+1,j}$ from the nonsolutions $N_i \subset E_i$. This is exactly the same regenerative cascade as in §9.4, except the startpoints at the top level are $W_X = X \cap \mathbf{V}(L)$ instead of $\mathbf{V}(L)$.

As usual, one must eliminate the junk points, J_i, and the nonsolutions, N_i, from E_i to obtain a witness point set for the pure $(\dim X - i)$-dimensional component of $\mathscr{X}$, then decompose these into irreducible components using monodromy and traces. As in all our intersection algorithms, isosingular deflation (§13.2.1) may be needed to complete the witness set of each irreducible component of the intersection by producing a polynomial system for which the component is an irreducible solution component [99].

12.3 ▪ Generalized Intersection

With the extension problem solved, we are one small step from computing a numerical irreducible decomposition for $(A \times B) \cap \mathbf{V}(f)$ when witness sets are given for irreducible multiplicity one sets A and B. All we need is a witness set for $X = A \times B$.

In a sense, we already have a witness set for $A \times B$ in the ordered set

$$\{\{f_A(x), f_B(y)\}, \{L_A(x), L_B(y)\}, W_A \times W_B\},$$

where by definition,

$$W_A \times W_B := \{(x,y) \mid x \in A, y \in B\}.$$

This is because $A \times B$ is an irreducible component of $\{f_A, f_B\}$ and

$$W_A \times W_B = (A \times B) \cap \mathbf{V}(L_A(x), L_B(y))$$

is a set of $\deg(A \times B) = \deg A \cdot \deg B$ isolated points. The only problem is that the system $\{L_A(x), L_B(y)\}$ is not a fully generic linear system. But this can be cured easily by choosing a fully generic linear system $L(x,y)$ and executing the homotopy $h_0(z,t)$ of (12.3).

Once the witness set for $A \times B$ is in hand, the regeneration extension method produces witness supersets for $(A \times B) \cap \mathbf{V}(f)$, and the usual techniques apply for decomposing these into witness sets for the irreducible components.

12.4 ▪ Fiber Products

Although fiber products sound like something to be found in a textile plant, it is also the name for a common construction in algebraic geometry. They are so natural, in fact, that engineers may have made use of the construction without knowing the mathematical name. Before giving a formal definition, let us sneak up on the construction by way of an example.

Let us suppose that a mechanical linkage, possibly a robot, has input variables $x \in \mathbb{R}^N$ (perhaps motor angles), output variables $y \in \mathbb{R}^M$ (perhaps a hand location), and parameters $q \in \mathbb{R}^K$ (link lengths and so on). These are related by the assembly conditions for the linkage, say, $f(x,y;q) = 0$. A typical analysis problem, called *forward kinematics*, is to give values of the inputs and the parameters, say, $x = x^*$ and $q = q^*$, and ask what outputs are obtained, that is, ask for the solution of $f(x^*, y,; q^*) = 0$ for the unknowns y. *Inverse kinematics* is similar, except outputs are given instead of inputs, so one must solve for x in $f(x, y^*; q^*) = 0$. In both these analysis problems, the linkage is known at the outset in terms of its parameters, q^*.

Another type of problem, one where fiber products show up, is *mechanism synthesis*. In such problems, one specifies something about the input/output relations that are desired in the mechanism, and the objective is to find linkage parameters that give this. Assume we have specified the mechanism type, so we have the equations $f(x,y;q)=0$, but we do not know the specific parameters q. Let's say we want to design a mechanism that can reach outputs $y_1^*,\ldots,y_\ell^*$. Once we build the mechanism, its parameters are fixed, but its inputs and outputs may vary. Clearly, the conditions to produce the outputs $y_1^*,\ldots,y_j^*$ are

$$\begin{bmatrix} f(x_1,y_1^*;q) \\ \vdots \\ f(x_\ell,y_j^*;q) \end{bmatrix} = 0, \tag{12.8}$$

where the unknowns are $(x_1,\ldots,x_\ell,q)\in\mathbb{C}^{\ell\cdot N+K}$. A solution to this system is a mechanism design, q, along with the inputs, $x_1,\ldots,x_\ell$, that make it reach the desired outputs, $y_1^*,\ldots,y_\ell^*$. Kinematicians will recognize this as a *precision point synthesis* problem, as we are requiring the mechanism to reach precisely the given outputs. To have maximal control of the output motion, one specifies as many outputs ℓ as possible, thereby determining a finite set of linkages to choose from. Giving fewer outputs leaves positive-dimensional solution sets, giving less control over output but allowing the designer larger choice to meet other considerations.

Where are the fiber products in this formulation? Each subsystem $f(x_i,y_i^*;q)=0$ has a solution set, say,

$$Z_i=\{(x,q)\in\mathbb{C}^{N+K}\mid f(x,y_i^*;q)=0\}.$$

In addition, we have the projection that for any point in Z_i gives just the parameters:

$$\pi: Z_i\to Q \text{ defined by } (x,q)\mapsto(q),$$

where in the development above we have $Q=\mathbb{C}^K$. The solution to the output synthesis problem (12.8) is the ℓ-fold fiber product

$$Z_1\times_Q\cdots\times_Q Z_\ell:=\{(z_1,\ldots,z_\ell)\in Z_1\times\cdots\times Z_\ell\mid \pi(z_1)=\cdots=\pi(z_\ell)\}.$$

More generally, if we have algebraic sets Z_1 and Z_2 with algebraic maps $\pi_1: Z_1\to Q$ and $\pi_2: Z_2\to Q$, then we define the fiber product as

$$Z_1\times_Q Z_2=\{(z_1,z_2)\in Z_1\times Z_2\mid \pi_1(z_1)=\pi_2(z_2)\}.$$

From this definition, it is clear that when a third set with a map enters the picture, $\pi_3:$ $Z_3\to Q$, associativity holds:

$$(Z_1\times_Q Z_2)\times_Q Z_3=Z_1\times_Q(Z_2\times_Q Z_3),$$

so we may write simply $Z_1\times_Q Z_2\times_Q Z_3$. This pattern extends indefinitely for more sets with maps. By the way, the name "fiber product" derives from the fact that for map $\pi: Z\to Q$ and $q^*\in Q$, mathematicians call the set

$$\pi^{-1}(q^*):=\{z\in Z\mid \pi(z)=q^*\}$$

the *fiber* over q^*. Most commonly, as in the output synthesis problem above, the maps π_1, π_2, etc., are simple projections, which is why the notation $Z_1\times_Q Z_2$ indicates the spaces involved but does not explicitly note the maps.

While fiber products arise naturally in problems such as output synthesis, the construction is also useful in finding sets having exceptional fiber dimension, a topic we discuss next.

12.4.1 ▪ Exceptional Loci

Exceptional loci are parameter sets where the fiber dimension of a solution set changes. Suppose that for a parameterized family of systems $f(z;q)$, as in Chapter 6, we have an irreducible set $A \subset \mathbf{V}(f(z,q^*))$ for generic q^*. If f describes the assembly conditions for a mechanical system, then $\dim A$ is the number of degrees of freedom of motion of the system having parameters q^*. It can be of interest to a kinematician to know what sets of parameters lead to more degrees of freedom than their generic neighbors. The resulting "overconstrained mechanisms" move quite differently than the ones having generic parameter values. While various exceptional mechanisms have been identified over time, these have largely been isolated discoveries with no overarching theory for finding new ones. The problem of finding exceptional loci is by no means limited to mechanisms, as it is a natural question to ask about any parameterized polynomial system. For example, see §13.1.2, where the problem of rank-dropping is treated as a case of exceptional loci.

A theory for systematically discovering exceptional loci, presented in [162], is based on fiber products. Suppose that X is an irreducible component of $\mathbf{V}(f(z;q))$. There is a natural projection $\pi : X \to Q$ given by $(z;q) \mapsto (q)$. For most $q \in Q$, the fibers $\pi^{-1}(q)$ all have the same dimension. It may happen though that there are some $q \in Q$ for which the fiber dimension increases.

How can the higher-dimensional fiber hide inside an irreducible component made up generically of lower-dimensional fibers? A simple example helps.

Example 12.3. Let $f(z;q) = q_1 z + q_2$, so that for general $q = (q_1, q_2)$, $\pi^{-1}(q)$ is a single point $(-q_2/q_1; q_1, q_2)$. We can anticipate from this that something unusual might happen if $q_1 = 0$, and indeed $\pi^{-1}(0,0)$ is a line, $(\lambda; 0, 0)$, $\lambda \in \mathbb{C}$. So the entire set $X = \mathbf{V}(f) \subset \mathbb{C}^3$ is two-dimensional, while nestled inside is the one-dimensional exceptional fiber $\pi^{-1}(0,0)$. One may visualize the situation by noting that for any given value of q_1, say, $q_1 = q_1^*$, the solution set $\mathbf{V}(f, q_1 - q_1^*) = X \cap \{(z; q_1, q_2) \in \mathbb{C}^3 \mid q_1 = q_1^*\}$ is a line of slope $-1/q_1^*$. This line becomes vertical, and thereby exceptional, when $q_1 = 0$. ∎

Though the details can seem quite technical, the basic idea is simple and comes down to merely counting dimensions. Suppose X is dimension d, and the dimension of $\pi(X) \subset Q$ is b (i.e., b is the base dimension). Then by the additivity of dimension, the dimension of $\pi^{-1}(q)$ must be dimension $h = d - b$ for generic $q \in \pi(X)$. (We may think of h as the height of X.) Consider then the fiber product of X with itself, $X \times_Q X$. Points in $X \times_Q X$ look like (z_1, q, z_2, q) and admit the projection map $\Pi : (z_1, q, z_2, q) \mapsto (q)$. For a generic point $q \in \pi(X)$, the dimension of $\Pi^{-1}(q)$ is $2h$ because it is the product of two h-dimensional fibers, and as q varies over $\pi(X)$, these fibers knit together to form a set of dimension $2h + b$. But what happens at the exceptional loci? Suppose that $Q^* \subset Q$ is an exceptional locus of dimension b^* with fiber dimension $h^* = \dim(\pi^{-1}(q^*)) > h$ for generic $q^* \in Q^*$. Then, letting q^* vary over Q^*, the fiber product $X \times_Q X$ must contain an exceptional set of dimension $2h^* + b^*$. Since $h^* > h$ by the definition of an exceptional fiber, there is a chance that $2h^* + b^* \geq 2h + b$. It might not be so, since $b^* < b$. But if $2h^* + b^* \geq 2h + b$ is true, then there is no way the exceptional set of dimension $2h^* + b^*$ can be contained inside the nonexceptional set of dimension $2h + b$. This implies that the fiber product operation has torn the exceptional set free of its capture within the set of nonexceptional cases.

One fiber product may not be enough to free the exceptional set. But for big enough k, we must have that $kh^* + b^* \geq kh + b$, so taking repeated fiber products,

$$\underbrace{X \times_Q \cdots \times_Q X}_{k \text{ times}},$$

will do the trick. Iterating the fiber product in this way guarantees that eventually the fiber product of the exceptional set becomes its own independent irreducible component, so it will show up in a numerical irreducible decomposition. The picture is complicated a bit by the fact that the fiber product can contain "side components" that mix up fibers of different types in the different directions, while the sets of interest are the "main components" that combine fibers all of the same type. Still, it can be shown that the exceptional sets will eventually be promoted to irreducibility. Moreover, it is easy to determine a k that suffices to reveal all possible exceptional sets, so the iterative process has a clear termination condition.

Example 12.4. Returning to Example 12.3, we have $b = 2$ and $h = 0$ for the base and fiber dimensions of X, while its exceptional set has $b^* = 0$ and $h^* = 1$. The fiber product $X \times_Q X$ has two irreducible components: the nonexceptional one with dimension $2h + b = 2$ and the exceptional one with dimension $2h^* + b^* = 2$. Since these are the same dimension, it is clear that the exceptional one cannot be contained in the nonexceptional one. ∎

12.4.2 ▪ Computing Fiber Products

The naive way to compute the numerical irreducible decomposition of a fiber product is to stack up the equations into one big system. For an ℓ-fold fiber product, $X_1 \times_Q \cdots \times_Q X_\ell$, where X_i is an irreducible component of $f_i(x_i;q)$ and the maps are $\pi_i : (x_i, q) \mapsto (q)$, the system is

$$F(x_1,\ldots,x_\ell;q) = \begin{bmatrix} f_1(x_1;q) \\ \vdots \\ f_\ell(x_\ell;q) \end{bmatrix} = 0. \tag{12.9}$$

Any of the general approaches, such as the regeneration cascade, can be applied to this system. However, stacking up systems in this way can lead to a high total degree for F, so this can quickly become a daunting computation.

The naive approach can sometimes be improved on. In particular, if $\mathbf{V}(f_1)$ has more than one component, say, both X_1 and Y_1, then $\mathbf{V}(F)$ contains not only $X_1 \times_Q X_2 \times_Q \cdots \times_Q X_\ell$, which we want, but also $Y_1 \times_Q X_2 \times_Q \cdots \times_Q X_\ell$, which perhaps we don't. This kind of trouble multiplies if other $\mathbf{V}(f_i)$ have several components. This suggests that an approach like the diagonal intersection method, which limits computation to the components of interest, should be followed.

Also, notice that in the two cases presented above, mechanism output synthesis and exceptional loci, the f_i are either identical or are from the same family of parameterized systems. So if we perform a numerical irreducible decomposition of $\mathbf{V}(f_1)$, we can use parameter homotopies to obtain decompositions of the other subsystems as well.

Once witness sets are available for each of $X_1,\ldots,X_\ell$, one can apply a modified version of the diagonal intersection technique. For $\ell = 2$, a more efficient approach to computing $X_1 \times_Q X_2$ is to use the regeneration extension technique to compute

$$\begin{aligned} X_1 \times_Q X_2 &= (X_1 \times X_2) \cap \mathbf{V}(\pi_1(X_1) - \pi_2(X_2)) \\ &= \{(x_1,q_1,x_2,q_2) \mid (x_1,q_1) \in X_1, (x_2,q_2) \in X_2, q_1 = q_2\}. \end{aligned}$$

This is almost the same as the diagonal intersection technique, except that only the projection coordinates are made equal instead of equating all coordinates.

In the case of finding exceptional loci, even more efficiencies can be garnered by taking advantage of the symmetries that arise because all the factors in the product are identical. Also, instead of taking general slices to find irreducible components, one may take slices that are aligned in the fiber and base directions. These ideas are the subject of ongoing research aimed at providing efficient, yet general, techniques for finding exceptional sets.

Chapter 13
Singular Sets

To motivate a study of singular sets, let's consider the polynomial $f(x,y) = x^2 - y^3$. It is easy to verify that $\mathbf{V}(f)$ is an irreducible curve of degree 3 such that $\frac{\partial f}{\partial x}$ and $\frac{\partial f}{\partial y}$ are both generically nonzero on $\mathbf{V}(f)$. The *singular points* of f are the points that satisfy

$$f(x,y) = \frac{\partial f}{\partial x}(x,y) = \frac{\partial f}{\partial y}(x,y) = 0,$$

which is just the origin. In particular, the Jacobian matrix $Jf(x,y) = \begin{bmatrix} 2x & -3y^2 \end{bmatrix}$ has rank 1 at all points of $\mathbf{V}(f)$ except the origin, where $Jf(0,0) = [0\ 0]$, which has rank 0.

In a slightly more general setting, suppose that $f(z)$ is a polynomial system and $V \subset \mathbf{V}(f) \subset \mathbb{C}^N$ is an irreducible component of dimension d such that the rank of Jf is $N-d$ generically on V. Hence, the tangent space of Jf has dimension d generically on V, and this being equal to the dimension of V implies that V has multiplicity one with respect to f. Consider the sets

$$\mathscr{S}_k(Jf) := \{z \in \mathbb{C}^N \mid \text{rank} Jf(z) \leq k\} \qquad \text{and} \qquad V_k := V \cap \mathscr{S}_k(Jf).$$

Each $\mathscr{S}_k(Jf)$ is defined by the vanishing of all the $(k+1)$-minors of $Jf(z)$, i.e., the determinants of all matrices formed by choosing $(k+1)$ rows and $(k+1)$ columns of $Jf(z)$. Since each determinant is an algebraic expression, each $\mathscr{S}_k(Jf)$ is an algebraic set. Since the rank of a matrix can decrease only at special points in V, we have $V = V_k$ for $k \geq N-d$. The *singular set* of V with respect to f is the set V_{N-d-1}, i.e., the set of points $z \in V$ such that rank $Jf(z) < N - d$. The set $V \setminus V_{N-d-1}$ is a nonempty Zariski open subset of V consisting of the *smooth points* of V with respect to f, which is a (path) connected set. Additionally, since V_{N-d-1} is defined as the intersection of V with $\mathscr{S}_k(Jf)$, we could use the ideas presented in Chapter 12 to compute a numerical irreducible decomposition for V_{N-d-1} starting with a witness set for V.

Considering where the Jacobian matrix drops rank is natural in some applications, especially in algebraic geometry. However, there are many other applications where we need to consider where other matrices drop rank, including augmented Jacobian matrices for computing the critical points of a projection (see §14.4) and the first-order critical points of an optimization problem (see §14.5), and Macaulay matrices (see §11.4) for computing a superset of the primary components [115]. Thus, in §13.1, we shall consider computing the numerical irreducible decomposition of points on an irreducible algebraic set where some matrix, whose entries are multivariate polynomials, drops rank. Since the minors defining the rank-dropping conditions can be of high degree and there could be an unwieldly number of them, this section uses a nullspace condition to compute the rank-dropping sets.

In §11.6, we used an iterative deflation approach with auxiliary variables to deflate irreducible components. By incorporating techniques of §13.1, we develop a more general deflation approach that permits the deflating of irreducible algebraic sets which are not irreducible components but have a common "singularity structure." For a particular example, consider the polynomial equation

$$f(x,y,z) = (x + y - z^2 + 1)(y^2 - xz - z) = 0,$$

whose solution set is the union of two surfaces. The irreducible curve defined by the intersection of the two surfaces is perforce singular, and it is not an irreducible component of $\mathbf{V}(f)$. Nevertheless, in Example 13.11, we show that this curve can be deflated. That is, we can construct a witness set for this curve, allowing us to perform additional numerical computations on it.

We remark that for any polynomial system $f(z)$, the "singularity structure" of $f^{-1}(0)$ is described by finitely many irreducible algebraic subsets, each called an *isosingular set* [98]. These sets can be deflated, thereby restoring the quadratic convergence of Newton's method locally near *any* point $x \in \mathbf{V}(f)$.

13.1 ▪ Rank-Dropping Conditions

This section looks at computing points where the dimension of the nullspace of a given matrix is exceptional, i.e., the rank is smaller than expected. However, before jumping into a nullspace formulation, we first review the algebraic representation of rank-dropping conditions based on determinants in §13.1.1.

In §12.4, we considered computing sets of exceptional dimension, such as mechanisms with an exceptional number of degrees of freedom, and found fiber products to be a useful construction for this. We explore using fiber products in §13.1.2 to find sets where the nullspace of a matrix of polynomials has exceptional dimension.

In §13.1.3, we describe an alternative approach from [18] which uses an explicit basis for the exceptionally large nullspaces. In this case, each linear space is considered a point in a *Grassmann variety*, the set of linear subsets of a given dimension in a Euclidean space of a given (at least as large) dimension.

In all three sections, we compute the rank-dropping sets on a Euclidean domain. However, one may consider restricting the domain to an irreducible algebraic set V. The computation of the rank-dropping sets on V given a witness set for V follows by using an intersection approach such as the diagonal approach of §12.1 or the regeneration extension method of §12.2.

13.1.1 ▪ Rank-Dropping via Determinants

For each $z \in \mathbb{C}^N$, suppose that $A(z)$ is an $r \times s$ matrix whose entries are polynomials in z. Since the ranks of $A(z)$ and $A(z)^T$ are equal, for simplicity we can assume that rank $A(z) \le s \le r$. For $k = 0, \dots, s$, consider the sets

$$\mathscr{S}_k(A) := \{z \in \mathbb{C}^N \mid \operatorname{rank} A(z) \le k\} \tag{13.1}$$

and let $F_k(z)$ be the polynomial system consisting of all the $(k+1)$-minors of $A(z)$, of which there are $\binom{r}{k+1} \cdot \binom{s}{k+1}$. Clearly, $\mathbf{V}(F_k) = \mathscr{S}_k(A)$, thereby showing that each $\mathscr{S}_k(A)$

is an algebraic set. In particular, $F_0(z)$ is simply the polynomial system consisting of the $r \times s$ entries of $A(z)$. The two drawbacks from using this determinantal approach are the potentially large number of polynomials in F_k and the fact that they may have high degrees. (If each entry of $A(z)$ has degree D, a k-minor of A has degree at most kD.)

Example 13.1. Consider the matrix

$$A(x,y) = \begin{bmatrix} x(2x-y) & 3y^2-2x-7y+4 \\ 2x^2-xy+2y-4 & xy+3y^2-x-8y+3 \end{bmatrix}.$$

We first compute the generic rank of $A(x,y)$ on $\mathbb{C}^2$ which, with probability one, is the rank of $A(x^*,y^*)$ at a random point (x^*,y^*). Even though $(0,0)$ is not random, we still know the generic rank of $A(x,y)$ is 2 since

$$A(0,0) = \begin{bmatrix} 0 & 4 \\ -4 & 3 \end{bmatrix}$$

is a 2×2 matrix of rank 2. Hence, $\mathscr{S}_2(A) = \mathbb{C}^2$.

For $k = 0, 1$, we aim to compute the algebraic sets $\mathscr{S}_k(A) = \mathbf{V}(F_k)$, where $F_1(x,y)$ is the single polynomial $\det A(x,y)$, and $F_0(x,y)$ is the system of four polynomials that are the entries of $A(x,y)$. Using the standard approach discussed in Part II, Bertini computes that $\mathscr{S}_1(A) = \mathbf{V}(F_1)$ is an irreducible curve of degree 4 and $\mathscr{S}_0(A) = \mathbf{V}(F_0)$ is the point $(1,2)$. ∎

13.1.2 ▪ Rank-Dropping Sets via Fiber Products

For an $r \times s$ matrix M, $r \geq s$, the rank-nullity theorem yields that rank $M \leq k$ if and only if dim null $M \geq s-k$. Thus, for our polynomial map $A : \mathbb{C}^N \to \mathbb{C}^{r\times s}$, we have

$$\mathscr{S}_k(A) = \{z \in \mathbb{C}^N \mid \operatorname{rank} A(z) \leq k\} = \{z \in \mathbb{C}^N \mid \dim \operatorname{null} A(z) \geq s-k\}. \tag{13.2}$$

By adding nullspace variables, we can set up a fiber product approach (see §12.4) to compute $\mathscr{S}_k(A)$. In particular, by considering fiber products over $\mathbb{C}^N$ of the set

$$X := \{(v;z) \in \mathbb{C}^s \times \mathbb{C}^N \mid A(z)\cdot v = 0\}, \tag{13.3}$$

we are able to identify where the matrix drops rank, i.e., has an exceptional nullspace. We demonstrate the approach on two examples, one simple enough to do by hand and one where we use Bertini.

Example 13.2. For an illustrative example, consider

$$A(x,y) = \begin{bmatrix} x & y \\ x & y \end{bmatrix}.$$

Clearly, the generic rank of $A(x,y)$ is 1, so $\mathscr{S}_1(A) = \mathbb{C}^2$. Let the set X be defined as in in (13.3), i.e., $X = \mathbf{V}(xv_1 + yv_2)$, and consider the projection $\pi : (v_1,v_2;x,y) \mapsto (x,y)$. Using the notation of §12.4.1, we have $b = 2$ and $h = 1$. The exceptional set $\mathscr{S}_0(A)$ has $b^* = \dim \mathscr{S}_0(A) \leq 1$ with fiber dimension $h^* = 2$. Clearly, $X = \mathbf{V}(xv_1 + yv_2)$ is irreducible. Any exceptional set contained in X must have $b^*+h^* < b+h$, which implies $b^* < 1$. Accordingly, $\mathscr{S}_0(A)$ is, at most, zero-dimensional, so if it is not empty, we have $b^* = 0$. It follows that one fiber product is enough to determine $\mathscr{S}_0(A)$ since the associated main component in the fiber product $X \times_{\mathbb{C}^2} X$ and that of $\mathscr{S}_1(A)$ have the same dimension: $b^* + 2h^* = b + 2h = 4$.

Specifically, the fiber product $X \times_{\mathbb{C}^2} X = \mathbf{V}(xv_1 + yv_2, xv_3 + yv_4)$ consists of two four-dimensional irreducible components V_0 and V_1:

$$V_0 = \{(v_1, v_2, v_3, v_4; 0, 0) \mid v_i \in \mathbb{C}\},$$
$$V_1 = \{(v_1, -v_1, v_3, -v_3; x, y) \mid x, y, v_1, v_3 \in \mathbb{C}\}.$$

We have $\mathscr{S}_1(A) = \mathbb{C}^2$ associated to V_1 and $\mathscr{S}_0(A) = \{(0,0)\}$ associated to V_0. ∎

Example 13.3. Reconsider the matrix $A(x,y)$ from Example 13.1 for which the generic rank on $\mathbb{C}^2$ is 2. Accordingly, the set $X = \mathbf{V}(A(x,y) \cdot v)$ contains a two-dimensional component $\{(0,0;x,y) \mid x,y \in \mathbb{C}\}$ associated to $\mathscr{S}_2(A)$. Since $\mathscr{S}_1(A)$ is given by a single nontrivial equation, $\det A(x,y) = 0$, we know it must be one-dimensional, and so adding in the one-dimensional nullspace, the associated component in X is two-dimensional. Furthermore, points in $\mathscr{S}_0(A)$ have two-dimensional nullspaces. According to this dimension accounting, none of the associated sets in X can contain the others, so it suffices to compute an irreducible decomposition of X. Decomposing X is accomplished using the following input file:

```
% RankDrop_FiberProduct.input
CONFIG
 TrackType:1;
END;
INPUT
 function f1,f2;
 variable_group x,y,v1,v2;

 f1 = x*(2*x-y)*v1 + (3*y^2-2*x-7*y+4)*v2;
 f2 = (2*x^2-x*y+2*y-4)*v1 + (x*y+3*y^2-x-8*y+3)*v2;
END;
```

Bertini computes that X has three irreducible components of dimension 2: two are linear, say, V_1 and V_2, and the other, say, V_3, has degree 7. In order to understand the components, we need to project each component onto the (x,y) variables. Even though projections are discussed in detail in Chapter 16, we will utilize the computations in Bertini here.

We define a projection in Bertini by constructing a `projection` file. In particular, the projection $\pi : (x,y,v_1,v_2) \mapsto (x,y)$ is defined in `projection` by the sequence

```
1 1 0 0
```

since we are projecting onto the first two coordinates. Keeping the `witness_data` file from the previous run, together with this `projection` file, we run Bertini using the same `input` file except for changing the `TrackType` configuration from 1 to 5. By running Bertini three times, selecting one of the irreducible components each time from the onscreen menu, we obtain the following information:

j	$\dim \overline{\pi(V_i)}$	fiber dim	$\deg \overline{\pi(V_i)}$	fiber degree
1	2	0	1	1
2	0	2	1	1
3	1	1	4	1

Hence, $\pi(V_1) = \mathscr{S}_2(A) = \mathbb{C}^2$, $\pi(V_2) = \mathscr{S}_0(A)$, which we can see in `main_data` is the point $(1,2)$, and $\pi(V_3) = \mathscr{S}_1(A)$ is a curve of degree 4. ∎

13.1.3 ▪ Rank-Dropping Sets via Grassmannians

For our polynomial map $A : \mathbb{C}^N \to \mathbb{C}^{r\times s}$ with $r \geq s$, we know, for $0 \leq k \leq s$, that $z \in \mathscr{S}_k(A)$ if and only if there exist linearly independent vectors $v_1,\ldots,v_{s-k} \in \mathbb{C}^s$ such that $A(z)\cdot v_i = 0$ for $i = 1,\ldots,s-k$. That is, $v_1,\ldots,v_{s-k}$ form a basis for some $(s-k)$-dimensional linear space L with $L \subset \text{null } A(z)$. By identifying L with the span of the columns of the matrix $[v_1 \ \cdots \ v_{s-k}]$, by a slight abuse of notation, we will write $A(z)\cdot L = 0$.

To simplify the presentation, we need to make one definition. The space of all $(s-k)$-dimensional linear spaces in $\mathbb{C}^s$ is called the *Grassmann variety*, denoted $\text{Grass}(s,s-k)$, which has dimension $k(s-k)$. Even though $\text{Grass}(s,s-k)$ is not typically parameterizable, we can easily parameterize a random Zariski dense subset of it as follows. Let I_{s-k} be the $(s-k)\times(s-k)$ identity matrix. For a random unitary[34] matrix $B \in \mathbb{C}^{s\times s}$, the set of all $(s-k)$-dimensional linear spaces generated by the columns of

$$B\cdot\begin{bmatrix} I_{s-k} \\ \Lambda \end{bmatrix}$$

for some $\Lambda \in \mathbb{C}^{s\times(s-k)}$ is a Zariski dense subset of $\text{Grass}(s,s-k)$. In particular, the number of entries of Λ, namely, $k(s-k)$, is equal to the dimension of $\text{Grass}(s,s-k)$.

Returning to the rank-dropping sets, consider the projection $\pi(z,\Lambda) = z$ and polynomial system

$$G_k(z,\Lambda) = A(z)\cdot B\cdot\begin{bmatrix} I_{s-k} \\ \Lambda \end{bmatrix}.$$

For a random unitary matrix B, with probability one, we have $\mathscr{S}_k(A) = \overline{\pi(\mathbf{V}(G_k))}$. In fact, the irreducible components of $\overline{\mathscr{S}_k(A)\setminus\mathscr{S}_{k-1}(A)}$, where $\mathscr{S}_{-1}(A) = \emptyset$, are exactly those with a generic zero-dimensional fiber.

Before looking at an example, we remark on the two extreme cases, namely, $k = 0$ and $k = s$, in which Λ has no entries. For $k = 0$, we have $G_0(z) = A(z)\cdot B\cdot I_s$. Since B and I_s are both invertible, solving $G_0(z) = 0$ is equivalent to solving $A(z) = 0$. For $k = s$, we have that $G_s(z)$ is the zero polynomial since, for every $z \in \mathbb{C}^N$, we always have $\dim\text{null } A(z) \geq 0$.

Example 13.4. Reconsider the matrix $A(x,y)$ from Example 13.1. Since $\mathscr{S}_0(A) = \mathbf{V}(A(z)) = \{(1,2)\}$ was solved using this formulation in Example 13.1, we need only consider computing $\mathscr{S}_1(A)$ using the Grassmannian approach. We use the following `input` file to compute the decomposition of $\mathbf{V}(G_1)$, where $B = \frac{1}{5}\left[\begin{smallmatrix} 3 & 4 \\ -4 & 3 \end{smallmatrix}\right]$:

```
% RankDrop_Grassmannian.input
CONFIG
 TrackType:1;
END;
INPUT
 function f1,f2;
 variable_group x,y,lambda;
```

[34] We use unitary matrices here for numerical stability, but this is not necessary theoretically.

```
    % L = B*[1;lambda]
    L1 = (3+4*lambda)/5;
    L2 = (3*lambda-4)/5;

    f1 = x*(2*x-y)*L1 + (3*y^2-2*x-7*y+4)*L2;
    f2 = (2*x^2-x*y+2*y-4)*L1 + (x*y+3*y^2-x-8*y+3)*L2;
  END;
```

Bertini computes that $\mathbf{V}(G_1)$ has two irreducible curves, one is linear, say, V_1, and the other, say, V_2, has degree 7.

We now compute $\overline{\pi(V_1)}$ and $\overline{\pi(V_2)}$, whose computation is discussed in detail in Chapter 16. To do this, we first need to construct a `projection` file for our projection $\pi(x,y,\lambda)=(x,y)$, which, since we are projecting onto the first two coordinates, contains simply the line:

```
  1 1 0
```

Keeping the `witness_data` file from the previous run, together with this `projection` file, we run Bertini using the same `input` file except for changing the `TrackType` configuration from 1 to 5. By running Bertini two times, selecting one of the irreducible components each time from the onscreen menu, we obtain the following information:

j	$\dim \overline{\pi(V_i)}$	fiber dim	$\deg \overline{\pi(V_i)}$	fiber degree
1	0	1	1	1
2	1	0	4	1

In particular, looking in `main_data`, we see that the first projects to the point $(1,2)$ which is in $\mathscr{S}_0(A)$. Hence, $\overline{\mathscr{S}_1(A)\setminus\mathscr{S}_0(A)}=\overline{\pi(V_2)}$, which is a curve of degree 4. In this specific case, algebraic geometry yields that $\mathscr{S}_1(A)$ is pure-dimensional, implying that $\mathscr{S}_1(A)$ is equal to just this curve of degree 4. ∎

13.2 ▪ Isosingular Sets*

Rather than dive into notation and present formal definitions, let's intuitively categorize into irreducible algebraic sets, which are as large as possible, the different types of "singularity structure" for a simple polynomial. Each such set is called an *isosingular set*.

Example 13.5. We consider the different types of "singularity structure" that the solutions of $f(x,y,z):=x^2yz=0$ could have. First, we have the irreducible components:

- $V_1:=\mathbf{V}(x)$,
- $V_2:=\mathbf{V}(y)$, and
- $V_3:=\mathbf{V}(z)$.

Here, V_1 has multiplicity 2 with respect to f, while the V_2 and V_3 are multiplicity one. In particular, from a local structure standpoint, general points on V_2 and V_3 are indistinguishable from each other.

The next sets that we need to consider are irreducible curves. Since, as a set, each V_i is smooth, the only places where the local "singularity structure" can be different is at the intersection of the irreducible components. Since each intersection is irreducible, we have the sets

- $V_4 := V_1 \cap V_2 = \mathbf{V}(x,y)$,
- $V_5 := V_1 \cap V_3 = \mathbf{V}(x,z)$, and
- $V_6 = V_2 \cap V_3 = \mathbf{V}(y,z)$.

From a local structure standpoint, general points on V_4 and V_5 are indistinguishable.

The last sets that we need to consider are singletons, i.e., sets consisting of a single point. Since, as sets, V_i are all smooth, we just need to consider the origin which arises as the intersection of the three irreducible components, that is,

- $V_7 := V_1 \cap V_2 \cap V_3 = \mathbf{V}(x,y,z)$.

For any point $(x,y,z) \in \mathbf{V}(f)$, we have a unique $i \in \{1,\ldots,7\}$ such that $(x,y,z) \in V_i$ and the "singularity structure" of $f^{-1}(0)$ at (x,y,z) is the same as a generic point on V_i. For example, the set associated to $(0,1,1)$ is V_1, while V_4 is the set associated to $(0,0,2)$. ∎

This illustrative example demonstrates two key properties of isosingular sets with respect to a given polynomial system $f(z)$:

1. there are finitely many isosingular sets $V_1,\ldots,V_k \subset \mathbf{V}(f)$, and
2. for each $z \in \mathbf{V}(f)$, there is a unique $i \in \{1,\ldots,k\}$ such that $z \in V_i$ and the "singularity structure" of z and a general point of V_i are equal.

In loose terms, the second property states that each point in $\mathbf{V}(f)$ is a "smooth" point on a unique isosingular subset of $\mathbf{V}(f)$.

There is one additional property, which will be discussed in detail in §13.2.1, that is the key link between isosingular sets and numerical computation:

3. each isosingular set is deflatable.

In particular, this means that given f and a general point $z \in V_i$, we can algorithmically construct a polynomial system g_i such that V_i is an irreducible component of $\mathbf{V}(g_i)$ that is multiplicity one as a component of $g_i^{-1}(0)$.

Example 13.6. Continuing with the setup from Example 13.5, we demonstrate the deflatability of isosingular sets for $f(x,y,z) = x^2yz$ and $V_6 = \{(x,0,0) \mid x \in \mathbb{C}\}$ using the sufficiently general point $(1,0,0) \in V_6$. To that end, since the Jacobian matrix of f evaluated at $(1,0,0)$ is the zero matrix, we know that $V_6 \subset \mathbf{V}(f')$, where

$$f'(x,y,z) = \begin{bmatrix} f(x,y,z) \\ \nabla f(x,y,z)^T \end{bmatrix} = \begin{bmatrix} x^2yz \\ 2xyz \\ x^2z \\ x^2y \end{bmatrix}, \quad \text{where } \nabla f = \begin{bmatrix} \frac{\partial f}{\partial x} & \frac{\partial f}{\partial y} & \frac{\partial f}{\partial z} \end{bmatrix}.$$

Next, we consider the Jacobian matrix of f' evaluated at $(1,0,0)$, which has a one-dimensional nullspace. Since $\dim V_6 = 1$, it follows that we can take $g_6 := f'$. ∎

We formalize this *isosingular deflation* in §13.2.1 and use it to encode the concept of "singularity structure" as a nonincreasing sequence of nonnegative integers that also arises in Thom–Boardman singularity theory [10, 71].

In §13.2.2, we return to the three key properties of isosingular sets enumerated above and revisit Example 13.5.

13.2.1 ▪ Isosingular Deflation*

The deflation approach discussed in §11.6 is called a *weak* deflation since it uses only a local *necessary* condition on the rank of a matrix. This is easily seen by considering the following example.

Example 13.7. Consider deflating the polynomial $f(x,y,z) = x^2+y^2+z^2$ at the origin. Since $Jf(0,0,0)$ is the zero matrix, every vector in $\mathbb{C}^3$ is in its nullspace. In this case, the weak deflation approach thereby selects a random vector and considers the set of points in $\mathbf{V}(f)$ for which this vector is in the nullspace of the Jacobian. This construction is based on the probability-one statement that a given matrix is zero if and only if a random vector is in its nullspace. However, in the weak deflation system, the random vector is selected first, turning this equivalency condition into a necessary condition, namely, that if a given matrix is zero, then a prescribed vector is in its nullspace. This idea is encapsulated by writing down the polynomial system generated by one deflation using intrinsic coordinates, namely,

$$g(x,y,z) = \begin{bmatrix} x^2+y^2+z^2 \\ \Box x + \Box y + \Box z \end{bmatrix}.$$

Upon filling each $\Box$ with a random number, $\mathbf{V}(g)$ defines a curve containing the origin such that $\operatorname{rank} Jf(x,y,z) = 1$ for every $(x,y,z) \in \mathbf{V}(g) \setminus \{(0,0,0)\}$. ∎

Isosingular deflation is considered a *strong* deflation in that it employs a *sufficient* condition on the rank of a matrix. For simplicity, we will state this using a determinantal approach, but an extrinsic and intrinsic linear slicing approach based on the Grassmannian approach in §13.1.3 is also described in [98]. Apart from its simplicity to state, the determinantal approach also does not utilize additional variables, meaning that we can avoid an extra projection computation. The downside, as discussed in §13.1.1, is the introduction of potentially many high-degree polynomials.

An isosingular deflation iteration takes a polynomial system $f : \mathbb{C}^N \to \mathbb{C}^n$ and a point $z^* \in \mathbf{V}(f)$ and returns the polynomial system

$$g(z) = \begin{bmatrix} f(z) \\ (N-d+1)-\text{minors of } Jf(z) \end{bmatrix}$$

with $d = \dim \operatorname{null} Jf(z^*)$, i.e., $\operatorname{rank} Jf(z^*) = N-d$. (If $d = \max\{N-n, 0\}$, i.e., $\operatorname{rank} Jf(z) = \min\{n, N\}$, we take $g(z) = f(z)$ since there are no $(N-d+1)$-minors of $Jf(z)$.) Clearly, the determinants ensure that $\mathbf{V}(g)$ consists of the points of $\mathbf{V}(f)$ for which the nullspace of the Jacobian is at least d. To simplify the notation later, we will think of an isosingular deflation iteration as an operator and simply write $(g, z^*) = \mathscr{D}(f, z^*)$. In particular, this permits easily writing the polynomial system obtained after $k \geq 2$ iterations via

$$(g_k, z^*) = \mathscr{D}^k(f, z^*) = \mathscr{D}(\mathscr{D}^{k-1}(f, z^*)).$$

Example 13.8. Let's return to deflating the polynomial $f(x,y,z) = x^2 + y^2 + z^2$ at the origin using isosingular deflation. Since $Jf(0,0,0)$ is the zero matrix, we have

$$g(x,y,z) = \begin{bmatrix} x^2+y^2+z^2 \\ 2x \\ 2y \\ 2z \end{bmatrix}.$$

Clearly, $\mathbf{V}(g) = \{(0,0,0)\}$, which is nonsingular with respect to g. ∎

With respect to deflating at generic points of irreducible components, which includes at isolated solutions, all the termination criteria for weak deflation extend to isosingular deflation. As demonstrated above in Examples 13.7 and 13.8, the difference is the firm control over the size of the nullspace of the Jacobian matrix.

Due to the intimate relationship between isosingular deflation and the nullspace dimension, it is natural to consider how the nullspace dimension changes as we perform additional isosingular deflation iterations. To simplify the notation, we define

$$\operatorname{dnull}(f,z^*) := \dim \operatorname{null} Jf(z^*).$$

The *deflation sequence* of f at $z^* \in \mathbf{V}(f)$ is the sequence $d_0(f,z^*), d_1(f,z^*), \ldots$ of nonnegative integers where

$$d_k(f,z^*) := \begin{cases} \operatorname{dnull}(f,z^*) & \text{if } k=0, \\ \operatorname{dnull}(\mathscr{D}^k(f,z^*)) & \text{otherwise.} \end{cases}$$

Since we are adding additional functions to f, the dimension of the nullspace cannot increase. Thus, the deflation sequence is a nonincreasing sequence of nonnegative integers which must reach its limit, i.e., *stabilize*, after finitely many iterations.

Example 13.9. Continuing with $f(x,y,z) = x^2 + y^2 + z^2$, let's consider the possible deflation sequences for points in $\mathbf{V}(f)$. For the origin, we can see from Example 13.8 that $d_0(f,0) = 3$ and $d_1(f,0) = 0$. Since deflation sequences are nonincreasing sequences of nonnegative numbers, the deflation sequence of the origin is

$$3,0,0,\ldots.$$

We remark that the limit of the deflation sequence is zero since the origin is an isolated singular point of the hypersurface $\mathbf{V}(f)$.

Since at every other point in $\mathbf{V}(f)$ except the origin the Jacobian has full rank, the operator $\mathscr{D}$ is simply the identity. In particular, the deflation sequence at any such point is

$$2,2,2,\ldots. \quad ∎$$

Due to upper semicontinuity of matrix ranks, it follows that for each irreducible algebraic set $V \subset \mathbf{V}(f)$, deflation sequences are constant generically on V. This means that we can define the deflation sequence of V to be the deflation sequence at a general point $z^* \in V$ with respect to f. In Example 13.9, $\mathbf{V}(f)$ was irreducible and every point except the origin had the same deflation sequence.

One remaining issue with deflation sequences is to determine when it has stabilized, i.e., reached its limit. In Example 13.9, this was straightforward, but this may not be the case in other situations. We will return to this question in the following section.

13.2.2 ▪ Isosingular Sets and Stabilization*

The definition of an isosingular set arises from the following question: for a polynomial system $f(z)$ and a nonincreasing, nonnegative sequence $\{d_r\}_{r=0}^{\infty}$, what points in $z \in \mathbf{V}(f)$ have $d_r(f,z) = d_r$ for all $r \geq 0$? Since such sets are often not closed, e.g., as seen in Example 13.9, a more natural algebraic geometric question is to consider the set consisting of the closure of such points. That is, if d is the sequence $\{d_r\}_{r=0}^{\infty}$, describe

$$X(f,d) := \overline{\{z \in \mathbf{V}(f) \mid d_r(f,z) = d_r \text{ for all } r \geq 0\}}. \tag{13.4}$$

For any sequence d, it turns out that $X(f,d)$ is a pure-dimensional algebraic set such that

$$\text{either } X(f,d) = \emptyset \text{ or } \dim X(f,d) = \lim_{r\to\infty} d_r.$$

The sets of interest are thus the irreducible components of $X(f,d)$ when $X(f,d) \neq \emptyset$, each of which we define to be an *isosingular set* with respect to f. It immediately follows that each irreducible component of $\mathbf{V}(f)$ is an isosingular set. Additionally, the third key property of isosingular sets discussed above, i.e., that each isosingular set is deflatable, follows since the dimension of an isosingular set will be equal to the nullspace dimension after finitely many deflation iterations.

Example 13.10. In Example 13.5 where $f(x,y,z) = x^2yz$, we listed seven irreducible algebraic subsets of $\mathbf{V}(f)$ based on an intuitive description of the "singularity structure." These are indeed the seven isosingular sets for f with deflation sequences:

- V_1: $3,2,2,\ldots,$
- V_2 and V_3: $2,2,2,\ldots,$
- V_4 and V_5: $3,3,1,1,\ldots,$
- V_6: $3,1,1,\ldots,$
- V_7: $3,3,3,0,0,\ldots.$ ∎

Suppose that $V \subset \mathbf{V}(f)$ is an irreducible component of dimension k that has multiplicity one with respect to f. The smooth points of V with respect to f are those points for which the Jacobian of f has a k-dimensional nullspace, while the singular points are those with a nullspace of dimension larger than k. In terms of deflation sequences, the smooth points of V have a deflation sequence $k,k,k,\ldots$, while the singular points must have a different deflation sequence since the first term is larger than k.

By using deflation sequences, we can naturally define the smooth and singular points of an isosingular set V with respect to f. The *smooth points* of V have the same deflation sequence as a general point of V, while the *singular points* have a different deflation sequence. The set of singular points of V is a nontrivial algebraic subset of V, so we can consider its irreducible components. If the set of singular points of V is nonempty, then each of its irreducible components is also an isosingular set with respect to f. In fact, each isosingular set for f either is an irreducible component of $\mathbf{V}(f)$ or arises in such a fashion. This classification immediately implies the first key property of isosingular sets from above, i.e., there are finitely many isosingular sets for f. Additionally, it provides an algorithm for computing each of them using rank-dropping sets discussed in §13.1. However, we will not explore this algorithm any further here.

For a point $z^* \in \mathbf{V}(f)$, $z^* \in X(f,\{d_r(f,z^*)\}_{r=0}^{\infty}) =: X$ so that there is an irreducible component of X containing z^*. In fact, the irreducible component must be unique since z^* is a smooth point of this component. This is the second key property of isosingular sets presented above. The deflation sequence at z^* has stabilized at d_k when z^* becomes a smooth point of $\mathscr{D}^k(f,z^*)$. This is clear because at stabilization the nullspace dimension of the Jacobian, d_k, is equal to the dimension d_∞ of the isosingular component on which z^* lies. By testing for smoothness, we can decide when a deflation sequence has stabilized. This test is described next.

Testing for stabilization

A necessary, but not sufficient, condition for the deflation sequence of f at point z^* to stabilize is that $d_k(f,z^*) = d_{k+1}(f,z^*)$. The deflation sequences of the isosingular sets V_4, V_5, and V_7 in Example 13.10 show that repetition alone is not sufficient to conclude stabilization, as these each begin with repetitions, $d_0 = d_1 = 3$, only to decrease later on.

For isolated solutions, the bound on the number of iterations needed for the deflation sequence to stabilize arises from the bounds for weak deflation, i.e., depth and multiplicity discussed in §11.6. Stabilization bounds are unknown in general, but we can test for stabilization based on the deflatability of isosingular sets. That is, the deflation sequence at z^* has stabilized when z^* is a smooth point on an irreducible component whose dimension is equal to the nullspace dimension of the Jacobian. This is a testable condition. If this test is successful, the nullspace dimension of the Jacobian is equal to the limit of the deflation sequence, which is called the *isosingular local dimension*.

The test works as follows. As above, let $g_k(z)$ be the system associated to $\mathscr{D}^k(f,z^*)$. We begin by randomizing g_k down to the same number of polynomials as $\operatorname{rank} J g_k(z^*)$; call this system $h_k(z)$. Since $h_k(z)$ is full rank at $z^* \in \mathbb{C}^N$, we know that z^* is a smooth point on an irreducible component of $\mathbf{V}(h_k)$ of dimension $d_k(f,z) = N - \operatorname{rank} J g_k(z^*)$. So we can sample this component by appending to h_k an appropriate linear system and moving the slice in a general direction (see §8.3). We need to generate just one general sample, say, z_{samp}, sufficiently far from z^*. If $g_k(z_{\text{samp}}) = 0$, then z^* is a smooth point of $\mathbf{V}(g_k)$ and the deflation has stabilized. Otherwise, the component of $\mathbf{V}(h_k)$ on which z^* lies is an artifact of the randomization, so we must increment k and try again.

Given a polynomial system $g(z)$ and a point $z^* \in \mathbf{V}(g)$ as above, Bertini can perform this isosingular stabilization test. In the event the point is indeed a smooth point on an irreducible component, the user has the option, controlled by the configuration `ConstructWitnessSet`, to ask Bertini to use monodromy loops with the trace test (see §10.2.2 and §10.2.1, respectively), to attempt to compute a witness set for this component. In particular, the default setting, namely, 0, simply exits after performing the test, while a setting of 1 will perform the additional computations.

At present, Bertini does not provide a function implementing the deflation step, $\mathscr{D}(f,z)$. However, the user could use symbolic differentiation software to do this step external to Bertini and then call Bertini to test for stabilization.

Example 13.11. Consider the polynomial system $f(x,y,z) = (x+y-z^2+1)(y^2-xz-2)$ at the point $\alpha = (1,-1,-1)$. Since $Jf(\alpha) = 0$, we know that at least one deflation iteration is needed. That is, we want to test whether α is a smooth point on an irreducible component for

$$g(x,y,z) = \begin{bmatrix} (x+y-z^2+1)(y^2-xz-2) \\ (y^2-xz-2)-z(x+y-z^2+1) \\ (y^2-xz-2)+2y(x+y-z^2+1) \\ -2z(y^2-xz-2)-x(x+y-z^2+1) \end{bmatrix}.$$

To do this, we need to create both an input file and a start file. The input file is

```
% IsosingularStabilization.input
CONFIG
 TrackType:6;
 ConstructWitnessSet:1;
END;
INPUT
 function g1,g2,g3,g4;
 variable_group x,y,z;

 a = y^2 - x*z - 2;
 b = x + y - z^2 + 1;

 g1 = a*b;
 g2 = a - z*b;
 g3 = a + 2*y*b;
 g4 = -2*z*a - x*b;
END;
```

while the start file is

```
1

 1 0
-1 0
-1 0
```

When Bertini is executed, it first computes the dimension of the nullspace of the Jacobian. The following is printed to the screen, which shows that this dimension is 1:

```
Testing for a component of dimension 1.
```

The test is then performed and the isosingular_summary file, which lists two integers, is created. The first is the computed dimension of the nullspace, in this case 1. The second describes the result of the stabilization test, which is either 0 for failure or 1 for success.

In this case, the test is successful so the value of ConstructWitnessSet instructs Bertini to perform monodromy loops with the trace test. The following screen output shows that four monodromy loops were needed along with the residual of the traces, which becomes numerically zero after computing four points:

```
Performing monodromy loops starting with 1 point
Trace residual for 2 points: 1.184999e-01
Performing monodromy loops starting with 2 points
Performing monodromy loops starting with 2 points
Trace residual for 3 points: 3.948950e-01
Performing monodromy loops starting with 3 points
Trace residual for 4 points: 1.817566e-15
```

The following onscreen output summarizes the result of the computation:

```
Dimension 1: 1 classified component
-----------------------------------------------------
   degree 4: 1 component
```

In particular, α is a smooth point (with respect to f) on the curve which lies at the intersection of the two hypersurfaces defined by f. ∎

We note that a general approach for deflating components which arise as the intersection of irreducible components is presented in [97].

13.3 ▪ Bertini Guide: Isosingular Stabilization Test

Isosingular Stabilization Test Summary

Input files
- `input` = run configuration & system definition
- `start` = a test point

Output files
- `isosingular_summary` = summarizes the test results

Key configurations:
- `TrackType:6;`
- `ConstructWitnessSet`

Execution:
```
>> bertini <input> <start>
```

Chapter 14

Real Solutions

In most applications, an engineer or scientist is interested in real solutions. Despite this, to apply continuation we routinely cast the problem in complex space. This simplifies matters because of the strong properties that come along with algebraic closure, not the least of which is that we can slice the solution set at random and be assured, with "probability one," that we will not miss. The same cannot be said for the real points of an algebraic set. Considering, for example, the unit sphere in $\mathbb{R}^3$, we see that real sets can start and stop, and so a random real slice might miss the set entirely!

This chapter begins with discussions on how to completely characterize the real solution set of a polynomial system. The treatment is organized by dimension, but not, as one might think, by the dimension of the real sets but rather by the dimension of the complex set that contains the real set. The dimensions are the same at regular real points, but there can be a difference in dimension for singular real sets. We start the chapter with a discussion of this issue (§14.1), followed by a discussion of the isolated case (§14.2).

When seeking positive-dimensional real solutions, there is a major decision to make about what kind of information is desired. One route is to try to fully capture the entire real set in a way that allows one to refine the representation to any degree of resolution. Our approach to this is based on cell decomposition, presented generally in §14.4.1, then applied specifically to curves (§14.4.2) and surfaces (§14.4.3). Another route is to limit the objective to obtain partial information about the set of real solutions. One approach is to compute a finite set of real solutions that contains at least one point on each connected component of the real solution set (§14.5). This, for example, can be used to decide if any real solutions exist as well as to compute a finite superset of the real isolated solutions. Another approach is to compute a numerical irreducible decomposition of the Zariski closure of the real solution set, which yields a real numerical irreducible decomposition (§14.6). For this decomposition, each irreducible component contains a real set of the same dimension and the real solutions that are isolated in the Zariski topology are enumerated.

Before commencing the discussion of our algebraic geometric approach to real sets, we should mention the alternative approach taken by exclusion methods. These approaches begin with an initial region of interest, a box in $\mathbb{R}^N$. Using a divide-and-conquer procedure, the algorithms exclude some regions as being devoid of solution points and divide the remaining regions into smaller boxes to be tested individually. The process terminates with a collection of tiny boxes that still contain all real solution points, thus providing a numerical approximation to the whole solution set in the initial region. The efficiency of

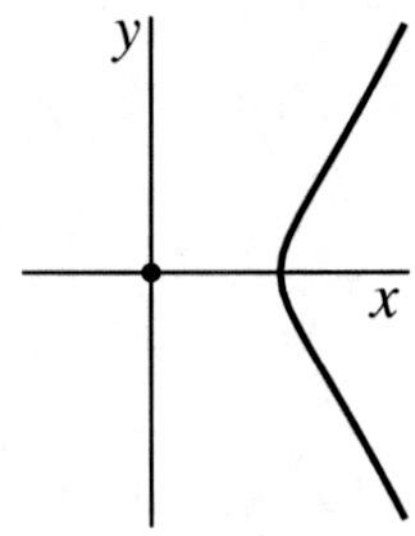

Figure 14.1. *The curve $y^2 = x^2(x-1)$, whose real points consist of a curve and an isolated point at the origin.*

such methods depends heavily on the effectiveness of the numerical tests used to exclude boxes. More effective methods are able to eliminate large empty regions early on, thereby avoiding further subdivision and testing. The size of the final boxes is governed by the resolution accuracy stipulated by the user, and the number of them grows exponentially with the dimension of the solution set. The methods for excluding boxes often depend on interval arithmetic to rigorously evaluate functions, while subdivision can rely on simple box bisection or use more sophisticated techniques such as interval Newton methods or linear relaxations. For more details, see [63, 64, 127, 128, 139, 145].

14.1 ▪ Real versus Complex Dimension

Suppose $f : \mathbb{C}^N \to \mathbb{C}^n$ is a polynomial system with real coefficients, so we may also consider f to be a map on the reals, $f : \mathbb{R}^N \to \mathbb{R}^n$. Recall that $\mathbf{V}(f)$ means the solution set of $f = 0$ in complex space:

$$\mathbf{V}(f) := \{z \in \mathbb{C}^N \mid f(z) = 0\}.$$

To talk about the real solution set of f, we introduce a subscript:

$$\mathbf{V}_{\mathbb{R}}(f) := \{z \in \mathbb{R}^N \mid f(z) = 0\} = \mathbf{V}(f) \cap \mathbb{R}^N.$$

When we speak of the dimension of $\mathbf{V}(f)$, we mean its complex dimension, whereas the dimension of $\mathbf{V}_{\mathbb{R}}(f)$ refers to real dimension. A locally n-dimensional complex set is locally biholomorphic to $\mathbb{C}^n$, whereas for a real set the local homeomorphism is to $\mathbb{R}^n$. The real dimension of $\mathbb{C}^n$ is $2n$, because each complex coordinate axis has a real and an imaginary part.

Let's begin with isolated roots. If p is an isolated solution in $\mathbf{V}(f)$ and p also happens to be real, it is clearly an isolated point of $\mathbf{V}_{\mathbb{R}}(f)$.

The situation is a bit more complicated for real points on positive-dimensional components of $\mathbf{V}(f)$. In some cases, such as $f = x^2 + y^2 - 1$, $\mathbf{V}(f)$ is a one-complex-dimensional set, and $\mathbf{V}_{\mathbb{R}}(f)$ is a one-real-dimensional set—the unit circle. But for $f = x^2 + y^2$, $\mathbf{V}(f)$ is still a one-complex-dimensional set, the pair of lines $x + iy = 0$ and $x - iy = 0$, while $\mathbf{V}_{\mathbb{R}}(f)$ is a single point, the origin, $(0,0)$, which we can think of as a circle of radius zero. Notice that the origin is singular, as $Jf = \begin{bmatrix} 2x & 2y \end{bmatrix}$ is $\begin{bmatrix} 0 & 0 \end{bmatrix}$ there. As we will explain in a moment, this is significant.

An even more surprising example is $f(x,y) = y^2 - x^2(x-1)$. This time, $\mathbf{V}(f)$ is an irreducible complex curve, while, as shown in Figure 14.1, $\mathbf{V}_{\mathbb{R}}(f)$ consists of a real curve that lies in the half-plane $x \geq 1$ and the singular isolated point at the origin. This is related to the former example in that the origin lies where the complex curve crosses itself. Indeed, for $|x| \ll 1$, we see that $f(x,y) \approx y^2 + x^2$.

These sorts of examples can be constructed in any dimension. For example, if $f(x,y,z) = x^2 + y^2 + z^2$, then $\mathbf{V}(f)$ is a surface, whereas $\mathbf{V}_{\mathbb{R}}(f)$ is just the origin. And for $f(x,y,z) = z^2 + (x-1)^2(y^2 - x^2(x-2))$, $\mathbf{V}(f)$ is an irreducible complex surface, but $\mathbf{V}_{\mathbb{R}}(f)$ has real solution components of 0, 1, and 2 dimensions:

- the origin, $(0,0,0)$;
- the line, $(1,\lambda,0)$, $\lambda \in \mathbb{R}$; and
- a surface that exists in the half-space $x \geq 2$.

The bottom line is that an n-dimensional complex algebraic set, even an irreducible one, can contain real sets of any dimension less than or equal to n.

There is one case where the relation between complex dimension and real dimension is simple. This is when $p \in \mathbb{R}^N$ is a smooth point of $f^{-1}(0)$. Recall that this means that the rank of the Jacobian at p is the same as the codimension of $\mathbf{V}(f)$ at p:

$$\text{rank} Jf(p) = N - \dim_p \mathbf{V}(f).$$

But since f has real coefficients, $Jf(p)$ has all real entries, and the real rank is the same as the complex one. This means that

$$\dim_p \mathbf{V}_{\mathbb{R}}(f) \geq N - \text{rank} Jf(p),$$

but also

$$\dim_p \mathbf{V}_{\mathbb{R}}(f) \leq \dim_p \mathbf{V}(f) = N - \text{rank} Jf(p),$$

so we must have

$$\dim_p \mathbf{V}_{\mathbb{R}}(f) = \dim_p \mathbf{V}(f).$$

So at smooth points of $f^{-1}(0)$, the complex and real dimensions are equal. Accordingly, it is no coincidence that in all the examples above, the points where the real dimension is smaller than the complex dimension are singular points. Be careful, however, not to conclude that singularity implies a difference in dimension; for $f(x,y) = (x^2 + y^2 - 1)^2$, both $\mathbf{V}(f)$ and $\mathbf{V}_{\mathbb{R}}(f)$ are one-dimensional, even though every point on both sets is singular.

14.2 ▪ Isolated Solutions

It is clear that isolated complex roots that happen to be real are also isolated in real space. Using the techniques of Part I, we can find a set of points that includes all the isolated complex solutions, and using a local dimension test, we can reliably sort out which of these are truly isolated. Then the real isolated points among these can be identified by checking the size of their imaginary parts. There is the minor practical issue of setting appropriate thresholds for declaring a root's imaginary part to be numerically zero, but this is generally not difficult to tune appropriately. The configuration setting for adjusting this is `ImagThreshold`. (See §E.7.)

The recipe above does not necessarily find all real isolated solutions, because there may be singular ones that are contained in positive-dimensional sets. One option is to use the approach presented in §14.5 to compute a finite superset of the real isolated solutions. However, the situation is simpler if the objective is limited to finding just all nonsingular isolated real roots, since these are also nonsingular isolated roots in complex space.

Bertini attempts to sort the nonsingular roots from a superset of the isolated roots using the threshold `CondNumThreshold`. (Also in §E.7.)

If one wishes for more certainty than the threshold heuristics provide, it is possible, if enough precision is used, to certify that a nonsingular solution is truly real and not a complex solution with a tiny imaginary part. That is our next subject.

14.3 ▪ Certification of Nonsingular Isolated Real Roots

Although in practical work it usually suffices to know that a solution has a very small imaginary part, there are instances when one might want more certainty in declaring that a real root exists. That is, when a numerical approximation to an apparently *nonsingular* isolated real root has been found, using Bertini or any other method, one might like to certify it. Certification in this context means the establishment of a computational proof that there is a real root nearby and that Newton's method will quadratically converge to it from the given approximation. This capability is provided by a separate module, called `alphaCertified` [95, 96].

For simplicity, we assume that our polynomial system f is square, i.e., $N = n$, with real coefficients. In this case, Smale's α-theory (see [36, Chapter 8] for a general overview) provides conditions to certify that Newton's method starting at a given point $z \in \mathbb{C}^N$ will quadratically converge to a point in $\mathbf{V}(f)$. Such a point z is called an *approximate solution* of $f(x) = 0$ and the corresponding limit point $\xi \in \mathbf{V}(f)$ is called the *associated solution* of z.

Since f has real coefficients, Newton's method maps real points to real points. With this property, α-theory can be used to certify the reality or nonreality of isolated nonsingular points in $f^{-1}(0)$ as shown in [96]. Clearly, if $z \in \mathbb{R}^N$ is an approximate solution with associated solution $\xi \in \mathbf{V}(f)$, then we must have $\xi \in \mathbf{V}_{\mathbb{R}}(f)$. Therefore, the reality of an associated solution can be certified by finding an approximate solution that is real. For this case in `alphaCertified`, this is obtained by using α-theory to certify that both z and its real part $(z + \mathrm{conj}(z))/2 \in \mathbb{R}^N$ have the same associated solution. Here, conj() means complex conjugation, that is, for real vectors a and b, $\mathrm{conj}(a+ib) = a-ib$ so that $(a+ib)+\mathrm{conj}(a+ib) = 2a$.

The reality assumption on f provides that if $z \in \mathbb{C}^N$ is an approximate solution with associated solution $\xi \in \mathbf{V}(f)$, then $\mathrm{conj}(z)$ is also an approximate solution with associated solution $\mathrm{conj}(\xi) \in \mathbf{V}(f)$. Therefore, the nonreality of ξ can be certified by showing that ξ and $\mathrm{conj}(\xi)$ are distinct points. For this case in `alphaCertified`, this is obtained using α-theory to compute upper bounds on

$$||z-\xi|| = ||\mathrm{conj}(z) - \mathrm{conj}(\xi)||$$

together with $||z - \mathrm{conj}(z)||$ and the triangle inequality.

In `alphaCertified`, the computations can be performed using either exact rational arithmetic or arbitrary precision floating-point arithmetic. In either case, the coefficients of f must be rational, whereas the presentation of the points depends on the choice of arithmetic. In the floating-point arithmetic case, some of the output files generated by Bertini including `nonsingular_solutions` and `finite_solutions` can be directly used by `alphaCertified`. In the rational arithmetic case, one must first convert each floating-point number into a rational number before calling `alphaCertified`. Nonetheless, all computations are *certifiable* in the rational arithmetic case. Since the bit length of rational numbers can substantially grow in the internal computations, the user has the option to use floating-point arithmetic in which the roundoff errors for the internal computations are not fully controlled but can be reduced by increasing the precision used in the computations.

A typical use is to combine parameter homotopies in Bertini (discussed in Chapter 6) with certification routines in `alphaCertified` to study the number of real roots of a parameterized polynomial system, e.g., [69, 96]. For example, suppose that $f(x;p) : \mathbb{C}^N \times \mathbb{C}^P \to \mathbb{C}^N$ is a polynomial system such that $f(x;p^*) = 0$ has exactly k solutions, all of which are nonsingular, for general $p^* \in \mathbb{C}^P$. If Bertini computes k points that `alphaCertified` has certified are all approximate solutions that correspond to distinct associated solutions, then one obtains that the number of real roots is the number of real associated solutions certified by `alphaCertified`.

14.4 ▪ Cell Decomposition of the Real Solution Set

A cell decomposition is a way of capturing an entire real algebraic set, much as a witness set allows us to represent a complex algebraic set. We have already noted that even if $A \subset \mathbb{C}^N$ is a pure-n-dimensional algebraic set, its real points, $A_{\mathbb{R}} := A \cap \mathbb{R}^N$, might break up into real components at several dimensions, although always with dimension no greater than n. Reflecting this, the cell decomposition of $A_{\mathbb{R}}$ may contain cells of various dimensions. Moreover, as we will explicate shortly, each k-dimensional cell, $k > 0$, is described in part by its boundary, which consists of $(k-1)$-dimensional cells. We begin our discussion with a general description of cell decompositions, and then proceed to more precise descriptions of how to construct cell decompositions for real algebraic curves and surfaces.

14.4.1 ▪ Cell Decomposition Definition

The objective of a cell decomposition is to break up a set into nonoverlapping pieces, called *cells*, that cover the whole set. Let's assume the set is pure N-dimensional. Then, the interior of each cell is also N-dimensional with a boundary that is $(N-1)$-dimensional. An N-dimensional cell is called simply an "N-cell," while 2-cells are often called "faces"; 1-cells, "edges"; and 0-cells, "vertices." If two cells touch, they do so only at their boundaries: no interior point is a member of more than one cell. Moreover, we require that the boundaries also have a cell decomposition, this property holding recursively all the way down until edges terminate in boundaries that are vertices.

Except for finite zero-dimensional sets, we cannot, of course, list all the points in a cell. Instead, for each N-cell, we list its boundary cells and give a method for constructing points in the interior to as fine a resolution as desired.

It may help to consider the case of polygons and polyhedra. A polygon can be given quite simply as a list of vertices, with their coordinates listed, and a list of edges, each of which is given merely by listing its two endpoints. For example, a triangle in $\mathbb{R}^3$ might be given as (the vertex data chosen here is arbitrary)

$$\begin{aligned} \text{Vertices:} &\quad v_1 = (1.1, 0, 1.3);\ v_2 = (5.2, 3.8, 2);\ v_3 = (2.2, 4, 5.6). \\ \text{Edges:} &\quad e_1 : 1,2;\ e_2 : 2,3;\ e_3 : 3,1. \end{aligned}$$

Notice that the vertex list contains numerical data, while the edge list is strictly symbolic, consisting of indices of the vertices ($e_1 : 1,2$ means edge 1 goes from v_1 to v_2, etc.). This suffices to describe the triangle, because it is clear how to reconstruct points on the line segment between two vertices:

$$e_1 = \{t v_1 + (1-t) v_2,\ t \in [0,1]\}, \text{ etc.} \tag{14.1}$$

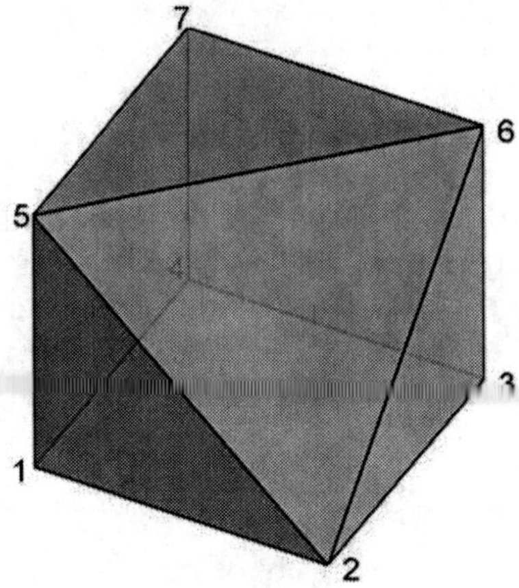

Figure 14.2. *A polyhedron with seven faces.*

It is even more instructive to consider a polyhedron, so let's list a cell decomposition of a cube with one corner lopped off, shown in Figure 14.2:

Vertices: $v_1 = (0,0,0)$; $v_2 = (1,0,0)$; $v_3 = (1,1,0)$; $v_4 = (0,1,0)$;
$v_5 = (0,0,1)$; $v_6 = (1,1,1)$; $v_7 = (0,1,1)$.
Edges: $e_1 : 1,2$; $e_2 : 1,4$; $e_3 : 1,5$; $e_4 : 2,3$; $e_5 : 2,5$; $e_6 : 2,6$; $e_7 : 3,4$;
$e_8 : 3,6$; $e_9 : 4,7$; $e_{10} : 5,6$; $e_{11} : 5,7$; $e_{12} : 6,7$.
Faces: $f_1 : 1,5,3$; $f_2 : 1,4,7,2$; $f_3 : 4,8,6$; $f_4 : 7,9,12,8$; $f_5 : 2,3,11,9$;
$f_6 : 6,10,5$; $f_7 : 10,12,11$.
3-Cells: $\text{cell}_1 : 1,2,3,4,5,6,7$.

Since the polyhedron in Figure 14.2 is convex, the list of vertices alone would suffice to describe it, with the 3-cell being the convex combination of the vertices. But the cells we will construct to represent real algebraic sets will not in general be convex, so we will need a more detailed enumeration like that above.

In fact, even the detailed listing as above is not quite enough for algebraic sets, because without convexity, more than one cell might share the same boundary. Consider, for example, the spherical surface illustrated in Figure 14.3, on which we have marked two vertices, v_1 and v_2, two edges (semicircles on the equator), and two faces (the north and south hemispheres). The two faces have the same boundary, so we need to differentiate between them somehow. We can do so by noting that one face contains the North Pole, b_1, and the other the South Pole, b_2. Intermediate points a_1 and a_2 can be inserted along the edges to differentiate them as well, since they also have the same boundary. So after listing the coordinates of v_1 and v_2, the edge and face data are

Edges: $e_1 : 1,2$ via a_1; $e_2 : 1,2$ via a_2;
Faces: $f_1 : 1,2$ via b_1; $f_2 : 1,2$ via b_2.

Of course, somewhere the data structure must store the coordinates of not only v_1 and v_2 but also those of a_1, a_2, b_1, b_2.

This now comes very close to the representation of cells that we will use in cell decomposition of real algebraic sets. What is missing is the rule for how to sweep out points

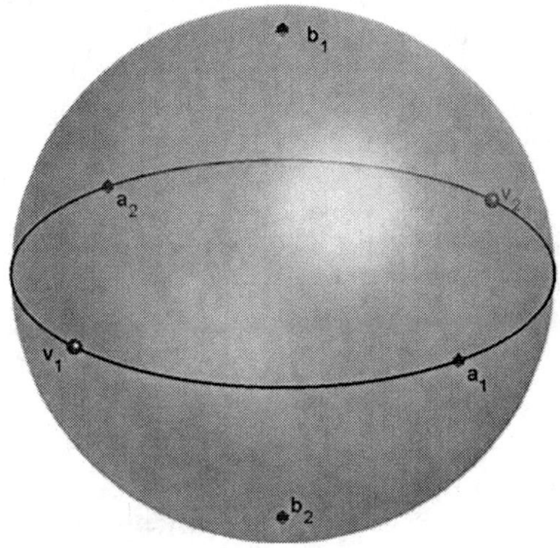

Figure 14.3. *Cell decomposition of a sphere.*

on each cell, the equivalent of (14.1). We will do so by specifying an implicit mapping for each positive-dimensional cell. The mapping extends out to the entire cell by continuation from an initial point given in the interior of the cell. Just as (14.1) defines an edge as a function of a local coordinate t, these mappings implicitly define an N-cell in terms of N local coordinates.

We have not yet specified a method for constructing cell decompositions. Let us do so next for curves and then for surfaces.

14.4.2 ▪ Real Points of a Complex Curve

We begin our treatment of complex curves with an observation about complex conjugates. Suppose A is a one-dimensional irreducible component of $\mathbf{V}(f)$, where $f:\mathbb{C}^N \to \mathbb{C}^n$ has all real coefficients. Then, for any point $p \in \mathbf{V}(f)$, we have $f(\mathrm{conj}(p)) = \mathrm{conj}(f(p)) = \mathrm{conj}(0) = 0$, so $\mathrm{conj}(p) \in \mathbf{V}(f)$, where conj() is complex conjugation. This implies that conj(A) is also an irreducible component of $\mathbf{V}(f)$. Accordingly, one of two conditions must hold:

- either A is self-conjugate, meaning $\mathrm{conj}(A) = A$; or
- there is a conjugate pair of irreducible components, A and $\mathrm{conj}(A) \neq A$.

If we have a witness set for A, we can easily distinguish between the two cases. All we need to do is choose one of the witness points, say, w, for A and test whether $\mathrm{conj}(w) \in A$ using a homotopy membership test from §8.4. If $\mathrm{conj}(w) \in A$, then A is self-conjugate; otherwise A and conj(A) are distinct. In many cases, the witness set for A has been obtained via a numerical irreducible decomposition, so if $\mathrm{conj}(w) \notin A$, further membership tests can be conducted for the other components of the same dimension as A until conj(A) is found. If the witness set $\{f_A, L_A, W_A\}$ for A was obtained by some means that does not also include one for conj(A), this is not a problem, because $\{f_A, L_A^*, \mathrm{conj}(W_A)\}$ suffices, where L_A^* is the linear system obtained from L_A by conjugating each coefficient:

$$c_0 + c_1 z_1 + \cdots + c_N z_N \longrightarrow \mathrm{conj}(c_0) + \mathrm{conj}(c_1) z_1 + \cdots + \mathrm{conj}(c_N) z_N.$$

In the case of a conjugate pair of irreducible components, $p \in A$ is real if and only if $p \in \text{conj}(A)$, so the set of real points of A is contained in $A \cap \text{conj}(A)$. Since neither set contains the other, $A \cap \text{conj}(A)$ must be smaller than one dimension and therefore it either is empty or is a finite set of points. We can compute $A \cap \text{conj}(A)$ using the intersection methods of Chapter 12 and extract the real points.

Example 14.1. For the polynomial $f(x,y) = (x^4-1)^2 + (xy+1)^2$, $\mathbf{V}(f)$ decomposes into a conjugate pair of irreducible components $A := \mathbf{V}((x^4-1) + i(xy+1))$ and $\text{conj}(A) = \mathbf{V}((x^4-1) - i(xy+1))$. Then,

$$A \cap \text{conj}(A) = \mathbf{V}(x^4-1, xy+1) = \{(1,-1), (-1,1), (i,i), (-i,-i)\}$$

so that $\mathbf{V}_{\mathbb{R}}(f) = \{(1,-1), (-1,1)\}$. ∎

The self-conjugate case takes more work, since $A \cap \mathbb{R}^N$ may have both zero- and one-dimensional pieces. The crux of the matter is how to represent the edges with a minimal data structure that allows reconstruction of the edge to any resolution. A natural approach in numerical algebraic geometry is to describe the edge using a homotopy function and a startpoint. Then, tracing the edge is just a matter of path tracking.

We construct homotopies for edges via a projection $\pi : \mathbb{C}^N \to \mathbb{C}$ that has real coefficients so that $\pi : \mathbb{R}^N \to \mathbb{R}$ as well. This is just a single linear function,

$$\pi(z) = a \cdot z, \qquad a = \begin{bmatrix} a_1 & \cdots & a_N \end{bmatrix} \in \mathbb{R}^{1\times N},$$

where we treat z as a column vector of coordinates. Now, $\mathbf{V}(\pi(z) - t)$, for some constant t, is just a hyperplane, and for generic $a \in \mathbb{R}^{1\times N}$ and $t \in \mathbb{R}$, $A \cap \mathbf{V}(\pi(z) - t)$ will consist of isolated points. As we move t along the real line keeping a constant, we sweep out a family of parallel hyperplanes, some hitting $A_{\mathbb{R}}$, some not. (All not if $A_{\mathbb{R}} = \emptyset$.)

Let us assume that we have a value of t so that the hyperplane cuts $A_{\mathbb{R}}$ transversely at p, so p is an isolated real point of $A \cap \mathbf{V}(\pi(z) - t)$. If A is a multiplicity one component of $f_A^{-1}(0)$, then p is a nonsingular root of

$$h(z,t) = \{f_A(z), \pi(z) - t\} = 0 \tag{14.2}$$

at $t = \pi(p)$. Then, we can clearly sweep out more points of $A_{\mathbb{R}}$ by following the homotopy starting from $(z,t) = (p, \pi(p))$ and continuing t in some open interval around $\pi(p)$, i.e., $t \in (t_0, t_1)$ with $t_0 < \pi(p) < t_1$.

What limits how far this homotopy can proceed in t? The path tracker proceeds without a hitch until

$$\text{rank}\left[\frac{\partial h}{\partial z}(z,t)\right] < N.$$

We may assume without loss of generality that $f_A : \mathbb{C}^N \to \mathbb{C}^{N-1}$, because if it consists of more than $(N-1)$ polynomials, we know (see §9.2) that we can replace them with $(N-1)$ random combinations and still preserve A as an irreducible component. Then,

$$M(z) := \frac{\partial h}{\partial z}(z,t) = \begin{bmatrix} \frac{\partial f_A}{\partial z}(z) \\ a \end{bmatrix} \tag{14.3}$$

is a square matrix, and the homotopy becomes singular at any $z \in A$ such that $\det M(z) = 0$. In other words, the singularities are $A \cap \mathbf{V}(\det M(z))$. But using the regeneration extension method of §12.2, we know how to extend a witness set for A into one for $A \cap \mathbf{V}(\det M(z))$.

Of course, the limits of the real edges that we trace with the homotopy are real points, so we solve for all isolated points of $A \cap \mathbf{V}(\det M(z))$ and sort out the real ones.

Notice that the singularity condition $\det M(z) = 0$ does not depend on the point p in any way, so the computation of $A \cap \mathbf{V}(\det M(z))$ suffices to find all the endpoints of all edges in the cell decomposition. Let A_{crit} be the set of real points in this set, i.e., $A_{\text{crit}} := [A \cap \mathbf{V}(\det M(z))] \cap \mathbb{R}^N$. We call A_{crit} the *critical points* of A.

Although A_{crit} tells us where all the edges start and end, we have not yet specified a computation to find an interior point on each edge. But this is simple; all we need to do is slice A with hyperplanes between the critical points or, more precisely, at values of t interspersed between $\pi(A_{\text{crit}})$. As it is also possible for $A_{\mathbb{R}}$ to go off to plus or minus infinity, we also need to slice at one value of t less than $\min(\pi(A_{\text{crit}}))$ and one greater than $\max(\pi(A_{\text{crit}}))$. (If $A_{\text{crit}} = \emptyset$, a single slice at $t = 0$ suffices.) Note that each such slice is $A \cap \mathbf{V}(\pi(z) - t)$ for some given t, and this amounts to just sampling A by moving the linear polynomial $L_A(z)$ to $L_t(z) := \pi(z) - t$ using the homotopy

$$H_t(z,u) = \begin{bmatrix} f_A(z) \\ uL_A(z) + (1-u)(\pi(z) - t) \end{bmatrix} = 0 \tag{14.4}$$

with startpoints $W_A = A \cap \mathbf{V}(L_A)$ at $u = 1$ and tracking to $u = 0$. (Note that t is constant and u is the homotopy parameter.)

One final note: If $A_{\mathbb{R}}$ contains some isolated points, these points must be singular, as we discussed in §14.1. Accordingly, they also appear in A_{crit}. After the edges have been traced out in the procedure outlined below, the isolated points of $A_{\mathbb{R}}$ are those in A_{crit} that are not the endpoints of any edge.

To summarize, the cell decomposition of curve A consists of

- a finite list of vertices, $V \subset A_{\mathbb{R}}$, and
- a finite list of edges, each given by
  - the projection, $\pi(z)$;
  - an interior point of the edge, $p \in A_{\mathbb{R}}$;
  - indices (ℓ, r) such that $v_\ell, v_r \in V$ are the endpoints of the edge, with $\pi(v_\ell) < \pi(v_r)$ (these are the "left" and "right" endpoints, respectively);
  - the edge homotopy (14.2).

The isolated points of $A_{\mathbb{R}}$ are those in V that are not endpoints of any edge. An edge can be traced out by initializing the homotopy at $(z,t) = (p, \pi(p))$ and tracking t left (negatively) to $\pi(v_\ell)$ and right (positively) to $\pi(v_r)$.

The decomposition is constructed by the following procedure:

1. Compute A_{crit} using regeneration extension. Initialize $V = A_{\text{crit}}$.
2. Let $T \subset \mathbb{R}$ be $\#(A_{\text{crit}}) + 1$ values that bracket and intersperse between $\pi(A_{\text{crit}})$.
3. Sample $A_{\mathbb{R}}$ by computing $A \cap \mathbf{V}(\pi(z) - t_i)$ for each $t_i \in T$ using (14.4). The real points in these slices are the set of interior points, say P.
4. For each $p \in P$, do the following:
   (a) Track the edge homotopy (14.2) from $(z,t) = (p, \pi(p))$ until t reaches nearest values in $\pi(A_{\text{crit}})$, left and right. Use a singular endgame to find the endpoints.

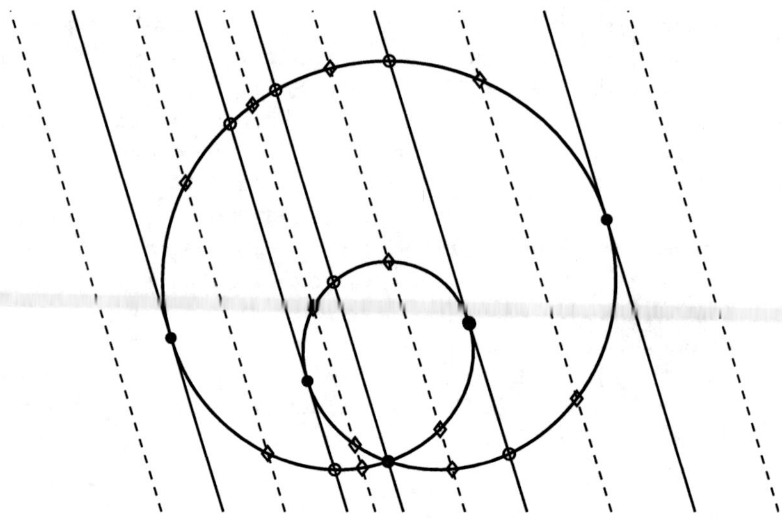

Figure 14.4. *Cell decomposition of a limaçon curve.*

(b) If either endpoint is not in V, append it to V.

(c) Record the edge by its interior point p and indices to its endpoints in V.

Note that if the slices at the bracketing values of $t_i \in T$—the ones for $t < \min(\pi(A_{\mathrm{crit}}))$ and $t > \max(\pi(A_{\mathrm{crit}}))$—give real points in P, then these edges continue to infinity. One can simply note that these are open edges, or one can cast the curve into projective space and track these edges to infinity using a projective transformation. More on the projective option can be found in [34].

Example 14.2 (Limaçon). Figure 14.4 illustrates the decomposition procedure applied to the limaçon defined by

$$(x^2+y^2-3y)^2-(x^2+y^2)=0.$$

The projection used here is given by $a=[1 \quad 0.3]$. The solid dots mark the critical points, four of which are at tangents of the family of sweeping lines, while one is a self-crossing of the curve. Dashed lines represent the slices at values bracketing and midway between the critical values. These give the interior points, P, of the edges, marked with diamond symbols. These slices sweep left and right to trace out the edges and reveal the endpoints of the edges. Some endpoints are critical points, while new vertices added to V at step 4(b) are marked with small circles. This cell decomposition has a total of 11 vertices and 12 edges. ∎

Some details*

The projection coefficients, $a \in \mathbb{R}^{1\times N}$, should be chosen at random. This avoids, with probability one, the possibility that the whole curve $A_{\mathbb{R}}$ lies in a hyperplane, $\pi(z)=c$, for some $c \in \mathbb{R}$. Equivalently, one could rotate A randomly and then project onto the first coordinate.

We assume in the above that A is multiplicity one, i.e., generically reduced. Otherwise, every point $z \in A$ will satisfy $\det M(z)=0$, and so if no compensatory measures are taken, the critical point calculation does not return the desired result. But if A is not multiplicity one, then it can be made so with a deflation operation (see §11.6, §13.2). This produces a new polynomial system to replace f_A such that generic points of A are smooth. A finite number of points of A may still be singular. This does not matter, as these points will simply be added to A_{crit}, and the procedure completes.

If step 4(b) above appends a new point to V, this means the edge is not singular at its endpoint and so $A_{\mathbb{R}}$ continues nonsingularly onto another edge. If desired, such edges can be merged into one, reducing the size of the final data structure. In the limaçon example, this would reduce the number of edges from 12 to six.

The limaçon example is a compact curve, as it does not have branches diverging to infinity. There are several alternatives for treating noncompact curves:

- mark diverging branches as going to infinity (there will be no more critical points beyond those already found);
- clip the curve at a finite boundary and trace branches to that boundary; or
- compactify the curve using projective coordinates, so branches can be traced all the way to infinity (see Chapter 4).

The essentials of the algorithm for cell decomposition of the real points of a complex curve first appear in [124] with a restatement closer to the description above given in [34].

14.4.3 ▪ Real Points of a Complex Surface

Cell decomposition of the real points in a complex surface follows a similar pattern as the curve case, but some new wrinkles come up. The main differences are the following:

- the implicit definition of the faces involves two local coordinates;
- in addition to left and right boundaries, faces have top and bottom ones; and
- because the boundary consists of curved edges instead of just points, the implicit mapping from local coordinates to the face is more complicated.

These differences notwithstanding, the same basic idea underlies the method; we project the surface onto two coordinates and establish where the surface reverses itself with respect to the projection. The set of points where this happens forms a "critical curve" that cells are not allowed to cross.

For starters, let's revisit the sphere we first considered in Figure 14.3, but now we project it down to a horizontal plane, as shown in Figure 14.5. We see that the shadow of the sphere is a circular disk, and the equator of the sphere maps to the outer boundary. The equator is the critical curve for this projection, and it forms the boundary of the faces. Let the surface coordinates be $x = (x_1, x_2, x_3)$ and let the projection coordinates be $(u, v) = (\pi_1(x), \pi_2(x))$, where in this illustration $\pi_i : x \mapsto x_i$, $i = 1, 2$. Solving for the criticality conditions yields a witness set for the equator. With this in hand, the method of the previous section gives a cell decomposition of the equator with respect to π_1 in terms of vertices v_1, v_2 and edge interior points a_1, a_2. Slicing the sphere halfway between the vertices, that is, slicing with the plane $\pi_1(x) = 0.5[\pi_1(v_1) + \pi_1(v_2)]$, gives a witness set for the dashed great circle. Cell decomposition of this set gives edges with endpoints a_1, a_2 and interior points b_1, b_2, but these edges do not form boundaries for the faces. Instead, these edges reveal how b_1, b_2 connect to the edges of the equator allowing the completion of the face descriptions. We assume in what follows that in the cell decomposition of the

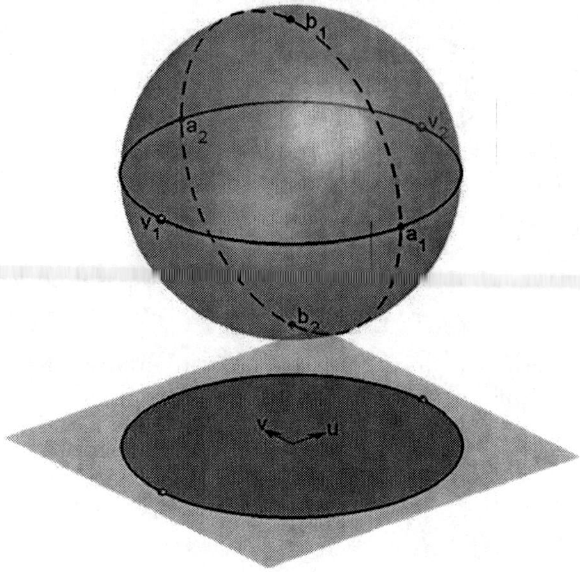

Figure 14.5. *Cell decomposition of a sphere with projection onto a plane.*

great circle slice, the interior points, b_1, b_2, of the edges are obtained by cutting the curve midway between its critical points, so that b_1, b_2 are the solutions to

$$\begin{bmatrix} f(x) \\ \pi_1(x) - 0.5[\pi_1(v_1) + \pi_1(v_2)] \\ \pi_2(x) - 0.5[\pi_2(a_1) + \pi_2(a_2)] \end{bmatrix} = 0.$$

Missing from this description is the implicit mapping from (u,v) to the face. Let's assume the equation of the sphere is $f(x) = x_1^2 + x_2^2 + x_3^2 - r^2$. We may let u run from $\pi_1(v_1)$ to $\pi_1(v_2)$, i.e., $u \in [-r, r]$, and then let the limits on v vary with u. Because the circle is so simple, we could write the limits on v as $\pm\sqrt{r^2 - u^2}$. But for more general surfaces where the critical curve is more complicated, we will not have a closed-form solution of this type. A method that works in general introduces tracer points on the critical curve that follow u. Let p_1 be the tracer point on edge 1 (marked by point a_1) and p_2 on edge 2. The criticality condition is, in general,

$$g(x) := \det\left(J \begin{bmatrix} f(x) \\ \pi_1(x) \\ \pi_2(x) \end{bmatrix} \right) = 0, \tag{14.5}$$

which for the sphere works out to $g(x) = 2x_3$. So as u varies, the tracer points are given by the homotopy

$$h(p_1, p_2, u) = \begin{bmatrix} f(p_1) \\ g(p_1) \\ \pi_1(p_1) - u \\ f(p_2) \\ g(p_2) \\ \pi_1(p_2) - u \end{bmatrix} \tag{14.6}$$

with startpoint $(p_1, p_2, u) = (a_1, a_2, \pi_1(a_1))$. Now, it is clear that v can range from $\pi_2(p_1)$ to $\pi_2(p_2)$. So introducing new local variables $(\hat{u}, \hat{v}) \in [0,1] \times [0,1]$ and functions $u(\hat{u})$ and $v(p_1, p_2, \hat{v})$, the full implicit mapping for either face is

$$H(x, p_1, p_2, \hat{u}, \hat{v}) = \begin{bmatrix} h(p_1, p_2, u(\hat{u})) \\ f(x) \\ \pi_1(x) - u(\hat{u}) \\ \pi_2(x) - v(p_1, p_2, \hat{v}) \end{bmatrix}, \tag{14.7}$$

where

$$u(\hat{u}) = (1 - \hat{u})\pi_1(v_1) + \hat{u}\pi_1(v_2), \tag{14.8}$$

$$v(p_1, p_2, \hat{v}) = (1 - \hat{v})\pi_2(p_1) + \hat{v}\pi_2(p_2). \tag{14.9}$$

To find the value of x corresponding to $(\hat{u}, \hat{v}) = (\alpha, \beta) \in [0,1]^2$ on the upper face, start at $(x, p_1, p_2, \hat{u}, \hat{v}) = (b_1, a_1, a_2, 0.5, 0.5)$ and follow any continuous path in $[0,1]^2$ from $(0.5, 0.5)$ to (α, β). The same works on the lower face, except starting at $x = b_2$.

This is almost as general as we need for any surface, except that the left and right boundaries of the faces on the sphere are points, v_1 and v_2, whereas for a more general surface, these may become curves—in fact, compound edges. For illustration, consider a torus:

$$f(x) = (x_1^2 + x_2^2 + x_3^2 + 3)^2 - 16(x_1^2 + x_2^2) = 0.$$

Figure 14.6 illustrates the cell decomposition of this torus with a projection whose kernel (the vector, say, $\mathbf{k}$ such that $\pi_1(\mathbf{k}) = \pi_2(\mathbf{k}) = 0$) is 20° from the z-axis. (The large arrow indicates the direction of the kernel.) Vertices are marked with small spheres, edge interior points with diamonds, and face interior points with cubes. Among the vertices, the ones that are critical points of the critical curve are marked in red; there are four of them in this projection. One view shows just the critical curve, another shows all vertices and edges, and two additionally show grid lines of constant $\hat{u}$ and $\hat{v}$ on the faces. One of these views is directed along the kernel of the projection, from which perspective the lines of constant $\hat{u}$ appear as straight lines.

The conformation of the cell decomposition can change depending on the projection used. Figure 14.7 shows another cell decomposition obtained when the projection direction is more canted, at 43° from the vertical. With this projection, there are eight critical points, six of which are on the inner hole of the torus. This results in 24 faces, some of which are so narrow that they are obscured in the drawing by the thickness of the edges that bound them.

To summarize, the cell decomposition of the real points of a complex surface follows these steps. Let $S \subset \mathbf{V}(f) \subset \mathbb{C}^N$ be the complex surface.

- Find a witness set for the critical curve, $S \cap \mathbf{V}(g)$, where g is as in (14.5) using the regeneration extension method of §12.2.
- Compute a cell decomposition for the real points of the critical curve, $\mathcal{K} := S \cap \mathbf{V}(g) \cap \mathbb{R}^N$, using the method of the previous section. The critical points of $\mathcal{K}$ are the critical points of S. The vertices and edges of $\mathcal{K}$ are also vertices and edges of S.
- Slice S at each critical point and compute a cell decomposition of the real points of the slice, merging edges that meet at regular points. The edges are added to the edges of S. (The vertices are also critical points and so have been already entered into the data structure at the previous step.)

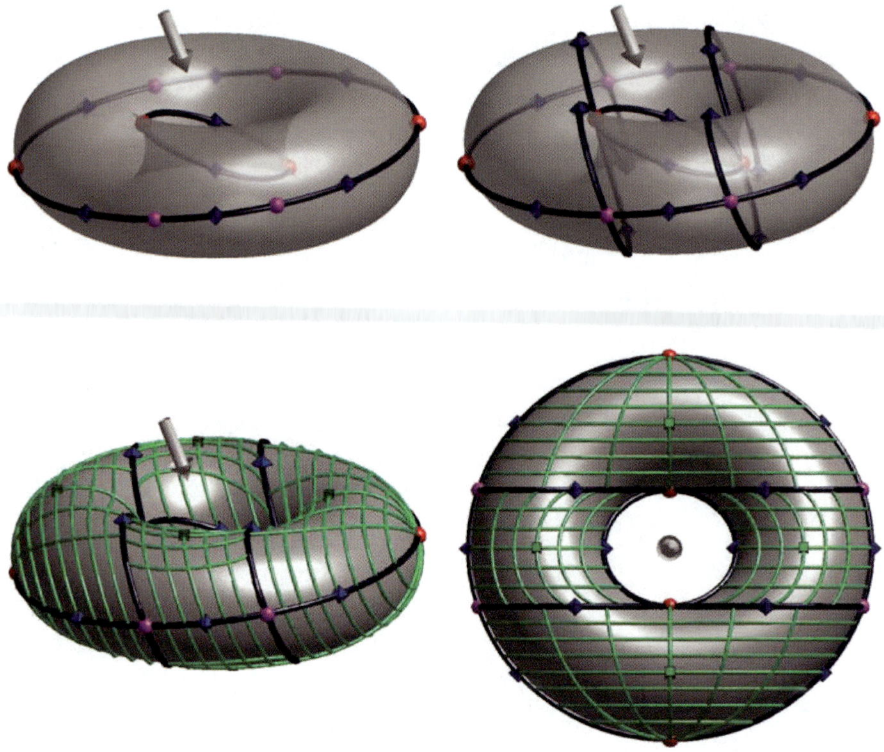

Figure 14.6. *Cell decomposition of a torus. Projection from high angle.*

- Slice S at the interior points of the edges of $\mathscr{K}$ and compute a cell decomposition of the real points of each slice, merging edges that meet at regular points. The interior points of these edges are the interior points of the faces, and the endpoints of the edges reveal the "up/down" edges of the faces, which are edges of $\mathscr{K}$.
- Track left/right from the interior point of each face to find its left and right edges. This is done in $(\hat{u},\hat{v})$-space in the interior of the square, $(\hat{u},\hat{v}) \in (0,1)^2$, using (14.7).
- Any real curve in S is composed of the edges of $\mathscr{K}$ that are not the boundary of any face.
- The isolated real points in S, if any, are the vertices of $\mathscr{K}$ (i.e., critical points) that are not adjacent to any edge.

The last two steps take care of cases where S, even though it is two-complex-dimensional, includes some lower-dimensional real pieces, as discussed in §14.1.

Some details*

The surface cell decomposition is subject to the same considerations as the curve case. The description above assumes the surface is multiplicity one; if it is not, it must be deflated to make it so. Moreover, even if S is multiplicity one, the critical curve on the surface might not be multiplicity one as a solution of (14.5). If that happens, it must also be deflated before it can be decomposed, and the deflated system must also be used in (14.6) to trace

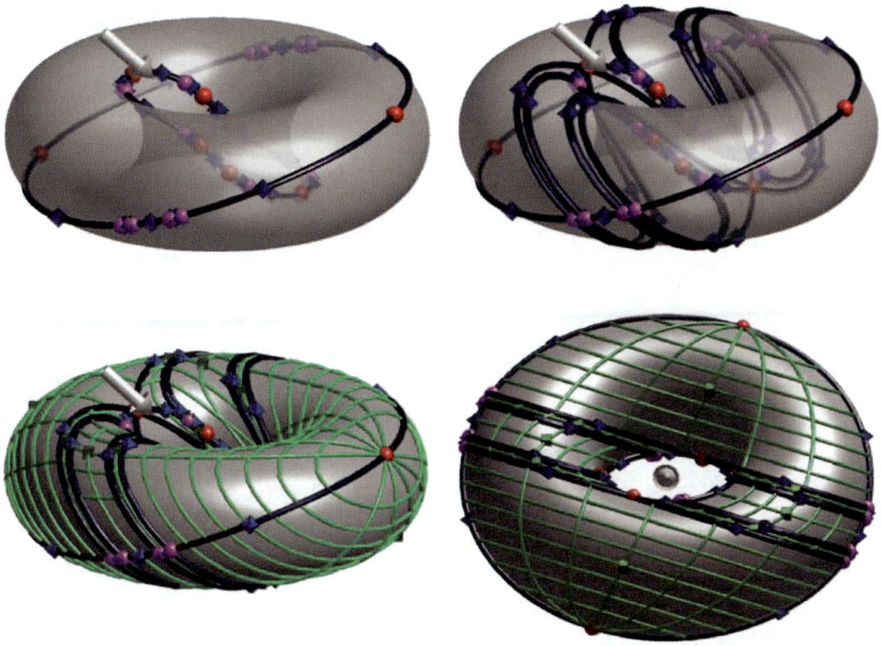

Figure 14.7. *Cell decomposition of a torus. Projection from low angle.*

points on the critical curve. The critical curve might also have more than one irreducible component, and these might have different multiplicities.

As in the curve case, surfaces can also extend to infinity. One has the same basic choices: mark open faces, clip them to a finite boundary, or compactify using projective coordinates. One method of compactifying is described in [34], which details the full surface algorithm in the "almost smooth" case, meaning a multiplicity one surface that contains no singular curve. (The isosingular deflation theory of §13.2 that is necessary for removing the almost smooth condition came later.)

14.5 ▪ A Point on Each Connected Component

For any real polynomial system f, the real solution set $\mathbf{V}_{\mathbb{R}}(f) \subset \mathbb{R}^N$ can be decomposed into finitely many path connected components $C_1, \ldots, C_k$ such that both C_i and $\mathbf{V}_{\mathbb{R}}(f) \setminus C_i$ are closed in the usual topology on $\mathbb{R}^N$. For example, the curve $y^2 = x^2(x-1)$ presented in Figure 14.1 has two connected components, namely, $C_1 = \{(0,0)\}$ and $C_2 = \{(x,y) \mid y^2 = x^2(x-1), x \geq 1\}$. In this section, we present a homotopy-based method for computing a finite set of points $S \subset \mathbf{V}_{\mathbb{R}}(f)$ such that $S \cap C_i \neq \emptyset$ for each $i = 1, \ldots, k$ following [87]. Clearly, $S = \emptyset$ if and only if $\mathbf{V}_{\mathbb{R}}(f) = \emptyset$. Moreover, since each real isolated point forms a connected component, the set S is a finite superset of the real isolated points.

We shall assume that $N > d > 0$ and $f : \mathbb{C}^N \to \mathbb{C}^{N-d}$ is a polynomial system with real coefficients such that $\mathbf{V}(f)$ is a pure d-dimensional algebraic set. Given $y \in \mathbb{R}^N$, following a classical approach of Seidenberg [152] that was further developed in [11, 150], we consider the points of $\mathbf{V}(f)$ which are critical points of the distance function

$$d_y(x) = (x-y)^T(x-y) = \sum_{i=1}^{N}(x_i - y_i)^2.$$

These are the points $x \in \mathbf{V}(f)$ that satisfy the following condition:

$$\operatorname{rank}\left[\begin{array}{cccc} x-y & \nabla f_1(x)^T & \cdots & \nabla f_{N-d}(x)^T \end{array}\right] \le N-d. \tag{14.10}$$

Since the matrix in (14.10) is $N \times (N-d+1)$, it has rank at most $N-d$ if and only if it has a nonzero null vector. That is, (14.10) holds at x if and only there exists $\lambda \in \mathbb{P}^{N-d}$ such that

$$\left[\begin{array}{cccc} x-y & \nabla f_1(x)^T & \cdots & \nabla f_{N-d}(x)^T \end{array}\right] \lambda = 0. \tag{14.11}$$

In fact, each singular point of $f^{-1}(0)$ is a critical point since $\operatorname{rank} Jf(x) < N-d$ implies (14.10).

With the aim of computing a finite set of points in $\mathbf{V}_{\mathbb{R}}(f)$, we shall consider ways in which the set of critical points is positive-dimensional. First, if the set of singular points is positive-dimensional, then the set of critical points will be as well. Another way is for y to be in a special position with respect to $\mathbf{V}(f)$, e.g., $f(x) = x_1^2 + x_2^2 - 1$ with $y = (0,0)$. Therefore, we shall select y at random and consider a family of deformations of $\mathbf{V}(f)$ limiting to $\mathbf{V}(f)$. Since we will only need to consider finitely many paths, we will dehomogenize $\mathbb{P}^{N-d}$, the space containing the null vectors, using a random patch as discussed in Chapter 4.

To summarize, we select $y \in \mathbb{R}^N$, $\epsilon \in \mathbb{C}^{N-d}$, and $\alpha \in \mathbb{C}^{N-d+1}$ randomly and consider the Fritz John homotopy[35]

$$H(x,\lambda,t) = \left[\begin{array}{c} f(x) - t\epsilon \\ \lambda_0(x-y) + \lambda_1 \nabla f_1(x)^T + \cdots + \lambda_{N-d} \nabla f_{N-d}(x)^T \\ \alpha_0\lambda_0 + \cdots + \alpha_{N-d}\lambda_{N-d} - 1 \end{array}\right]. \tag{14.12}$$

Let $A \subset \mathbb{C}^N \times \mathbb{C}^{N-d+1}$ be the set of solutions of $H(x,\lambda,1) = 0$, let $E \subset \mathbb{C}^N \times \mathbb{C}^{N-d+1}$ be the set of endpoints defined by H with startpoints A, and let $\pi(x,\lambda) = x$. Then,

$$S = \pi(E) \cap \mathbb{R}^N \subset \mathbf{V}_{\mathbb{R}}(f)$$

is a finite set of points containing a point on each connected component of $\mathbf{V}_{\mathbb{R}}(f)$. A word of caution is that S is determined after ignoring the λ coordinates, whereas Bertini's postprocessor will test for reality using both the x and λ coordinates.

The only remaining issue is to solve $H(x,\lambda,1) = 0$, i.e., to compute the set of startpoints A. Based on genericity, $H(x,\lambda,1) = 0$ has finitely many solutions, each of which is nonsingular. Moreover, the system $H(x,\lambda,1)$ is naturally 2-homogeneous. These two facts make possible many solving options including using a standard 2-homogeneous homotopy (see §5.1), a 2-homogeneous regeneration (see §5.4), or even a regeneration extension approach (see §12.2) where one first computes a witness set for the pure d-dimensional smooth algebraic set $\mathbf{V}(f - \epsilon)$ and uses this witness set to compute A.

Example 14.3. As an illustrative example, consider the sextic plane curve defined by

$$f(x_1,x_2) = x_2^2 + x_1^2(x_1-1)^3(x_1-2).$$

[35]The name arises from the first-order Fritz John conditions [107] used to construct this homotopy.

For illustrative purposes, we take $y = (0.5, 0.8)$, $\epsilon = 1.1 + 0.7i$, and $\alpha = (-1.3, 1.7)$. Our first task is to solve $H(x, \lambda, 1) = 0$, where H is defined as in (14.12). For this, we simply use a standard 2-homogeneous homotopy by constructing the following `input` file:

```
% FritzJohnInitial.input
 function F,G1,G2,patch;
 variable_group x1,x2;
 variable_group l0,l1;
 constant y1,y2,eps,a0,a1;

 y1 = 0.5;
 y2 = 0.8;
 eps = 1.1 + 0.7*I;
 a0 = -1.3;
 a1 = 1.7;

 F = x2^2 + x1^2*(x1-1)^3*(x1-2) - eps;
 G1 = l0*(x1-y1) + l1*x1*(x1-1)^2*(6*x1^2-13*x1+4);
 G2 = l0*(x2-y2) + 2*l1*x2;
 patch = a0*l0 + a1*l1 - 1;
END;
```

Solving using Bertini produces 16 finite solutions, all of which are nonsingular. We rename `finite_solutions` as `start` since these points are the startpoints for our Fritz John homotopy, which we construct using a user-defined homotopy (see §7.5):

```
% FritzJohnHomotopy.input
CONFIG
 UserHomotopy:2;
END;
INPUT
 function F,G1,G2,patch;
 variable_group x1,x2;
 variable_group l0,l1;
 constant y1,y2,eps,a0,a1;
 pathvariable T;
 parameter t;

 t = T;
 y1 = 0.5;
 y2 = 0.8;
 eps = 1.1 + 0.7*I;
 a0 = -1.3;
 a1 = 1.7;

 F = x2^2 + x1^2*(x1-1)^3*(x1-2) - t*eps;
 G1 = l0*(x1-y1) + l1*x1*(x1-1)^2*(6*x1^2-13*x1+4);
 G2 = l0*(x2-y2) + 2*l1*x2;
 patch = a0*l0 + a1*l1 - 1;
END;
```

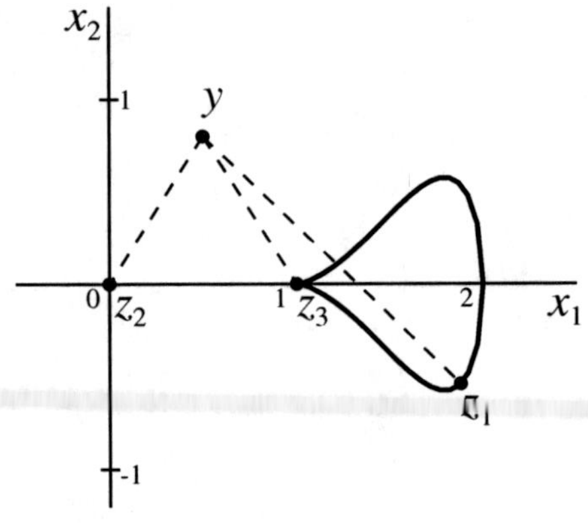

Figure 14.8. *Real part of a sextic curve with critical points.*

For this homotopy, all 16 paths converge. For $\pi(x,\lambda)=x$, the set $S=\pi(E)\cap\mathbb{R}^2$, where E is the set of endpoints, contains three points, namely (rounded to four decimal places),

$$z_1=(1.8796,-0.5380),\quad z_2=(0,0),\quad \text{and}\quad z_3=(1,0),$$

where z_j is the endpoint of exactly j paths. The set $\pi(E)$ also contains 10 nonreal points which arise as the endpoint of a unique path.

Figure 14.8 plots $\mathbf{V}_{\mathbb{R}}(f)$ with dotted lines connecting y to each z_j. Clearly, $\mathbf{V}_{\mathbb{R}}(f)$ consists of two connected components, an isolated point and a curve. In this case, S contains the isolated point as well as two points on the curve, one of which is the cuspidal point with the other maximizing the distance between y and the connected curve. ■

14.6 ▪ Real Numerical Irreducible Decomposition*

An algebraic approach for understanding the real solution set of a polynomial system is to compute the set of polynomials which vanish on the real solution set, called the *real radical ideal.* A *real numerical irreducible decomposition* is a numerical irreducible decomposition for the solution set of the real radical ideal, which is simply the Zariski closure of the real solution set. In this case, each irreducible component of dimension d contains a real set of dimension d, which need not be the case in general.

Example 14.4. To illustrate, consider $f(x,y,z)=(x+y-z)^2+(x+y-2)^2$ which defines a quadratic hypersurface $\mathbf{V}(f)$, but $\mathbf{V}_{\mathbb{R}}(f)=\{(t,2-t,2)\mid t\in\mathbb{R}\}$ is a real line. Thus, the Zariski closure of $\mathbf{V}_{\mathbb{R}}(f)$ is simply the complex line $\{(t,2-t,2)\mid t\in\mathbb{C}\}$, which is defined by $x+y-z=x+y-2=0$. ■

From an algebraic standpoint, one challenge for computing the real radical ideal is that it may require computations outside of the field of coefficients of the original polynomial system. For example, $f(x)=x^3-2$ has rational coefficients with only real point, namely, $\sqrt[3]{2}$. In this case, the real radical ideal is generated by the polynomial $x-\sqrt[3]{2}$ which has coefficients in an extension field. When using numerical approaches, this challenge is alleviated. For example, when $\mathbf{V}_{\mathbb{R}}(f)$ consists of finitely many points, the numerical approach of [111, 112] uses a semidefinite characterization to numerically approximate a set of generators for the real radical ideal. For the general case, our current aim is to only compute a witness set for each irreducible component defined by the real radical ideal rather than explicitly defining equations for each component or the whole real radical ideal.

As we seek witness sets, another challenge arises, which is constructing a polynomial system associated to each witness set. That is, for each irreducible component V of the real radical ideal, we need to construct a polynomial system f_V for which V is an irreducible component of $\mathbf{V}(f_V)$. The key for constructing such a polynomial system f_V for V is to use isosingular theory (see §13.2) given a sufficiently general point on V.

Example 14.5. Reconsider $f(x,y,z)=(x+y-z)^2+(x+y-2)^2$ from Example 14.4 for which $\mathbf{V}_{\mathbb{R}}(f)=\{(t,2-t,2) \mid t\in\mathbb{R}\}$. Let $V=\{(t,2-t,2)\mid t\in\mathbb{C}\}$ be the Zariski closure of $\mathbf{V}_{\mathbb{R}}(f)$. Given the sufficiently general point $(1,1,2)\in V$, the isosingular set associated with $(1,1,2)$ with respect to f is V with deflation sequence $3,1,1,\ldots$ and deflated system

$$f_V(x,y,z)=\begin{bmatrix} f(x,y,z) \\ \nabla f(x,y,z)^T \end{bmatrix}=\begin{bmatrix} (x+y-z)^2+(x+y-2)^2 \\ 2(x+y-z)+2(x+y-2) \\ 2(x+y-z)+2(x+y-2) \\ -2(x+y-z) \end{bmatrix}.$$

In this simple case, we have $V=\mathbf{V}(f_V)$. ∎

14.7 ▪ Software Summary

As of this writing, the cell decomposition and real numerical irreducible decomposition algorithms described in this chapter have been tested on sample problems but not yet publicly released. In the tests, some steps rely on Bertini user-defined homotopies driven by external interfaces. However, we have explicitly shown the two-step process needed to compute a point on each connected component of the real solution set using Bertini.

The `alphaCertified` module mentioned in §14.3 is publicly released and available at [95]. A users guide and an algorithmic description of the software in [96] can be found at that same location.

Chapter 15

Applications to Algebraic Geometry

In addition to the various computations discussed earlier in this book (numerical irreducible decomposition, finding real algebraic sets, etc.), algebraic geometers might be interested in several recent applications of numerical algebraic geometry to problems from algebraic geometry. This chapter provides just a sampling of the various numerical techniques that have been used for various algebraic geometric problems, whereas the next chapter focuses more in depth on the particular topic of images of algebraic sets. In particular, this chapter includes a discussion on computing invariants in §15.1, computations related to secant varieties in §15.2, and the use of numerical methods to develop and test conjectures in §15.3. Algebraic geometers (and others!) might wonder how the numerical methods of Bertini compare to symbolic methods in other packages, such as Macaulay2 [66] or Singular [67]. This is discussed briefly in §15.4.

The point of this chapter is to introduce algebraic geometers to tools that might be useful in their own work, not to provide details that already appear elsewhere in the literature. Thus, we focus throughout this brief chapter on how methods from numerical algebraic geometry have proved to be useful for algebraic geometry problems, leaving details regarding algorithms and justification entirely to the cited articles.

15.1 ▪ Invariants

Invariants are used to classify or otherwise study algebraic sets. Symbolic methods, particularly those based on Gröbner bases, have traditionally been the standard choice for computing such invariants. Recently, however, progress has been made in developing numerical and hybrid symbolic-numeric methods for the computation of such quantities. In this section, we briefly discuss two different methods for computing certain invariants.

15.1.1 ▪ Geometric Genus of a Curve

Three invariants of a curve are the degree, the arithmetic genus, and the geometric genus. Computing the degree of a curve using numerical algebraic geometry was discussed in §8.1.2. We will highlight here, and revisit in §16.5.2, the numerical approach for computing the geometric genus that was presented in [27]. The key to this approach is the Hurwitz formula [129].

Hurwitz formula. Let $\pi : X \to Y$ be a generically d-to-1 map from a smooth irreducible compact curve X of genus g_X onto a compact smooth curve Y of genus g_Y. Then

$$2g_X - 2 = d(2g_Y - 2) + \rho,$$

where ρ is the sum of the cycle numbers at each ramification point of π minus the number of ramification points.

We already encountered ramification points and branch points in our discussion of basic homotopies in §2.1.2 and again in the treatment of endgames in §3.3.1. In those cases, the curve under consideration is the one defined by a homotopy of the form $h(x,t) = 0$, i.e., $X = \mathbf{V}(h) \subset \mathbb{C}^N \times \mathbb{C}$, and the projection is $\pi : (x,t) \mapsto t$, so Y in the Hurwitz formula is just $\mathbb{C}$. Recall that ramification points are the points of X where sheets of the d-sheeted cover touch each other, and the branch points are the images of the ramification points under the projection π, i.e., the values of t where ramification occurs.

The smoothness of X in the above statement of the Hurwitz formula may easily be dropped. If we do this, then we can take any set $\mathcal{D}$ containing all the singular points of X and all ramification points. Let $\widehat{B} = \pi(\mathcal{D})$. Then, for each point $y \in \widehat{B}$, choose a small closed disk Δ containing y, but no other points of $\widehat{B}$. Note that the degree d of the map is the number of points in the set $\pi^{-1}(z)$ for any $z \in \partial\Delta$. The contribution of y to ρ is the sum of the cycle numbers on X as we go around $\partial\Delta$ minus the number of distinct cycles, which is just the degree of the map $\pi : X \to Y$ minus the number of distinct cycles (because the cycle numbers must sum to the degree of the map).

Considering just the affine case,[36] let $X \subset \mathbb{C}^N$ be a curve. Let $Y := \mathbb{C}$ and consider the projection $\pi : \mathbb{C}^N \to Y$ defined by $\pi(x) = A \cdot x$, where $A \in \mathbb{C}^{1\times N}$ is general. Since $g_Y = 0$, all that remains in the Hurwitz formula to compute g_X is ρ since $d = \deg X$ is obtained from a witness set for X.

A superset $\widehat{B}$ of the branch points can be computed by using rank-dropping conditions (see §13.1) based on X and π. We need to allow for ∞ to be a branch point. So we formally add the point ∞ to the set $\widehat{B}$. Then, for each $b \in \widehat{B}$ with $b \neq \infty$, we prescribe a small loop D_b in $\mathbb{C}$ encompassing b which does not pass through or contain any other point in $\widehat{B}$. For convenience, this loop can be either circular or polygonal, such as the polygonal loops we use in the Cauchy endgame (see §3.3.3). For ∞ we choose $D_b := \partial\Delta$, where $\Delta \subset \mathbb{C}$ is a closed disk containing all $b \in \widehat{B} \setminus \{\infty\}$.

Fix $b \in \widehat{B}$ with loop D_b. Then, for each $c \in D_b$, we know that $\pi^{-1}(c) \cap X$ consists of d distinct points, where d is the degree of the map π. Upon fixing a starting point $v_0 \in D_b$, one tracks the paths of $\pi^{-1}(v) \cap X$ as v moves around the loop D_b and returns to v_0. Then one observes the monodromy action, similar to the one displayed in Figure 3.4, and in particular, one counts the number of distinct cycles. This gives the contribution of b to ρ in the Hurwitz formula. For example, if all the paths end at their respective startpoints, then the number of cycles is the same as the degree, and the contribution of b to ρ is zero.

After obtaining the contribution to ρ for each $b \in \widehat{B}$, using $g_Y = 0$, and rearranging the Hurwitz formula, the geometric genus is simply

$$g_X = \rho/2 + 1 - d.$$

We invite the reader to explore more details and examples in [27].

---

[36]We leave the projective case to [27].

15.1.2 ▪ Characteristic Classes

The *Chern classes* of a smooth irreducible scheme $Z \subset \mathbb{P}^r$ are fine invariants from which many other invariants may be computed. In fact, the degrees of the Chern classes are of particular interest. In [15], a probability-one numerical algorithm is presented to compute these invariants. Since any satisfying description of this method would entail numerous definitions (e.g., Chow ring, tangent bundle, and polar class), we only describe this technique from a very high level.

To avoid technical details, we remark that the degrees of the Chern classes can be obtained by computing the degrees of various projective schemes via *polar classes*. Therefore, this reduces to a standard computation in numerical algebraic geometry that is readily available within Bertini. Once these degrees are in hand, they can be organized into a triangular system of linear equations whose solution gives the degrees of the Chern classes.

The discussion in [15] provides further details on this computation. It also provides a lengthy list of formulas relating degrees of Chern classes to well-known invariants. In fact, [15] is a generalization of [49] that was generalized to Segre classes [54] which has a Macaulay2 package named `CharacteristicClasses`.

15.2 ▪ Secant Varieties

Secant and join varieties are useful tools in a number of fields, including algebraic geometry, combinatorics, (algebraic) statistics, and physics, mostly due to the fundamental connection between secant varieties and tensor rank. This section includes two recent methods using numerical algebraic geometry to analyze these varieties.

15.2.1 ▪ Terracini's Lemma

A fundamental tool in the numerical algebraic geometry toolbox discussed in Part II of this book is the ability to numerically compute arbitrarily many points that are arbitrarily close to points on algebraic sets. By using this tool with Terracini's lemma, one can compute the dimensions of join and secant varieties [26].

The join variety of two algebraic sets X and Y is the closure of the union of the linear span of a point in X and a point in Y. For example, if $x \in X$ and $y \in Y$ are distinct, then the linear span of x and y is the unique line passing through these points. The join variety of k algebraic sets $X_1,\ldots,X_k$, denoted $J(X_1,\ldots,X_k)$, is the closure of the union of the linear span of $x_1,\ldots,x_k$, where $x_i \in X_i$.

Secant varieties are join varieties where the algebraic sets are identical. That is, the kth secant variety of X, denoted $\sigma_k(X)$, is $J(X_1,\ldots,X_k)$, where $X_i := X$.

The key realization in [26] is that numerical algebraic geometry can be used to obtain a numerical approximation of a general point on a join variety, because such a point is just a general point in the linear span of general points sampled on each algebraic set. This, together with Terracini's lemma, provides a useful tool for studying join and secant varieties since it provides a probability-one approach for computing the dimension of the space being studied. In short, Terracini's lemma describes the tangent space to $J(X_1,\ldots,X_k)$ at a generic point in terms of the span of the tangent spaces at generic points on each X_i.

15.2.2 ▪ Algebraic Phylogenetics

In recent years, there has been a move to study *phylogenetics* (the study of evolutionary relationships between species) using algebraic geometry. An introduction to this topic can be found in [143]. One variety of particular interest in phylogenetics is the secant variety

$\sigma_4(\mathbb{P}^3 \times \mathbb{P}^3 \times \mathbb{P}^3)$. As discussed in [8], invariants for a statistical model of any binary evolutionary tree with any number of states can be computed from invariants of this variety. Due to this reduction, the so-called *salmon conjecture* was developed. In particular, the first person who could describe the set of polynomials that vanish on this secant variety would be awarded a fresh-caught Alaskan salmon by E. Allman.

In [25], details of work toward solving this problem culminated in a *numerical proof* of the conjecture. The foundation for this numerical proof came from insight related to the irreducible components of a polynomial system, computed using Bertini via a numerical irreducible decomposition. The use of numerical algebraic geometry was necessary since the irreducible decomposition could not be computed using the symbolic methods available at that time. The insight gained from this numerical computation helped lead to a nonnumerical proof in [58].[37]

15.3 ▪ Developing and Testing Conjectures

The goal of mathematical inquiry is typically the proof of—or discovery of counterexamples to—conjectures. For problems involving polynomial systems, numerical algebraic geometry methods provide a means for testing many examples relatively rapidly. The results of these examples can either lead to the development of new conjectures, provide computational evidence for a conjecture, or even yield a counterexample. In this section, we present three instances of using numerical algebraic geometry in this context.

15.3.1 ▪ Maximum Likelihood Degree

Maximum likelihood estimation is a fundamental method in statistics used for estimating parameters of a model. By considering algebraic models and likelihood functions, one obtains a statistical problem that can be formulated in terms of algebraic geometry. See [52] for an introduction to the connection between statistics and algebraic geometry, a field known as *algebraic statistics*.

In [89], the model of interest is the set $\mathcal{M}_r := \mathcal{V}_r \cap \Delta_{m,n}$, where $\mathcal{V}_r$ is the algebraic set consisting of the $m \times n$ matrices of rank at most r and $\Delta_{m,n}$ is the set of $m \times n$ matrices whose entries are nonnegative and sum to 1. With this model, given a nonnegative integer $m \times n$ matrix U, one aims to maximize the *likelihood function* ℓ_U on $\mathcal{M}_r$, where

$$\ell_U(P) := \frac{\prod_{i=1}^m \prod_{j=1}^n P_{i,j}^{U_{i,j}}}{P_{++}^{U_{++}}} \quad \text{with } P_{++} := \sum_{i=1}^m \sum_{j=1}^n P_{i,j} \text{ and } U_{++} := \sum_{i=1}^m \sum_{j=1}^n U_{i,j}.$$

We note that $P_{++} = 1$ on $\mathcal{M}_r$, but this is included in the denominator so that ℓ_U is a rational function defined on the projective space $\mathbb{P}^{mn-1}$. When U is generic, the function ℓ_U has finitely many, say, $d(m,n,r)$, critical points on $\mathcal{V}_r \subset \mathbb{P}^{mn-1}$. The number $d(m,n,r)$ is called the *maximum likelihood degree* of $\mathcal{V}_r \subset \mathbb{P}^{mn-1}$, which is a fundamental algebraic invariant used to understand algebraic properties of the maxima of ℓ_U.

Clearly, $\mathcal{V}_r$ is a determinantal variety described by the vanishing of all $(r+1)$-minors. The critical points of ℓ_U can also be described using determinants. Typically, such a determinantal formulation results in a polynomial system that is overdetermined. One key contribution of [89] is a local kernel formulation, developed using techniques discussed

[37]To finish the story, S. Friedland was awarded the salmon for his work and, based on the picture on his webpage, he seems to have enjoyed it!

in §13.1, which is a square system. Working with this square system, numerical algebraic geometry has two distinct uses for this problem: to compute $d(m,n,r)$ and to use parameter homotopies (see Chapter 6) to compute the critical points of ℓ_U for many different random U. In particular, once an ab initio phase has been completed, thereby computing $d(m,n,r)$, each parameter homotopy requires tracking $d(m,n,r)$ paths.

Using the square system formulation, regeneration (see §5.4) in Bertini was used to compute various values of $d(m,n,r)$ for the larger values of m, n, and r presented in [89], as well as the critical points for various U. As an illustration of the values, $d(4,4,2) = d(4,4,3) = 191$, $d(4,5,2) = d(4,5,3) = 843$, and $d(5,5,2) = d(5,5,4) = 6776$. For $m \leq n$, not only did the computations show that $d(m,n,r) = d(m,n,m-r+1)$ for the various m, n, and r computed, but there was a bijection between the critical points. It was observed that if $d = d(m,n,r) = d(m,n,m-r+1)$, $P_1,\ldots,P_d$ are the critical points for ℓ_U on $\mathbf{V}_r$, and $Q_1,\ldots,Q_d$ are the critical points for ℓ_U on $\mathbf{V}_{m-r+1}$, then there is a reordering so that

$$P_1 \star Q_1 = \cdots = P_d \star Q_d,$$

where $\star$ is the Hadamard (entrywise) product. These statements, which were first conjectured from computations in the original version of [89] along with their analogue for symmetric matrices, were proved in [51].

One open conjecture from [89] is that $d(3,n,2) = 2^{n+1} - 6$ for $n \geq 3$. This statement was developed by using Bertini to compute for $3 \leq n \leq 10$.

15.3.2 ▪ Galois Groups of Schubert Problems

Our final two examples explore the use of numerical algebraic geometry to develop conjectures concerning problems in enumerative geometry. In this section, we discuss using monodromy (see §10.2.2) for computing Galois groups of Schubert problems, while §15.3.3 focuses on a problem in real enumerative geometry. A nice reference for this and related problems is [164].

A Schubert problem is a type of enumerative geometry problem based on linear subspaces. For example, a famous Schubert problem asks us to determine the number of lines that meet four given general lines in space.[38] Following [86], the algebraic Galois group is the same as the geometric monodromy group. An element in the monodromy group, which is a permutation in a symmetric group, can be obtained by performing a monodromy loop using techniques in numerical algebraic geometry. For example, by using random monodromy loops, a subgroup of the monodromy group is obtained from the finitely many elements computed. This generated subgroup provides insight into the monodromy group.

In [116], numerical algebraic geometry was applied to the study of Galois groups of Schubert problems, primarily using the software package PHCpack [172]. The numerical computations of this article showed that the Galois group of a particular Schubert problem of degree 6006 is the full symmetric group on 6006 characters.

15.3.3 ▪ Lines and Conics

A real enumerative geometry problem is an enumerative geometry problem where one considers both real parameters and real solutions. For example, [61] provides computational evidence for real Schubert problems. In this section, we focus on another problem

[38] We won't spoil the answer here: see [116].

in enumerative geometry, which is not a Schubert problem. Over $\mathbb{C}$, the number of plane conics that intersect eight general lines in space is 92. Over $\mathbb{R}$, the question is to determine how many of the 92 plane conics can be real when the eight lines are real. Here, "real" simply means that the coefficients of the equations defining the plane conic and the lines are all real. A summary of solving over 15 million random instances of this problem was presented in [96]. For each instance, eight real lines were selected at random, then Bertini was used to run a parameter homotopy (see Chapter 6) to approximate the 92 solutions, and finally `alphaCertified` [95, 96] was used to determine how many of the solutions were real (see §14.3). The results showed that any even number[39] of real solutions between 0 and 90 was possible. Thus, the only remaining case was to determine whether there exists a set of eight real lines in general position so that all 92 plane conics which intersect them are real.

In [69], so-called gradient descent homotopies together with an approach based on Dietmaier's local linearization method [50] were used to move the eight lines in such a way that a conjugate pair of nonreal solutions merged into a double real root and then split into two distinct real solutions. This produced a collection of eight real lines for which Bertini suggested that all of the 92 conics intersecting them were real. A rigorous proof that this was indeed the case was then developed using `alphaCertified` and the triangle inequality, thereby completely settling the question.

15.4 ▪ Symbolic versus Numerical Computation

For many decades, standard techniques for solving polynomial systems relied on the computation of Gröbner bases. These, and other symbolic methods [168], offer results together with a proof of their correctness over the field used in the exact computation. This section highlights some topics to help an analyst decide how to approach their specific problem, e.g., using only a symbolic or numerical approach or using a hybrid approach. For a more complete discussion together with results using Bertini and `Singular` [67] on a variety of benchmark problems, see [14].

Symbolic methods provide a direct approach to produce certified results in a wide variety of problems with exact coefficients. Two notable examples discussed in [14] are solving binomial systems and the highly structured overdetermined system arising from a Sudoku puzzle. Additionally, finite field computations can be used to produce some insight into the solutions over $\mathbb{Q}$.

As demonstrated throughout this book, numerical methods provide another approach to study solutions of polynomial systems. Both a posteriori certification methods (e.g., see §14.3) as well as certified tracking approaches (e.g., [29, 30, 31, 32, 36, 40]) are increasing the ability to use numerical approaches to develop certified results. Even without rigorous certification of results, probability-one numerical approaches are also useful for experimentation in algebraic geometry, as demonstrated throughout this chapter. In particular, experiments based on solving over many points a parameter space using parameter homotopies (see Chapter 6) is a notable example discussed both here and in [14].

In recent years, a growing amount of research has been in the direction of developing hybrid algorithms which combine advantages of both symbolic and numerical approaches. One example is to use a symbolic preprocessor to simplify the system that is then solved using numerical methods. As a final example, consider the hybrid approach developed in [17] for computing exact results. In this approach, one first solves

[39] Recall that nonreal solutions to a polynomial system with real coefficients must come in conjugate pairs.

the system using numerical techniques discussed in this book. The ability to compute approximations of points to arbitrary accuracy then enables lattice basis reduction methods to compute explicit polynomials that vanish on the algebraic sets of interest. Afterward, symbolic methods can (rapidly) check the correctness of the polynomials.

In summary, both numerical and symbolic approaches have distinct advantages in various situations, and the development of more algorithms for combining these advantages is an area ripe for further inquiry.

Chapter 16

Projections of Algebraic Sets

A fundamental question which arises in a variety of topics is how to compute the projection of an algebraic set. For example, §13.1 computes sets based on rank conditions of a matrix of multivariate polynomials using projections. Other classical problems involving projections include computing discriminant varieties and solving implicitization problems. Algebraically, the answer to understanding the projection of an algebraic set is through *elimination theory*, which involves computing polynomials that vanish on the image. However, obtaining such polynomials can be extremely difficult. In this chapter, which might have been titled *Numerical Elimination Theory*, we discuss how to numerically manipulate the images of algebraic sets without requiring explicit defining equations for the image. For more details and applications, we refer the reader to [90, 91, 88].

Throughout, we have stated that a key component of a witness set for an irreducible variety V is a polynomial system f such that V is an irreducible component of $\mathbf{V}(f)$. Without defining equations for the projection of an algebraic set, we are unable to construct a true witness set. Nonetheless, the goal of this chapter is to describe both a structure and how it is used to perform similar computations as one could do with a witness set. This structure, called a *pseudowitness set*, is described, with a procedure for constructing it, in §16.2.

Given an irreducible algebraic set $V \subset \mathbb{C}^N$ and a linear projection map $\pi : \mathbb{C}^N \to \mathbb{C}^M$, Chevalley's theorem (Theorem 16.1) states that $Y = \pi(V)$ is constructible. In particular, the closure of Y in both the usual topology and the Zariski topology is equal, which we will denote as $\overline{Y}$. We provide more details on constructible sets in §16.1.

In §16.3, we describe using witness sets and pseudowitness sets to develop a membership test for both the constructible set Y and the irreducible algebraic set $\overline{Y}$. The key to this and related computations involving pseudowitness sets is that computations are actually performed on the algebraic set V.

Building from membership tests, §16.4 presents another variant of the numerical irreducible decomposition, namely, the decomposition of the algebraic set $\overline{\pi(\mathbf{V}(f))}$, where f is a polynomial system.

There are two related topics that we highlight in this chapter, both presented in §16.5, which focus on developing a deeper understanding of Y and $\overline{Y}$. The first considers the codimension 1 components in $\overline{Y} \setminus Y$, while the second considers the geometric genus of a general curve section of $\overline{Y}$ when $\dim \overline{Y} \geq 1$.

Even though the rest of this chapter focuses on projections, please note that these topics apply to any image of an algebraic set under an algebraic map. To see this, suppose that $V \subset \mathbf{V}(f)$ is an irreducible component and π is an algebraic map defined on $\mathbf{V}(f)$. Consider the polynomial system

$$g(x,y) = \begin{bmatrix} x - \pi(y) \\ f(y) \end{bmatrix}$$

with projection map $\gamma(x,y) = x$. Then,

$$W = \{(\pi(y), y) \mid y \in V\}$$

is an irreducible component of $\mathbf{V}(g)$ with $\gamma(W) = \pi(V)$. By using, for example, the regeneration extension approach of §12.2, a witness set for W can be computed starting with a witness set for V.

16.1 ▪ Technical Details on Constructible Sets and Slicing*

Let Z be an irreducible k-dimensional algebraic set, and let $p(z)$ be a polynomial on $\mathbb{C}^N$. By [72, Theorem III.C.14] (see also [161, §12.2.1]), either $Z \subset \mathbf{V}(p)$ or $Z \cap \mathbf{V}(p)$ is pure-dimensional of dimension $\dim Z - 1$. In this section, we consider similar statements assuming some mild genericity.

A slight generalization of an algebraic set is a constructible set. (See [161, §12.5] for more details.) A subset of $\mathbb{C}^N$ is said to be *constructible* if it is in the set of subsets of $\mathbb{C}^N$ constructed from affine algebraic sets by finite sequences of the operations of union, intersection, and complementation. A typical example of a constructible set is $(\mathbb{C}^2 \setminus \mathbf{V}(z_1)) \cup \{(0,0)\}$, which arises as the image of the map from $\mathbb{C}^2 \to \mathbb{C}^2$ given by $(z_1, z_2) \to (z_1, z_1 z_2)$. As the next theorem shows, constructible sets often arise as in this example.

Theorem 16.1 (Chevalley's theorem). *Let $f : X \to Y$ be an algebraic map between algebraic sets and $Z \subset X$ be constructible. Then, $f(Z)$ is a constructible set.*

The following shows that constructible sets are very close to algebraic sets.

Lemma 16.2. *Let $X \subset \mathbb{C}^N$ be a constructible set. Then, the closure $\overline{X}$ of X in the usual topology is an affine algebraic set and there is a Zariski-open dense set $U \subset \overline{X}$ such that $U \subset X$.*

The last result means that it makes sense to define $\dim X$ as $\dim U = \dim \overline{X}$.

The following consequence of [153, Theorem 3.42] (itself a consequence of vanishing theorems of Picard–Kodaira type) will be sufficient for our needs. Seidenberg's theorem is a special case of it.

Theorem 16.3. *Let $Z \subset \mathbb{C}^N$ be an irreducible algebraic set. Let $g : \mathbb{C}^N \to \mathbb{C}^K$ be an algebraic map. If $L \subset \mathbb{C}^K$ is a general r-codimensional affine linear space with* $\dim g(Z) \geq r+1$, *then $g^{-1}(L) \cap Z$ is irreducible of dimension* $\dim Z - r$.

We note that g needs to be defined on a Zariski-open dense set $U \subset \mathbb{C}^N$ only as long as $U \cap Z$ is nonempty.

16.2 ▪ Pseudowitness Sets

A witness set for an irreducible algebraic set V contains the fundamental information needed to perform numerical algebraic geometric computations with V. A pseudowitness set aims to perform the same role for the image of an algebraic set when defining equations for the image are unknown.

This section defines a pseudowitness set, first described in [90], followed by a description of computing a pseudowitness set. As mentioned above, for simplicity, we restrict our exposition to projections of algebraic sets.

Let $f : \mathbb{C}^N \to \mathbb{C}^n$ be a polynomial system, $V \subset \mathbf{V}(f)$ be an irreducible component, $\pi : \mathbb{C}^N \to \mathbb{C}^M$ be a projection map, and $Y = \pi(V)$. Without loss of generality, we assume that

1. V has multiplicity one with respect to f and
2. $\pi(x_1, \ldots, x_N) = (x_{\sigma_1}, \ldots, x_{\sigma_M})$, where $1 \le \sigma_1 < \sigma_2 < \cdots < \sigma_M \le N$.

The former assumption may be achieved using isosingular deflation (see §13.2), while the latter assumption can be achieved upon renaming variables.

With this setup, let's first consider a witness set of $\overline{Y}$ assuming we have a polynomial system g defined on $\mathbb{C}^M$ such that $\overline{Y}$ is an irreducible component of $\mathbf{V}(g)$. If $d = \dim \overline{Y}$, this is the triple

$$\{g, L, W\},$$

where $L : \mathbb{C}^M \to \mathbb{C}^d$ is a system of general linear polynomials and $W := \overline{Y} \cap \mathbf{V}(L)$ with $\deg \overline{Y} = \# W$. In particular, since L is general, we know that $W = Y \cap \mathbf{V}(L)$.

Since we do not have access to such a system g, we instead work on $\mathbb{C}^N$ with V. In particular, consider the set $A_1 := V \cap \mathbf{V}(\mathcal{L}_1)$, where $\mathcal{L}_1(x) := L(\pi(x))$. We have

$$A_1 = \bigcup_{w \in W} \pi^{-1}(w) \cap V$$

so that A_1 consists of $\# W$ irreducible components, each of which has the same dimension, namely, $\dim_{gf}(V, \pi) := \dim V - d$, and degree, say, $\deg_{gf}(V, \pi)$. Thus, we need to consider using an additional $\dim_{gf}(V, \pi)$ general slices to yield finitely many points. In particular, if $\mathcal{L}_2 : \mathbb{C}^N \to \mathbb{C}^{\dim_{gf}(V,\pi)}$ is a system of general linear polynomials and $A_2 := V \cap \mathbf{V}(\mathcal{L}_2)$, then π defines a generically $\deg_{gf}(V, \pi)$-to-one map from A_2 to Y. With this setup, a *pseudowitness set* for Y with respect to V is the quadruple

$$\{f, \pi, \mathcal{L}, \mathcal{W}\}, \text{ where } \mathcal{L} := \begin{bmatrix} \mathcal{L}_1 \\ \mathcal{L}_2 \end{bmatrix} \text{ and } \mathcal{W} := V \cap \mathbf{V}(\mathcal{L}). \tag{16.1}$$

The pseudowitness point set $\mathcal{W}$ consists of $\deg \overline{Y} \cdot \deg_{gf}(V, \pi)$ points with $\pi(\mathcal{W}) = W$. In particular, $\# \pi(\mathcal{W}) = \deg \overline{Y}$ and $\# \left(\mathcal{W} \cap \pi^{-1}(w)\right) = \deg_{gf}(V, \pi)$ for each $w \in \mathcal{W}$.

The following section shows how to compute a pseudowitness set for Y from a witness set for V.

Computing a pseudowitness set Let $\{f, \mathcal{L}_V, \mathcal{W}_V\}$ be a witness set for V, which has multiplicity one with respect to f, where $\mathcal{L}_V : \mathbb{C}^N \to \mathbb{C}^{\dim V}$ is a system of general linear polynomials and $\mathcal{W}_V = V \cap \mathbf{V}(\mathcal{L}_V)$. In order to compute a pseudowitness set for $Y = \pi(V)$, we need to compute $\dim_{gf}(V, \pi)$ and $\dim \overline{Y} = \dim V - \dim_{gf}(V, \pi)$. With probability one, for every $z \in \mathscr{W}_V$

$$\dim_{gf}(V,\pi) = \dim \operatorname{null} \begin{bmatrix} Jf(z) \\ J\pi(z) \end{bmatrix},$$

where $Jf(z)$ and $J\pi(z)$ are the Jacobians of f and π, respectively, evaluated at z.

Example 16.4. To illustrate, consider computing the discriminant locus of a univariate quadratic polynomial. We have

$$f(a,b,c,x) = \begin{bmatrix} ax^2+bx+c \\ 2ax+b \end{bmatrix}, \quad V = \mathbf{V}(f), \text{ and } \pi(a,b,c,x) = (a,b,c).$$

For simplicity, we compute $\dim_{gf}(V,\pi)$ by evaluating at $(a,b,c,x) = (1,-2,1,1)$:

$$\dim_{gf}(V,\pi) = \dim \operatorname{null} \begin{bmatrix} x^2 & x & 1 & 2ax+b \\ 2x & 1 & 0 & 2a \\ 1 & 0 & 0 & 0 \\ 0 & 1 & 0 & 0 \\ 0 & 0 & 1 & 0 \end{bmatrix} = \dim \operatorname{null} \begin{bmatrix} 1 & 1 & 1 & 0 \\ 2 & 1 & 0 & 2 \\ 1 & 0 & 0 & 0 \\ 0 & 1 & 0 & 0 \\ 0 & 0 & 1 & 0 \end{bmatrix} = 0.$$

Hence, $\dim \overline{\pi(V)} = 2 - 0 = 2$. ∎

After computing $\dim \overline{Y}$ and $\dim_{gf}(V,\pi)$, the construction of the linear system $\mathscr{L}$ can commence as described above. That is, $\mathscr{L}$ consists of $\dim \overline{Y}$ general linear polynomials involving only the variables in $\pi(x)$ and $\dim_{gf}(V,\pi)$ general linear polynomials. All that remains is to compute the pseudowitness point set $\mathscr{W} = V \cap \mathbf{V}(\mathscr{L})$. This is accomplished using a homotopy that deforms $\mathbf{V}(\mathscr{L}_V)$ to $\mathbf{V}(\mathscr{L})$ on V with startpoints $\mathscr{W}_V$. In order to construct a square homotopy, we randomize (see §9.2) f down to $c := N - \dim V$ functions as shown in the following:

$$H(x,t) = \begin{bmatrix} [I_c \ A] \cdot f(x) \\ (1-t)\mathscr{L}(x) + t\mathscr{L}_V(x) \end{bmatrix}, \tag{16.2}$$

where I_c is the $c \times c$ identity matrix and A is a general $c \times (n-c)$ matrix. Due to the specialized construction of $\mathscr{L}$, some of these homotopy paths may diverge with the pseudowitness point set $\mathscr{W}$ being the set of finite endpoints. In fact, if n_∞ is the number of paths that diverged, then

$$\deg V = \#\mathscr{W}_V = n_\infty + \#\mathscr{W} = n_\infty + \deg \overline{Y} \cdot \deg_{gf}(V,\pi).$$

Example 16.5. Continuing with the discriminant locus from Example 16.4, it is easy to verify that $\deg V = 3$. Since $\overline{Y} = \mathbf{V}(b^2 - 4ac)$ has degree 2 and

$$\pi^{-1}(a,b,c) \cap V = \left\{ \left(a,b,c,\frac{-b}{2a} \right) \right\} \text{ for general } (a,b,c) \in \overline{Y},$$

we have $\deg_{gf}(V,\pi) = 1$, yielding $n_\infty = 1$. Thus, one of the three paths tracked using (16.2) will diverge whereas the other two will converge to distinct points that map to distinct points under π. ∎

The following example uses Bertini to compute a pseudowitness set.

Example 16.6 (Enneper surface). Consider the implicitization problem of computing the degree of the Enneper surface $S \subset \mathbb{C}^3$ given parametrically by

$$\begin{bmatrix} x(u,v) \\ y(u,v) \\ z(u,v) \end{bmatrix} = \begin{bmatrix} u(1-u^2/3+v^2)/3 \\ -v(1-v^2/3+u^2)/3 \\ (u^2-v^2)/3 \end{bmatrix}.$$

That is, $S = \overline{\pi(V)}$, where $\pi(x,y,z,u,v) = (x,y,z)$ and $V = \mathbf{V}(f)$ with

$$f(x,y,z,u,v) = \begin{bmatrix} 3x - u(1-u^2/3+v^2) \\ 3y + v(1-v^2/3+u^2) \\ 3z - (u^2-v^2) \end{bmatrix}.$$

By algebraic geometry, V is an irreducible surface of multiplicity one with respect to f. We compute a witness set for V using the following `input` file:

```
% Enneper.input
CONFIG
 TrackType:1;
END;
INPUT
 function f1,f2,f3;
 variable_group x,y,z,u,v;

 f1 = 3*x - u*(1 - u^2/3 + v^2);
 f2 = 3*y + v*(1 - v^2/3 + u^2);
 f3 = 3*z - (u^2 - v^2);
END;
```

Running Bertini yields that $\deg V = 9$. Apart from maintaining `witness_data`, we also need to change `TrackType` in `input` from 1 to 5 and define π in the file `projection`. The `projection` file consists of a list of integers, either 0 or 1, one for each variable in the original system. The ith integer is 1 if π maps onto the ith variable; otherwise it is 0. So, for $\pi(x,y,z,u,v) = (x,y,z)$, the projection file simply contains the list

```
1 1 1 0 0
```

Now, rerunning Bertini displays a menu that prompts the user to first select the dimension and then select the component number of the component to project. After navigating the menu to select the unique irreducible component, the following is displayed on the screen:

```
Dimensions
  Projection: 2
       Fiber: 0

Degrees
  Projection: 9
       Fiber: 1
```

This summary states $\dim S = 2$, $\dim_{gf}(V,\pi) = 2$, $\deg S = 9$, and $\deg_{gf}(V,\pi) = 1$ so that $n_\infty = 0$. We remark that S is defined by the following degree 9 polynomial:

$$\begin{aligned} g(x,y,z) \;=\;& 64z^9 - 128z^7 + 144(y^2-x^2)z^6 + 16(4-27(x^2+y^2))z^5 \\ & + 240(y^2-x^2)z^4 + 3(16x^2 - 2y^2(117x^2-8) - 45(x^4+y^4))z^3 \\ & + 162(y^4-x^4)z^2 + 9(x^2-y^2)^2 z - 27(x^2-y^2)^3. \end{aligned}$$ ∎

In the previous example, we computed a pseudowitness set by first computing a witness set for V. However, if $\dim \overline{Y}$ and $\dim_{gf}(V,\pi)$ are already known and a witness set for V has not been computed, one can construct the linear polynomials for the pseudowitness set and use techniques from Part I to directly compute the associated pseudowitness point set. Such an approach, for example, was used in [35] to verify the degree of some parametrically defined hypersurfaces related to Hilbert's sums of squares cones.

16.3 ▪ Membership Tests Using Pseudowitness Points

One use of defining equations for $\overline{Y} = \overline{\pi(V)}$ is to determine whether a given point lies in $\overline{Y}$. In this section, we describe homotopy membership tests to decide whether a given point lies in the constructible set Y or the algebraic set $\overline{Y}$ following [91].

Let's first start with testing for membership in $\overline{Y}$. To that end, suppose that $V \subset \mathbf{V}(f) \subset \mathbb{C}^N$ is an irreducible component of multiplicity one with respect to f and $\{f,\pi,\mathscr{L},\mathscr{W}\}$ is a pseudowitness set for $Y = \pi(V) \subset \mathbb{C}^M$. As above, we shall consider the case where π is a linear projection map. Following (16.1), let $\mathscr{L} = [\mathscr{L}_1,\ \mathscr{L}_2]^T$, where $\mathscr{L}_1(x) = L(\pi(x))$ with L consisting of $\dim \overline{Y}$ general linear polynomials on $\mathbb{C}^M$ and $\mathscr{L}_2$ consists of $\dim_{gf}(V,\pi) = \dim V - \dim \overline{Y}$ general linear polynomials on $\mathbb{C}^N$. Then, to determine whether a given point is in $\overline{Y}$, we need to simply deform $\mathscr{L}_1$ to an appropriate system of linear polynomials and track the points in the pseudowitness point set $\mathscr{W}$ during this deformation. As above, the construction of a square homotopy may require randomizing (see §9.2) f down to $c := N - \dim V$ functions. To be explicit, given a test point $y \in \mathbb{C}^M$, we will simply move $\mathscr{L}_1(x)$ to $\mathscr{L}_1(x) - L(y) = L(\pi(x)) - L(y)$ using the following homotopy:

$$H(x,t) = \begin{bmatrix} [I_c\ A]\cdot f(x) \\ \mathscr{L}_1(x)(1-t+\gamma t) - L(y)(1-t) \\ \mathscr{L}_2(x) \end{bmatrix}, \tag{16.3}$$

where I_c is the $c\times c$ identity matrix, A is a general $c\times(n-c)$ matrix, and $\gamma\in\mathbb{C}$ is general. For $w \in \mathscr{W}$, let $w(t)$ be the path starting at w defined by H and consider

$$E := \left\{ z \in \mathbb{C}^M \;\middle|\; z = \lim_{t\to 0} \pi(w(t)) \text{ for some } w \in \mathscr{W} \right\}. \tag{16.4}$$

Then, with probability one, $y \in E$ if and only if $y \in \overline{Y}$. We can simplify down to tracking at most $\deg \overline{Y}$ paths since if $w_1, w_2 \in \mathscr{W}$ such that $\pi(w_1) = \pi(w_2)$, then

$$\lim_{t\to 0} \pi(w_1(t)) = \lim_{t\to 0} \pi(w_2(t)).$$

In fact, computing E amounts to running an endgame (see §3.3) on just the coordinates appearing in $\pi(x)$.

Since the path $w(t)$ may diverge even while $\pi(w(t))$ converges, we have $\pi(\mathscr{E}) \subset E$, where

$$\mathscr{E} := \left\{ \xi \in \mathbb{C}^N \;\middle|\; \xi = \lim_{t\to 0} w(t) \text{ for some } w \in \mathscr{W} \right\}. \tag{16.5}$$

In fact, $y \in \pi(\mathscr{E})$ is a sufficient condition for concluding that $y \in Y$. It is also a necessary condition when $\pi(\mathscr{E}) = E$ or $\dim \overline{Y} = 1$. Therefore, when $y \in \overline{Y}$, $\pi(\mathscr{E}) \subsetneq E$, and $\dim \overline{Y} > 1$, we need another test to determine whether $y \in Y$. Since $y \in Y$ if and only if

$V \cap \pi^{-1}(y) \neq \emptyset$, the additional test is to use a witness set for V and a diagonal homotopy approach (see §12.1) to compute a witness point superset for this intersection.

Example 16.7. To illustrate the basic test for deciding membership in $\overline{Y}$, reconsider the quadratic discriminant locus from Example 16.4. Since $\dim V = \dim \overline{Y}$, for simplicity, we can take $\{f, \pi, \mathcal{L}, \mathcal{W}\}$ as a pseudowitness set for Y, where

$$\mathcal{L}(a,b,c,x) = \begin{bmatrix} a-3c-4 \\ b+3c+2 \end{bmatrix} \text{ and } \mathcal{W} = \left\{(-2,4,-2,1), \left(6,-4,\frac{2}{3},\frac{1}{3}\right)\right\}.$$

By using user-defined homotopies (see §7.5), we will employ the membership test at

$$P_1 = (1,2,1), \quad P_2 = (0,0,1), \quad \text{and} \quad P_3 = (1,3,2).$$

There are various ways to set up the space where we perform the homotopy. In order to accurately compute the set E from (16.4), we first treat the space $\mathbb{C}^4$ as $\mathbb{C}^3 \times \mathbb{C}$ with the two different variable groups representing the image variables of π, namely, a, b, and c, and the fiber variables of π, namely, x. Second, we partially compactify the space to $\mathbb{C}^3 \times \mathbb{P}^1$ so that the convergent paths in this space are exactly those that yield points in E. This requires homogenizing f with respect to the fiber variable and casting the points in W to this space. The following `input` file defines a homotopy for testing point P_1:

```
% Discriminant.input
CONFIG
 UserHomotopy:2;
END;
INPUT
 function f1,f2,l1,l2;
 variable_group a,b,c;
 hom_variable_group x,z;
 constant a1,b1,c1;
 random gamma;
 pathvariable T;
 parameter t;

 t = T;
 a1 = 1;
 b1 = 2;
 c1 = 1;

 f1 = a*x^2 + b*x*z + c*z^2;
 f2 = 2*a*x + b*z;
 l1 = (a - 3*c - 4)*(1 - t + gamma*t) - (a1   3*c1 - 4)*(1 - t);
 l2 = (b + 3*c + 2)*(1 - t + gamma*t) - (b1 + 3*c1 + 2)*(1 - t);
END;
```

The following `start` file contains the two startpoints represented in $\mathbb{C}^3 \times \mathbb{P}^1$:

```
2

-2 0
 4 0
-2 0
 1 0
 1 0
```

```
 6 0
-4 0
 0.666666666666666666667 0
 0.333333333333333333333 0
 1 0
```

Using the notation from `input`, the points in $\mathscr{E}$ are the endpoints for which $z \neq 0$. Moreover, to test points P_2 and P_3, we simply need to update `a1`, `b1`, and `c1` in `input` accordingly. We summarize the results of the three runs in the following table:

i	$\#E$	$P_i \in E$	$\#\pi(\mathscr{E})$	$P_i \in \pi(\mathscr{E})$	Summary
1	2	Yes	2	Yes	$P_1 \in Y \subset \overline{Y}$
2	2	Yes	1	No	$P_2 \in \overline{Y}$, inconclusive for Y
3	2	No	2	No	$P_3 \notin \overline{Y}$

∎

16.4 ▪ Decomposing Images of Algebraic Sets

Until now, we have been focusing on the image of one irreducible component $V \subset \mathbf{V}(f)$ under π. By using the membership test based on pseudowitness sets described in §16.3, we can compute a numerical irreducible decomposition of $\overline{\pi(\mathbf{V}(f))}$ where each irreducible component is described by at least one pseudowitness set.

Let $V_1, \ldots, V_k$ be the irreducible components of $\mathbf{V}(f)$. From a witness set for each V_i, we can compute a pseudowitness set for $\pi(V_i)$ using the method in §16.2. Then, the irreducible components of $\overline{\pi(\mathbf{V}(f))}$ are the inclusion-maximal elements of

$$\left\{ \overline{\pi(V_i)} \;\middle|\; i = 1, \ldots, k \right\}.$$

That is, $\overline{\pi(V_i)}$ is an irreducible component of $\overline{\pi(\mathbf{V}(f))}$ if and only if there does not exist j such that $\overline{\pi(V_i)} \subsetneq \overline{\pi(V_j)}$. With probability one, $\overline{\pi(V_i)} \subset \overline{\pi(V_j)}$ if and only if, for general $w \in \overline{\pi(V_i)}$, $w \in \overline{\pi(V_j)}$. Hence, using the membership test of §16.3 and considering dimensions, we can identify the irreducible components of $\overline{\pi(\mathbf{V}(f))}$. We note that we may have more than one pseudowitness set to describe the same irreducible component since it may happen that there exists $j \neq i$ such that $\overline{\pi(V_i)} = \overline{\pi(V_j)}$. The additional pseudowitness sets for the same irreducible component of $\overline{\pi(\mathbf{V}(f))}$ can add flexibility in subsequent computations.

Example 16.8. As an illustrative example, consider the polynomial system

$$f(x,y,s,t) = \begin{bmatrix} (s-1)x - st \\ (s-1)y - t + 1 \\ (s-2)(t-2) \end{bmatrix} \quad \text{with projection map } \pi(x,y,s,t) = (x,y).$$

It is easy to verify that $\mathbf{V}(f)$ has two irreducible components, namely,

$$V_1 = \{(2y+2, y, 2, y+1) \mid y \in \mathbb{C}\} \text{ and } V_2 = \{(2y+2, y, s, 2) \mid sy = y+1 \text{ and } s, y \in \mathbb{C}\}.$$

In particular, $\overline{\pi(V_1)} = \overline{\pi(V_2)} = \overline{\pi(\mathbf{V}(f))} = \mathbf{V}(2y+2-x)$. ∎

16.5 ▪ Further Applications of Pseudowitness Sets*

In §16.3, we described approaches for deciding whether a given point is in $Y = \pi(V)$ and $\overline{Y}$. Even though one can test specific points, it would nice to be able to fully understand the difference between $\overline{Y}$ and Y by computing a numerical irreducible decomposition of $\overline{\overline{Y} \setminus Y}$. Even though we presently do not know of any numerical approach to compute this decomposition, we describe in §16.5.1 an approach for computing a numerical irreducible decomposition of the codimension 1 components of $\overline{\overline{Y} \setminus Y}$. This codimension 1 information often suffices in applications such as computing the geometric genus of a general curve section of $\overline{Y}$. The computation for the geometric genus of a curve given by a witness set is presented in §15.1.1, so in §16.5.2 we provide only a brief description of how this computation is adapted for use with pseudowitness sets.

16.5.1 ▪ Codimension 1 components of $\overline{\overline{Y} \setminus Y}$

Let $V \subset \mathbb{C}^N$ be an irreducible algebraic set and, for simplicity, we assume that $\pi : \mathbb{C}^N \to \mathbb{C}^M$ is the projection onto the first M coordinates, where $0 < M < N$, i.e., $\pi(x_1,\ldots,x_N) = (x_1,\ldots,x_M)$. Thus, we can write $\pi : \mathbb{C}^M \times \mathbb{C}^{N-M} \to \mathbb{C}^M$ and consider the product projection $\overline{\pi} : \mathbb{C}^M \times \mathbb{P}^{N-M} \to \mathbb{C}^M$ extending π. Note that we already used this construction in Example 16.7 when performing membership tests.

Let $\overline{V}$ be the closure of V in $\mathbb{C}^M \times \mathbb{P}^{N-M}$ and $Y := \pi(V)$. The map $\overline{\pi}$ is proper, implying that $\overline{\pi}(\overline{V})$ is an algebraic set. Since $Y = \pi(V)$ is dense in $\overline{\pi}(\overline{V})$, it follows that $\overline{Y} = \overline{\pi}(\overline{V})$.

The next step is to consider $\overline{V} \cap \mathscr{P}$, where $\mathscr{P} := \mathbb{C}^M \times \left(\mathbb{P}^{N-M} \setminus \mathbb{C}^{N-M}\right)$. Given a witness set for V, it is trivial to construct a witness set for $\overline{V}$. Then, one can compute witness sets for the $\dim V - 1$ components, say, $D_1,\ldots,D_k$, of $\overline{V} \cap \mathscr{P}$. This computation is aided by the fact that $\mathscr{P}$ is linear. We then use the tests from §16.3 to compute $\mathscr{I} \subset \{1,\ldots,k\}$ consisting of the indices i such that $\dim \overline{\pi(D_i)} = \dim \overline{Y} - 1$ and $\overline{\pi(D_i)} \subset \overline{\overline{Y} \setminus Y}$. In the end, each set $\overline{\pi(D_i)}$ where $i \in \mathscr{I}$ is a codimension 1 irreducible component of $\overline{\overline{Y} \setminus Y}$.

Example 16.9. Consider the polynomial systems $f_1(x,y) = xy - 1$ and $f_2(x,y) = x(xy-1)$ with projection map $\pi(x,y) = x$. Let $V_i := \mathbf{V}(f_i)$ and $Y_i := \pi(V_i)$. Since $\dim V_i = \dim \overline{Y}_i = 1$, we know that $\overline{Y}_i = \mathbb{C}$. Thus, we are aiming to compute the points, if any, in $\mathbb{C}$ outside of Y_i.

Let $\overline{V}_i$ be the closure of V_i in $\mathbb{C} \times \mathbb{P}^1$, which is $\mathbf{V}(g_i)$, where $g_1(x,y,z) = xy - z$ and $g_2(x,y,z) = x(xy-z)$. Since $\mathscr{P} := \mathbb{C} \times (\mathbb{P}^1 \setminus \mathbb{C}) = \mathbf{V}(z)$, it follows that $\overline{V}_i \cap \mathscr{P} \subset \mathbb{C} \times \mathbb{P}^1$ has a unique irreducible component, namely, $D := \{(0,1,0)\} \subset \mathbb{C} \times \mathbb{P}^1$. Since $\pi(D) = \{0\}$, all that remains is to determine whether $0 \in Y_i$. Clearly, $0 \notin Y_1$ implying $\overline{Y}_1 \setminus Y_1 = \{0\}$, while $0 \in Y_2$ implying $Y_2 = \overline{Y}_2 = \mathbb{C}$. ∎

16.5.2 ▪ Geometric Genus of a General Curve Section

In §15.1.1, given a witness set for an irreducible curve C, an approach for computing the geometric genus of C, which is the genus of the unique smooth compactification of the desingularization of C, is presented. If $V \subset \mathbb{C}^N$ is an irreducible algebraic set of dimension at least 1, the geometric genus of a general curve section of V is the geometric genus of the irreducible curve $V \cap \mathscr{K}$, where $\mathscr{K}$ is a general linear space of $\mathbb{C}^N$ of dimension $N - \dim V + 1$, which is independent of $\mathscr{K}$. This is an important enumerative invariant of V, e.g., it arises in calculations involving the Riemann–Roch theorem.

Rather than using a witness set, we would like to perform this computation using pseudowitness sets. We simply highlight the key differences here and refer to [91] for more details. Suppose that $\overline{Y} = \overline{\pi(V)} \subset \mathbb{C}^M$ is an irreducible algebraic set of dimension at least 1. Naturally, the geometric genus of a general curve section of $\overline{Y}$ is the geometric genus of the irreducible curve $\overline{Y} \cap \mathscr{K}$, where $\mathscr{K} \subset \mathbb{C}^M$ is a general linear space of dimension $M - \dim \overline{Y} + 1$. Let $Z := Y \cap \mathscr{K}$ and thus $\overline{Z} = \overline{Y} \cap \mathscr{K}$. A pseudowitness set for $\overline{Z}$ can be easily computed from a pseudowitness set for Y. Then, using §16.5.1, one is able to compute the finitely many points in $\overline{Z} \setminus Z$. Apart from the addition of analyzing these finitely many points and using pseudowitness sets in place of witness sets, the geometric genus algorithm presented in §15.1.1 can be used to compute the geometric genus of $\overline{Z}$, i.e., the geometric genus of a general curve section of $\overline{Y} = \overline{\pi(V)}$.

16.6 ▪ Bertini Guide: Computing Pseudowitness Sets

Computing a Pseudowitness Set Summary

Input files
- `input` = run configuration & system definition
- `projection` = defines projection
- `witness_data` = contains witness sets for system

Output files
- `witness_data_projection` = pseudowitness set for selected component with respect to the given projection

Key configurations:
- `TrackType:5;`

Notes:
- Component must have multiplicity one with respect to the input system.

Execution:
- `>> bertini <input>`

Final note. As of this writing, Bertini is unable to directly perform additional computations with the computed pseudowitness sets. However, by printing the pseudowitness set (see §11.1), the additional algorithms in this section can be and have been performed on sample problems, using user-defined homotopies in Bertini.

Chapter 17

Big Polynomial Systems Arising from Differential Equations

For the applications discussed previously, it is reasonable for the practitioner to expect to be able to compute a complete numerical description of the solution set of the corresponding polynomial systems with enough computational power and time. In this chapter, we explore "solving" polynomial systems which could have more isolated solutions (e.g., $2^{1000} \approx 10^{300}$) than atoms in the universe (say, 10^{83}). This chapter provides only a brief snapshot of this rapidly changing area, much of which is outside the scope of this book. The interested reader can further explore the referenced articles as well as the survey [81].

The systems of particular interest in this chapter are those which arise via the discretization of systems of partial differential equations (PDEs). These systems pose many challenges since

1. they consist of a large number of variables and equations;
2. computing all solutions is usually impractical;
3. the number of solutions to the corresponding differential equations may be none, one, a finite number, or even infinitely many; and
4. the tools such as continuation, which are inexpensive for most of the systems of this book, are costly even for a single path.

Such systems of differential equations are arising out of engineering and science at an ever-increasing rate with, typically, little guidance from theoretical differential equations. To put this in perspective, in §17.1.3 we present a set of differential equations arising from a simple tumor-growth model. A toy version of this already simplified problem reduces to the classical Hele–Shaw problem with surface tension, which has its own extensive literature. In addition to this model, which is defined on a two-dimensional domain with free boundary, we illustrate some of the issues as they arise in two simpler situations: a two-point boundary value problem on a one-dimensional domain (§17.1.1) and a predator-prey model on a two-dimensional domain with fixed boundary (§17.1.2).

In §17.2, we discuss various methods of arriving at solutions. In particular, §17.2.1 discusses *bootstrap methods*. The *filtering method*, first introduced in [4], attempts to solve a polynomial system arising through discretization by incrementally adding variables to systems from smaller grids and filtering out at each stage unreasonable solutions. The *domain decomposition method*, first introduced in [78], may be considered as a multigrid approach as it builds up to a system of polynomials arising from a full discretization by using simpler systems of polynomials suggested from a domain decomposition. Finally, a discussion of using continuation and analysis of bifurcations to construct new solutions from old solutions occurs in §17.2.2.

Endgames, discussed in §3.3, estimate the endpoint of a path by building a local model of the path near the endpoint. In §17.2.3, we summarize the work of [80] which combines endgames with a WENO-based discretization scheme to solve some steady-state problems coming from hyperbolic conservation laws.

17.1 ▪ Polynomial Systems from Discretizing Differential Equations

The following provides some examples illustrating certain key features of systems arising from differential equations. Since much is beyond the scope of this book, we refer the reader to [4, 5, 6, 74, 75, 76, 79, 77, 78, 80, 82, 81, 84] for many more examples and details.

17.1.1 ▪ Two-Point Boundary Value Problems

To illustrate constructing a discretization, consider the two-point boundary value problem [4, §3.1] on the interval [0, 1]:

$$y'' = 2y^3, \quad y(0) = \frac{1}{2}, \quad y(1) = \frac{1}{3}. \tag{17.1}$$

This problem has a unique solution, namely, $y(x) = (x+2)^{-1}$.

For $N \geq 2$, consider splitting the interval $[0,1]$ into N equal subintervals $[x_i, x_{i+1}]$, $i = 0, \ldots, N-1$, where $x_i := i \cdot h_N$ and $h_N := N^{-1}$. In order to approximate a potential solution $y(x)$, we let y_i be an approximation of $y(x_i)$. Since $y(0)$ and $y(1)$ are given, we have $y_0 := 1/2$ and $y_N := 1/3$. For $i = 1, \ldots, N-1$, we approximate $y''(x)$ using a central second-order finite difference scheme, namely,

$$y(x_i) \approx \frac{y_{i+1} - 2y_i + y_{i-1}}{h_N^2}.$$

The resulting discretized polynomial system is thus

$$F_N(y_0, y_1, \ldots, y_N) = \begin{bmatrix} y_0 - 1/2 \\ y_2 - 2y_1 + y_0 - 2h_N^2 y_1^3 \\ \vdots \\ y_N - 2y_{N-1} + y_{N-2} - 2h_N^2 y_{N-1}^3 \\ y_N - 1/3 \end{bmatrix}. \tag{17.2}$$

Though for even small N a solution of $F_N = 0$ provides a good approximation of the unique solution of (17.1), the system $F_N = 0$ has 3^{N-1} solutions. This is a fact of life in the numerical solution of systems of differential equations: deciding which solutions to a discretization correspond to actual solutions is not obvious. Moreover, it is rare (unfortunately) that the number of solutions of the system of ordinary differential equations (ODEs) is known a priori, while for systems of PDEs, the a priori knowledge about solutions is even sparser. Furthermore, as finer discretizations are considered, it rapidly becomes impossible to compute all solutions. Thus, as discussed in more detail in §17.2, a practical approach must use heuristics to find a promising subset of the full solution set of the discretization and then perform further checks to decide which of these approximate an actual solution of the differential equations. Using solutions for a coarse discretization to spawn solutions for a finer grid, one may hope to find many, if not all, of the solutions that can be approximated at a final target level of discretization.

Merely changing the boundary conditions can significantly alter the solution set. For example, consider the problem [4, §3.3] on $[0,1]$:

$$y'' = -y^3, \quad y(0) = 0, \quad y(1) = 0. \tag{17.3}$$

In contrast to (17.1), which has a unique solution, (17.3) has infinitely many solutions. Of these, only a finite number are approximated by solutions for a given discretization. Creating a discretization of (17.3) similar to (17.2), the discretized system corresponding to $N = 7$ has 729 solutions (41 of which are real), while the system corresponding to $N = 8$ has 2187 solutions (79 of which are real). Each real solution approximates an actual solution of (17.3).

17.1.2 ▪ A Predator-Prey Model

The following Lotka–Volterra population model with diffusion [123, 175] was investigated using numerical algebraic geometry techniques by the second and third authors together with B. Hu:

$$-\Delta u = u(1-v), \tag{17.4}$$
$$-\Delta v = -v(1-u). \tag{17.5}$$

Here, u and v are functions defined on $[0,1]\times[0,1]$ and Δ is the Laplacian. The system also includes some nonzero boundary condition, such as

$$u = \begin{cases} 0 & \text{if } x = 0, \\ 1 & \text{if } x = 1, \\ x & \text{if } y = 0, \\ \sin(x)/\sin(1) & \text{if } y = 1, \end{cases} \quad \text{and } v = \begin{cases} 1 & \text{if } x = 0, \\ 0 & \text{if } x = 1, \\ 1-x & \text{if } y = 0, \\ \sin(1-x)/\sin(1) & \text{if } y = 1. \end{cases}$$

Choosing a positive integer N and discretizing using central differences with the grid-points $(x_i, y_j) = \left(\frac{i}{N+1}, \frac{j}{5}\right)$ with $0 \le i \le N+1$ and $0 \le j \le 5$ yields a polynomial system consisting of $8N$ quadratic equations,

$$\frac{u_{i+1,j} - 2u_{i,j} + u_{i-1,j}}{25} + \frac{u_{i,j+1} - 2u_{i,j} + u_{i,j-1}}{(N+1)^2} + \frac{u_{i,j}(1-v_{i,j})}{25(N+1)^2} = 0, \tag{17.6}$$
$$\frac{v_{i+1,j} - 2v_{i,j} + v_{i-1,j}}{25} + \frac{v_{i,j+1} - 2v_{i,j} + v_{i,j-1}}{(N+1)^2} + \frac{v_{i,j}(u_{i,j}-1)}{25(N+1)^2} = 0, \tag{17.7}$$

with variables $\{u_{i,j}, v_{i,j} \mid 1 \le i \le N, 1 \le j \le 4\}$ and constants

$$u_{0,j} = v_{N+1,j} = 0, \quad u_{N+1,j} = v_{0,j} = 1,$$

$$u_{i,0} = i/5,\ u_{i,5} = \frac{\sin(i/5)}{\sin(1)}, \quad v_{i,0} = 1 - i/5,\ v_{i,5} = \frac{\sin(1-i/5)}{\sin(1)}.$$

In [93, §9.4], a similar polynomial system was used as an example of the regeneration algorithm discussed in §5.4. Although the total degree is 2^{8N}, the system has only 2^{4N}

solutions (all isolated and nonsingular). As shown in [93], Bertini took under 8 minutes for $N = 4$ and under 4 hours for $N = 5$ running in parallel on 64 cores (8 nodes each with two 2.33-GHz quad-core Xeon 5410 processors). Though this grid is fine enough to get some useful information, the computing times make clear that even with more cores, a brute force approach is limited.

17.1.3 ▪ Free Boundary Problems

In a pair of articles [59, 60], Friedman and Hu developed bifurcation theories and stability theorems for some free boundary problems arising in tumor growth. Numerical algebraic geometry techniques have been applied successfully to free boundary problems of this sort [79, 76, 75, 77, 78] and also to related biological models [82, 84]. For expediency, we consider only one of the simplest two-dimensional models here.

Suppose that the tumor is a simply connected open set $\mathscr{D}(t) \subset \mathbb{R}^2$ with appropriately smooth boundary $\partial\mathscr{D}(t)$ that varies with time t. A simple system of PDEs describing the evolution of $\mathscr{D}(t)$ and $\partial\mathscr{D}(t)$ is

$$\begin{array}{rcll} \sigma_t &=& \Delta\sigma - \sigma & \text{on } \mathscr{D}(t), \\ \sigma &=& \bar{\sigma} & \text{on } \partial\mathscr{D}(t), \\ -\Delta p &=& \mu(\sigma - \bar{\sigma}) & \text{on } \mathscr{D}(t), \\ p &=& k & \text{on } \partial\mathscr{D}(t), \\ \dfrac{\partial p}{\partial n} &=& -V_n(t) & \text{on } \partial\mathscr{D}(t). \end{array} \tag{17.8}$$

This system arises by considering the tissue to be a porous medium, where Darcy's law (i.e., velocity is $-\nabla p$, where p is the pressure) holds. The mean curvature of the boundary is k, V_n is the velocity of the boundary in the normal direction, and σ is the density of the nutrient which is assumed to be at a constant level $\bar{\sigma}$ outside the cell. The derivative (in the outward normal direction) of the pressure p along the boundary is denoted $\frac{\partial p}{\partial n}$. Finally, the number μ is called the *tumor-aggressiveness* factor, which is a measure of the aggressiveness of the tumor.

We consider the steady-state solutions of (17.8) which satisfy

$$\begin{array}{rcll} \sigma &=& \Delta\sigma & \text{on } \mathscr{D}, \\ \sigma &=& \bar{\sigma} & \text{on } \partial\mathscr{D}, \\ -\Delta p &=& \mu(\sigma - \bar{\sigma}) & \text{on } \mathscr{D}, \\ p &=& k & \text{on } \partial\mathscr{D}, \\ \dfrac{\partial p}{\partial n} &=& 0 & \text{on } \partial\mathscr{D}. \end{array} \tag{17.9}$$

Given μ, Friedman and Hu [59, 60] computed the radially symmetric solutions. They showed there were certain values of μ, which they computed, for which the branch of radially symmetric solutions met the branch of nonradial solutions.

A discretization of this problem presented in [74, 76] was designed for domains that are close to convex. We assume, without loss of generality, that at the start of our computations the origin of $\mathscr{D}$ is the centroid of the region. We choose N rays coming out from the origin at the angles $\theta = 0, \frac{2\pi}{N}, \ldots, \frac{2(N-1)\pi}{N}$. For the ray with $\theta = \frac{2j\pi}{N}$, we let R_j be the distance from the origin to the point where the ray meets $\partial\mathscr{D}$. For gridpoints we take (in polar coordinates) the points $\left(\frac{iR_j}{M}, \frac{2j\pi}{N}\right)$, where $j = 0, \ldots, N-1$ and i goes from 0 to M. We choose discretizations of the equations giving approximate values of the functions σ and p at the gridpoints.

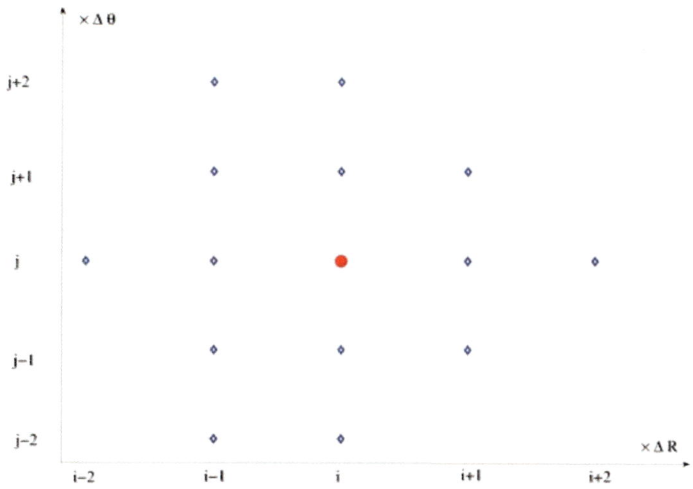

Figure 17.1. *Stencil of points.*

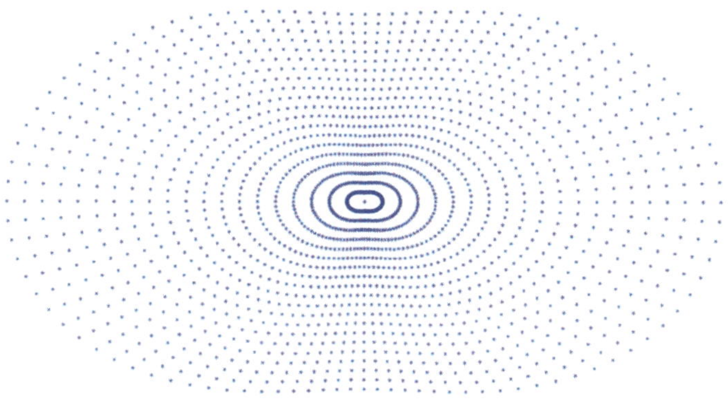

Figure 17.2. *Grid with* 1901 *gridpoints:* $(N,M) = (100,20)$.

In the interior, the discretization uses a stencil of points of the form presented in Figure 17.1, that is, the value at the (i,j) gridpoint is calculated in terms of the values at the stencil points. See [74, 76] for full details on the numerical discretization of this problem, which is outside the scope of this book. In [83] an alternate version of the grid is investigated. A sample grid is given in Figure 17.2.

To summarize, the variables in this discretization are

1. the distances R_j from the origin to the N gridpoints on the boundary $\partial\mathscr{D}$, and
2. the approximate values of σ and p at the distinct $N(M-1)+1$ gridpoints.

As the boundary gridpoints change, the interior gridpoints change continuously with the boundary points in the same ways as above. The only exception is that the origin is replaced with the average value of the boundary gridpoints. In particular, there are $2NM+2$ variables so that, for $(N,M) = (100,20)$ as used in Figure 17.2, the discretization uses 1901 gridpoints with the system depending on 4004 variables.

17.1.4 ▪ Steady-State and Hyperbolic Conservation Laws

Consider the one-dimensional hyperbolic conservation laws

$$u_t + \big(f(u)\big)_x = g(u,x).$$

By setting u_t to zero, the steady-state problem becomes

$$\big(f(u)\big)_x - g(u,x) = 0. \tag{17.10}$$

Certain solutions satisfying the strong entropy condition (see [80]) are of particular interest and theory provides that they may be obtained using the homotopy of differential equations:

$$H(u,\epsilon) = \Big(\big(f(u)\big)_x - g(u,x) - \epsilon u_{xx}\Big)(1-\epsilon) + \epsilon(u - u^0) \equiv 0 \tag{17.11}$$

as $\epsilon \to 0$ given initial conditions u^0. A discretization of (17.11) using a WENO scheme is presented in [80] together with several example systems of PDEs. We revisit this discretized homotopy in §17.2.3.

17.2 ▪ Computing Solutions to Big Polynomial Systems

The previous section discussed discretizing systems of nonlinear differential equations to produce systems of nonlinear equations. In this section, we consider ways to compute solutions to the discretized system with the goal that the computed solutions are physically reasonable since there are often too many solutions to compute all of them. A solution to a discretized system may fail to be physically reasonable in several ways:

- it may be far from real,
- it may oscillate more than is consistent with the system of differential equations, or
- it may approximate an unstable solution of the differential equations.

If we could generate all solutions to the discretized system, then we could sift out all the meaningful ones by testing against these criteria. In practice, we may try to arrive at the meaningful ones while computing fewer of the unreasonable ones by solving a sequence of discretizations, proceeding from coarse to fine, and testing for reasonableness as we proceed. Alternatively, in a one-parameter family of systems, we may use continuation to follow a solution path as the parameter varies and seek to generate new solution branches where bifurcations are discovered. Either way, stability testing is best done after accurate solutions have been found, and although we do not discuss this question further here, numerical algebraic geometry has been applied to it elsewhere [74, 75, 76, 77, 79, 82].

At any stage where a solution to a discretized system appears to be a likely candidate as a solution of the original system of differential equations, a further check is warranted. Typically, one takes a finer discretization, interpolates the discretized solution to the finer grid, and then checks (typically with Newton's method) whether the interpolated solution is sufficiently near a solution of the finer discretization. A striking example of this procedure is given in [82]. The starting point for finding solutions of the complicated system of PDEs in [82] is finding the radial symmetric solutions. In this case the system specializes to a system of ODEs. Solving the polynomial system arising by discretization with six gridpoints took about a day with Bertini running on 160 cores. There were

148 real solutions. The 148 solutions were interpolated to 12 gridpoints, which was followed by a use of Newton's method to check whether they were close to solutions of the polynomial system arising by discretization with 12 gridpoints. Only 24 of the 148 solutions survived. Table 3 of [82] shows the outcomes of this procedure as the gridpoints were further increased to 24, then 48, then 96, and finally 192 gridpoints. Only one of the original 148 real solutions survived to 96 gridpoints. That solution persists and is considered to be the discretization of a solution of the original system of ODEs.

In §17.2.1, we discuss filtering and domain decomposition methods, which are used to build up from small polynomial systems to the solutions of the polynomial systems coming from discretization. Each resulting solution is a good candidate to be an approximation of a solution of the original system of differential equations.

In §17.2.2, we discuss how to use bifurcation analysis to generate interesting new solutions from known solutions of systems of differential equations.

Finally, in §17.2.3, we discuss a class of problems where endgames give an effective method for computing a desired solution. For many examples, this method appears to be superior to the standard methods.

17.2.1 ▪ Solutions from Filtering and Domain Decomposition

We discuss two bootstrap methods, which in practice we usually combine together.

Filtering To give an illustration of filtering, consider (17.1) and its discretization $F_N(y_0,\ldots,y_N) = 0$ from (17.2). The filtering starts with a finite set S_N of solutions of $F_N(y_0,\ldots,y_N) = 0$ and ends with a finite set S_{N+1} of solutions of $F_{N+1}(y_0,\ldots,y_{N+1}) = 0$. Typically, we start by finding the set A_{N_0} of all isolated finite solutions of $F_{N_0} = 0$ for some small N_0. Then, starting with $N = N_0$, we bootstrap to finer grids by alternating between a filtering heuristic that cuts A_N down to a smaller set S_N of solutions of $F_N = 0$ and then uses these to spawn a new set of solutions A_{N+1} to $F_{N+1} = 0$. Iterating this process, a final application of the filtering heuristic produces candidate solutions at a target level of discretization.

Filtering heuristics can be of several types, possibly used simultaneously. For example, since we are interested in real solutions, we might throw out all solutions with the absolute value of the imaginary parts above some positive number. For another example, if we knew that the solutions of interest was everywhere greater than some constant, we might throw out solutions that dipped too far below that constant. A third example is obtained via differentiation of the original ODE. For example, differentiating (17.1) produces $y''' = 6y^2y'$, which the solutions of (17.1) must satisfy. Thus, one could consider throwing out solutions for which the absolute value of some discretization of $y''' - 6y^2y' = 0$ is greater than some positive constant. Note that some looseness should be allowed in all these tests, as it is possible that in refining the discretization, a solution in A_{N+1} may improve its test score compared to its parent solution in A_N. The purpose of the filtering is to remove solutions that are unlikely to spawn reasonable solutions at any subsequent stage.

One approach to computing S_{N+1} from S_N is as follows.

1. Add a new node, e.g., $x_{N-0.5} = (N-0.5)\cdot h_N$, and a new equation involving a new variable $y_{N-0.5}$, e.g.,

$$g(y_{N-1}, y_{N-0.5}, y_N) = y_N - 2y_{N-0.5} + y_{N-1} - 2\left(\frac{h_N}{2}\right)^2 y_{N-0.5}^3 = 0;$$

2. for each $y^* \in S_N$, solve $g(y^*_{N-1}, y_{N-0.5}, y^*_N) = 0$: since g is cubic in the example, we expect a finite set B_N consisting of at most $3 \cdot \#S_N$ solutions of the expanded system consisting of F_N and g;

3. construct a homotopy moving from the system consisting of F_N and g to the system F_{N+1} (where we rename the variables as needed);

4. use the homotopy to track the solutions B_N to a finite set of solutions A_{N+1} of $F_{N+1} = 0$; and

5. apply filtering rules to produce our new finite set of solutions S_{N+1}.

There are many choices of where to put a new node (or nodes) and many choices of homotopies. We refer to [4, §3.1] for one set of choices applied to this problem.

Domain decomposition A domain decomposition method [78] provides an alternative method for generating solutions on a finer grid from ones on a coarser grid. This approach has proved to be a flexible and powerful technique to find many solutions of systems of nonlinear differential equations, e.g., to find many new solutions to tumor-growth models such as that of §17.1.3 with μ a given positive constant. We again take the model problem (17.1).

As with the filtering approach above, the first step is to compute all solutions of $F_{N_0} = 0$ for some small N_0. Let S_{N_0} denote the set of isolated solutions where, possibly, the solutions with "large" imaginary part have been removed.

Assume that we have S_N consisting of solutions of $F_N = 0$. Recall that this discretization defines h_N and $x_0, \ldots, x_N$. For each $y^* = (y^*_0, \ldots, y^*_N) \in S_N$, consider the N variants of (17.1) for $j = 0, \ldots, (N-1)$ on the interval $[x_j, x_{j+1}]$:

$$y'' = 2y^3, \quad y(x_j) = y^*_j, \quad y(x_{j+1}) = y^*_{j+1}. \tag{17.12}$$

Fixing a small integer $M \geq 1$, we solve each variant following a similar discretization presented in (17.2), namely,

$$G_{j,M}(z_0, z_1, \ldots, z_M) = \begin{bmatrix} z_0 - y^*_j \\ z_2 - 2z_1 + z_0 - 2h_M^2 z_1^3 \\ \vdots \\ z_M - 2z_{M-1} + z_{M-2} - 2h_M^2 z_{M-1}^3 \\ z_M - y^*_{j+1} \end{bmatrix} = 0. \tag{17.13}$$

We note that parameter homotopies (see Chapter 6) can be used to efficiently solve $G_{j,M} = 0$, where the parameters are y^*_j and y^*_{j+1}. After solving these for $j = 0, \ldots, (N-1)$, each of which may have multiple solutions, we combine one solution from each to form an approximate solution of the form

$$w = (w_i, i = 0, \ldots, (NM)),$$

where w_{jM+k}, $j = 0, \ldots, (N-1)$, $k = 0, \ldots, M$, is the value of z_k in the selected solution to $G_{j,M} = 0$. That is, w_i for integer multiples of M coincide with the coarse discretization, i.e., $w_{j \cdot M} = y^*_j$, while the rest satisfy the fine discretization between these. Although the endpoints of the fine discretization segments match, the finite difference approximation of the differential equations at these nodes is not satisfied to the fine resolution, so an additional homotopy moves this mixed discretization to a fine discretization defined by $F_{NM} = 0$. By this process, different combinations of solutions to $G_{j,M} = 0$ produce various solutions to the fine discretization $F_{NM} = 0$. We may filter the results and use the most promising solutions to repeat the procedure, producing an even finer discretization, as needed.

17.2.2 ▪ Solutions from Continuation and Bifurcation

In §17.2.1, we discussed two techniques to construct solutions of nonlinear systems. A third technique is to take a solution dependent on a parameter and use continuation as the parameter changes to construct new solutions. The particularly interesting situation is when a bifurcation happens. This technique was applied in the articles on the tumor-growth model of §17.1.3 and the more complicated models cited in that subsection to obtain new solutions on different solution branches.

By monitoring the condition number of the Jacobian of the polynomial system associated to (17.9) as we use parameter continuation with the tumor-aggressiveness factor μ changing, we can find bifurcation values of μ which occur where different solution branches meet. In the cases where these values are known via theoretical computation, this approach computes the same values. Once a bifurcation value is known, the tangent cone to the solution curve at the bifurcation point provides the directions needed to track the solutions along the crossing branch. The tangent cone at a point is computable from the higher partial derivatives of the curve.

17.2.3 ▪ Hyperbolic Conservation Law Problems

Consider the polynomial system associated to the discretization of the differential equation (17.11) from §17.1.4. It is often the case that the solution arising as the limit of the homotopy (17.11) is either singular or ill-conditioned. Using Bertini with an endgame (the Cauchy endgame was used in [80]) yields a good approximation of the sought-after solution. For the examples of [80], this method is over an order of magnitude faster than the usual time-marching approach using (17.10). Moreover, the advantage increases as the size of the problem grows.

17.3 ▪ Software Summary

As of this writing, the algorithms described in this chapter have been applied to several research problems using code, which has not yet been publicly released, built using tools in Bertini.

Part IV

Bertini Users Manual

Appendix A
Bertini Quick Start Guide

The goal of this short appendix is to help new users learn how to run some simple first examples in Bertini. Specific syntax and formatting requirements for the numerous modes of running Bertini are provided in the subsequent appendices.

A.1 ▪ Installing Bertini

Bertini is currently freely available at the website

`bertini.nd.edu`

A link to this site, along with files for the examples contained in this book, is available at `www.siam.org/books/se25`. The permanent `doi` for the Bertini home page is

`dx.doi.org/10.7274/R0H41PB5`

Users may download an executable or the source code. For beginners, the proper executable should suffice.

No installation is necessary for the executable, though the user may find it useful to move Bertini onto the path, e.g., into a `bin` directory in the case of Linux. Windows users can run Bertini in Cygwin, a free Linux emulator available at `www.cygwin.com`.

A basic `Makefile` is provided with the source code. The user will need to have a C compiler (e.g., `gcc`), `GMP`, `MPFR` [57], `bison`, and `flex` installed in order to compile Bertini. Further details regarding compilation may be found in the `README` file downloaded with the source code.

A.2 ▪ The First Run

This section is intended to help the first-time user understand how a simple Bertini run works. It is assumed that the reader already has the executable available.

Bertini input and output are based primarily on files. For most sorts of runs, the user must create an input file, call Bertini, then read either the output summary on the screen or an output file. Many examples and types of runs are described later in this appendix and the subsequent appendices.

For this first run, create a new directory and either put Bertini on your path (preferred) or move the Bertini executable to this new directory. Make it your current working directory, as well. This can all be done on the command line.

Using your favorite text editor (emacs, vim, etc.), create a file named `input` with exactly the following content:

```
 variable_group x,y;
 function f,g;

 f = x^2 - 1;
 g = y^2 - 1;
END;
```

Obviously, this system will have four solutions, $(\pm 1, \pm 1)$.

Once `input` has been created, it is time to call Bertini. On the command line, just type:

```
» bertini input
```

and watch the output fly by. Bertini should run this system extremely quickly.

All forms of the output are described in §D.1. For now, try opening `main_data` and skimming through the four blocks of information, one block per solution. Once you have spotted the four solutions, you know the basics of running Bertini and are prepared to read ahead to discover the tools that are now available to you.

A.3 ▪ Basic Format for Input Files

For most runs, Bertini needs one file, the `input` file. The `input` file consists of an optional list of configurations (e.g., the desired accuracy of the solutions), declarations for almost all names used in the definition of the polynomials (e.g., all variables), and the target polynomial system itself. The exact syntax for the `input` file is given in §B.1. This file need not be named *input*; any legal Unix file name will suffice. The name `input` is used here for convenience.

For example, the following `input` file could be used to find the isolated complex solutions of $x^2 - 1$ with all configurations set to the default values:

```
% Sqrt1.input
 function f;
 variable_group x;

 f = x^2 - 1;
END;
```

Note that each statement must end with a semicolon and that the entire file is terminated with the keyword `END`. All white space (spaces, tabs, and blank lines) is ignored by Bertini. The `%` character is the comment marker; all text after it to the end of the line is ignored. Here and throughout this book, we use a comment to note the file name under which the input can be found in the website for this book. Since no configuration settings appear in this example, Bertini will execute using all default settings (see Appendix E).

Suppose that you would like to have solutions computed to 20 digits of accuracy, rather than the default of 11. Since we want to change a configuration, we must add a "`CONFIG` block" to the beginning of the `input` file. Thus, the `input` file would have the form

```
% MoreAccuracy.input
CONFIG
 FinalTol:1e-20;
END;
INPUT
 function f;
 variable_group x;

 f = x^2 - 1;
END;
```

In this case, the beginning of the input section was specified by including the keyword INPUT just before the declarations. This was necessary because of the inclusion of the configuration section at the top of the file, starting with the keyword CONFIG and ending with END;. The configuration statements are listed between those keywords.

A.4 ▪ Declaring Variables and Subfunctions

Most names used in a Bertini input file are declared as described above, with a line consisting of a Bertini data type followed by a list of names. However, variables and subfunctions follow special sets of rules, described in this section.

Notice that the type for the variables in the above example is variable_group and not just variable. In fact, Bertini will internally solve a homogenized version of the input system on projective space but provide output in the user-provided affine coordinates. (See Chapter 4 for more on projective spaces and homogenization.)

Alternatively, if a group of variables appears homogeneously in the polynomial system, it means that only their ratios matter. Such a group is declared using the hom_variable_group keyword. In this case, Bertini solves for the ratios (in projective space) and reports the results scaled so that one of the variables is exactly 1. Usually, hom_variable_group is used only when the variables of the problem naturally appear homogeneously, such as the elements of an eigenvector (see Example 4.1). A simple example of a homogeneous system is

```
% Homogeneous.input
 function f;
 hom_variable_group x,y;

 f = x^2 - y^2;
END;
```

Bertini will find two roots: $[x,y] = [1,1]$ and $[x,y] = [1,-1]$. About half the time, the second of these will be reported as $[x,y] = [-1,1]$ instead, which amounts to the same thing, since only the ratio of the variables matters.

When using hom_variable_group, Bertini checks that the system is properly homogenized. If not, Bertini displays an error message and quits.

When seeking only the isolated solutions of a system, the user may declare multiple variable groups using several variable_group or hom_variable_group statements, or a mix of both. These groupings determine how Bertini forms a homotopy, and different ways of grouping of the variables for the same polynomial system typically result in different numbers of solution paths in the homotopy, which affects the amount of computer time consumed in solving the problem. (See Chapter 5 for more on this.)

Here is an example, adapted from [68], of an input file in which two groups of variables are declared:

```
% TwoGroups.input
 variable_group z1;
 variable_group z2;
 function f1,f2;

 f1 = (29/16)*z1^3 - 2*z1*z2;
 f2 = z2 - z1^2;
END;
```

For this problem, splitting the variables into two groups results in a homotopy with five paths instead of the six paths that one gets with both variables in the same group. The difference is slight here, but for some problems, choosing the right groupings can have a powerful effect.

Bertini also allows for the use of subfunctions when defining functions. If some expression appears repeatedly throughout a polynomial system, it may be named, and all instances of that expression can then be replaced with that name. In this way, the expression, i.e., subfunction, will be evaluated only once per polynomial system evaluation. Subfunctions are the only names in an input file that do not need to be declared before being used. The following is an example with three subfunctions, S, T, and U, adapted from [155]:

```
% Subfunctions.input
 variable_group x,y,z;
 function f1,f2,f3;

 S = x^2+y^2+z^2-1;
 T = y-x^2;
 U = x*y-z;

 f1 = S*T*(x-2);
 f2 = S*U*(y-2);
 f3 = S*T*U*(z-2);
END;
```

A.5 ▪ Positive-Dimensional Examples

A "positive-dimensional run" is one in which Bertini is asked to find not just isolated points but also solution curves, surfaces, etc., if they exist. The input files for positive-dimensional runs are almost identical to those for zero-dimensional runs. In fact, other than a few configuration settings, it is impossible to distinguish between the two. The key differences are as follows:

- To specify that Bertini should do a positive-dimensional run (rather than the default zero-dimensional run), the user must change the `TrackType` setting in the `CONFIG` block of the input file. A basic positive-dimensional run is `TrackType:1;` with more options described in §E.1.1.
- The user can specify only one variable group, i.e., the variables must all be declared in a single statement using `variable_group` or `hom_variable_group`.

For example, a witness set for a curve known as the twisted cubic could be computed by Bertini using the following `input` file:

```
% TwistedCubic.input
CONFIG
  TrackType:1;
END;
INPUT
  variable_group x,y,z;
  function f1,f2;

  f1 = x^2-y;
  f2 = x^3-z;
END;
```

Using this `input` file, Bertini will produce three points on the twisted cubic curve. In fact, it will display to the screen that it has discovered a degree 3 curve, and the points themselves may be found in various output files, including one named `main_data`. More details on the output for positive-dimensional runs may be found in §D.2. Further details on what happens behind the scenes for such a run may be found in Part II of the book.

Suppose more points on the twisted cubic are needed, say, 200. Then, simply change the `TrackType` line to

```
TrackType:2;
```

Assuming the `witness_data` file from the previous run is still intact, this run will cause a catalog of all components in all dimensions to appear on the screen. Bertini will prompt the user to enter a codimension, a component number, and a number of points (to which you would respond 2, 0, and 200, respectively, in this case). The final prompt asks where to print the sample points. To print the points to a file, enter 0 followed by the name of the file. Otherwise, enter 1 to print the points to the screen. Bertini then produces the appropriate number of points and reports them as requested. Sampling is discussed briefly in §8.3. Further details regarding this and other interactive modes of Bertini may be found in §C.2.

Finally, suppose you have a point in $\mathbb{C}^3$ that you think might be on the twisted cubic curve. You could place that point into a file named `member_points` (with syntax identical to the `start` file described in §B.2), change `TrackType` to 3, and again ensure that the `witness_data` file is intact. Bertini would then perform the component membership test on that point, along with any other points listed in the `member_points` file, and report which component the point appears to lie on, up to the accuracy specified in `FinalTol`. Membership testing is described in §8.4.

A.6 ▪ User-Defined and Parameter Homotopies

For most types of runs, Bertini reads in the target system from the `input` file and then automatically generates and solves a compatible start system. However, two types of runs—user-defined homotopies and parameter homotopies—are a bit different. User-defined homotopies (see §7.5) require the user to fully specify the homotopy to be used, including the start system and the solutions of the start system. Parameter homotopies, discussed in full detail in Chapter 6, involve a regular run, followed by one or more runs at different points in a parameter space.

A.6.1 ▪ User-Defined Homotopies

User-defined homotopies are included in Bertini to allow users to create their own homotopies. This means that Bertini will simply track paths, as it is up to the user to create the homotopy and to provide Bertini with a list of startpoints. The important configuration keyword to provide your own homotopy is `UserHomotopy`.

There are two types of user-defined homotopies, the first being a user-defined homotopy in affine coordinates. In this case, the user must specify the variables using the keyword `variable` (since there are no groups involved as far as Bertini is concerned). It is also necessary to specify the name of exactly one path variable (i.e., the variable that marches from 1 to 0) and at least one parameter that depends on the path variable. The path variable may not appear in the definition of the functions, so if no parameter is actually desired, it is necessary to declare one parameter and set it equal to the path variable. Here is a simple `input` file to solve $x^2 = 0$ by moving from $x^2 - 1$ via a linear homotopy:

```
% UserHomotopy_Affine.input
CONFIG
 UserHomotopy:1;
END;
INPUT
 variable x;
 function f;
 pathvariable t;
 parameter s;
 constant gamma;

 gamma = 0.8 - 1.2*I;
 s = t;

 f = (x^2-1)*gamma*s + (x^2)*(1-s);
 % Note: t does not explicitly appear in f
END;
```

This homotopy explicitly employs the gamma trick (see Chapter 6) since Bertini will always move the path variable from 1 to 0 through $\mathbb{R}$. We note that the parameters need not be linear functions of the path variable. In fact, the dependence of the parameters on the path variable does not even have to be polynomial; for an example, see §2.3.

The second type of user-defined homotopy variable is declared using the keywords `variable_group` and `hom_variable_group`. (This causes Bertini to homogenize the affine groups and run the homotopy in a product of projective spaces, dehomogenizing at the end to report the affine groups in affine space.) In this case, the `UserHomotopy` configuration is set to 2, as shown in the following `input` file.

```
% UserHomotopy_Projective.input
CONFIG
 UserHomotopy:2;
END;
INPUT
 variable_group x;
 function f;
 pathvariable t;
 parameter s;
 constant gamma;
```

```
    gamma = 0.8 - 1.2*I;
    s = t;

    f = (x^2-1)*gamma*s + (x^2)*(1-s);
    % Note: t does not explicitly appear in f
  END;
```

With user-defined homotopies, the user must also provide the list of starting points in a file called `start`. For both of the above examples, the following is a valid `start` file:

```
  2
-1.0 0.0
 1.0 0.0
```

Further details regarding the `start` file may be found in §B.2.

A.6.2 ▪ Parameter Homotopies

As discussed in Chapter 6, the parameter homotopy can be a very powerful tool if one needs to find the nonsingular isolated solutions of a parameterized polynomial system for several values of the parameters. Bertini has a special run type for parameter homotopies, controlled by the configuration setting `ParameterHomotopy`. To get started, this is set to 1, causing Bertini to solve the parameterized polynomial system specified by the user in the `input` file at a random point in parameter space; that is, Bertini sets all the parameters to random complex numbers. These are stored for future use in a file called `start_parameters`. Bertini solves the problem just as in any other zero-dimensional run. This initial run is called the ab initio run to distinguish it from the true parameter runs that follow.

The power of parameter homotopies comes when one wishes to solve several problems in the same parameterized family. In this case, the user provides the target parameters in a file named `final_parameters` and sets `ParameterHomotopy` to 2. This causes Bertini to use a homotopy that moves the parameters in a straight line from the stored values in `start_parameters` to the target values in `final_parameters`. The solution paths of the homotopy begin at the points listed in a file named `start`. Usually, this file is obtained by renaming the file `nonsingular_solutions` that was produced in the ab initio run. (The sole purpose of the ab initio run is to provide this file.) The savings come about when the number of these solutions is less than the number of paths tracked in the ab initio run.

As a simple example, consider the `input` file

```
% ParameterHomotopy.input
CONFIG
 ParameterHomotopy:1;
END;
INPUT
 variable_group x;
 function f;
 parameter s;

 f = x^2-s;
END;
```

Running this file through Bertini, the file `start_parameters` is populated with the randomly chosen value of s, while `finite_solutions` is populated with the two square roots of that value of s. After building the file `final_parameters`, renaming `finite_solutions` to `start`, and changing `ParameterHomotopy` to 2, we are ready for our parameter run. Here's an example of `final_parameters`:

```
1

2 0
```

This corresponds to $s = 2 + 0i$. Running Bertini again, we find the square roots of 2 in the various output files; see Appendix D for more on output. (Parameter homotopy provides no savings in this simple example, which we discuss here merely to show the correct procedure. See Chap. 6 for examples that do result in computational savings.)

The user needs to execute the ab initio run just once, after which any number of parameter homotopies can be launched. Each such homotopy requires its own copy of `final_parameters`, and each is launched in a separate call to Bertini. If instead one would like to automate solving the problem for a grid of parameter values, the software module `Paramotopy`, described in Appendix H, provides this capability, making use of Bertini for performing the path tracking.

Finally, it is possible to "trick" Bertini into doing parameter homotopies without first doing the ab initio run. It is adequate to create the `input`, `start`, `start_parameters`, and `final_parameters` files. This could be useful if, for example, you know the solutions to one general polynomial system in a parameterized family and do not want to take the time to have Bertini recompute them.

At this point, you should be able to carry out several different sorts of Bertini runs. Many further details about the use of Bertini are provided in the appendices that follow.

Appendix B
Input Format

Bertini receives most of its user-provided input via files. In this chapter, we detail the various forms of input that may be provided to Bertini, along with the pertinent syntax. Section B.1 covers the `input` file. The subsequent sections cover the `start` file and other sorts of input files, and say a few words about interactive modes.

B.1 ▪ Syntax for Input Files

This section details the syntax of the Bertini `input` file. After going over the basic structure of the input file in §B.1.1, we provide restrictions on the choice of names (§B.1.2) and functions (sin(), cos(), etc.) that may be used in the `input` file to specify constants (§B.1.3). Care has been taken to provide some syntax checking in Bertini together with some information to help pinpoint the error. Some common syntax errors are recorded in §B.1.4.

B.1.1 ▪ The `input` File

The `input` file has two parts, grouped as follows (where the `%` symbol is the comment character in the `input` file):

```
CONFIG

% Lists of configuration settings (optional)

END;
INPUT

% Name declarations

% Optional assignments (parameters, constants, etc.)

% Function definitions

END;
```

The upper portion of the file consists of a list of configuration settings. Any configuration that is not listed in the `input` file will be set to its default value. A table of all configuration settings that may be changed, along with their default settings and acceptable ranges, may be found in Appendix E.

The syntax for the configuration lines is straightforward. Each line consists of the name of a setting (case does not matter), followed by a colon, the setting, and a semicolon. For example, to change the tracking type to 1 (the default is 0), simply include the following line somewhere in the `CONFIG` portion of the `input` file:

```
TrackType:1;
```

The lower portion of the `input` file begins with a list of symbol declarations (for the variables, functions, constants, and so on). All such declarations have the same format,

```
KEYWORD a1,a2,...,aN;
```

where `KEYWORD` depends upon the type of declaration. All symbols used in the input file must be declared, with the exception of subfunctions, as described in §A.4. Here are details regarding each type of symbol that may be used in the input file:

- **FUNCTIONS**

  Regardless of the type of run, all functions must be named, and the names must be declared using the keyword `function`.

  Here's an example:

  ```
  ...
  function f,g;

  f = 3*x^2-x*y+5;
  g = 8*y^2-2*x+2;
  ....
  ```

  Functions (or subfunctions, below) can use the basic operations +, - (unary or binary), *, /, ^, and parentheses in any sentence that is mathematically valid. For polynomial solving, the denominators must be constant and the exponents must be nonnegative integers. Some other basic functions are available, e.g., sin() and exp() (see §B.1.3).

  The only names that may be used to define functions or subfunctions are constants, variables, parameters, and subfunctions. Any parameters or subfunctions that appear in a function definition must be defined above the function. Path variables should not appear in functions or subfunctions. See the item CONSTANTS below for details on allowable formats for numbers.

- **VARIABLES**

  In all cases *except* user-defined homotopies on affine space, the variables are declared by group with one group per declaration statement. These declarations begin with either the keyword `variable_group` (for affine variable groups against which the polynomials have not been homogenized) or the keyword `hom_variable_group` (for variable groups against which the polynomials have been homogenized). Only one variable group may be used for a positive-dimensional run. For example, if there are two affine variable groups, the appropriate syntax would be

  ```
  variable_group z1,z2;
  variable_group z3;
  ```

In the case of affine user-defined homotopies, the keyword is `variable`.

- **PATH VARIABLES**

  If the user is specifying a user-defined homotopy, then a path variable must be declared in the `input` file. A path variable should not be declared in any other context. The path variable, often denoted by the letter "t", is the independent variable that is controlled during homotopy continuation. In Bertini, the homotopy always moves from the start system at $t = 1$ to the target system at $t = 0$. In a user-defined homotopy, it is also necessary to declare at least one parameter, as described in the next item. The keyword for path variables is `pathvariable`.

- **PARAMETERS**

  Homotopy continuation relies on the ability to cast a given polynomial system as a member of a parameterized family of polynomial systems. Sometimes there is only one parameter involved, but sometimes there are several. Parameters should be used only in the case of user-defined or parameter homotopies.

  In the case of user-defined homotopies, each parameter depends directly upon the path variable. The user must both declare each parameter and assign to it an expression depending only upon the path variable. Here is an example:

```
...
parameter p1,p2;
pathvariable t;
...
p1 = t^2;
p2 = t^3;
....
```

  For user-defined and parameter homotopies, Bertini assumes that there is at least one parameter. In the case that the user wishes to build a homotopy depending only upon the path variable, it is still necessary to declare a parameter, set it to the path variable in the assignments section, and then use only that parameter (and *not* the path variable) in the functions. Here is an example:

```
...
function f;
pathvariable t;
parameter s;
...
s = t;
f = x^2 - s;
....
```

  In the case of parameter homotopies, a path variable is automatically constructed by Bertini, so the user should not define a path variable. The parameters are declared but not defined (since the starting and ending parameters are handled in other files—see §A.6.2). The homotopy that Bertini constructs linearly interpolates between the given starting and ending parameters.

  No parameters should appear in the `input` file unless the user is doing a user-defined homotopy or a parameter homotopy, and the path variable should never appear directly in any polynomial definition.

- **CONSTANTS**

  Since it is sometimes useful to have constants gathered in one location (rather than scattered throughout the functions), Bertini has a `constant` type. If a name of constant type is to be used, it must be both declared and assigned a constant expression (see below). Here is an example:

```
...
constant g1,g2;
constant phi;
...
g1 = 1.25;
g2 = 0.75 - 1.13*I;
phi = exp(2*Pi*I/3);
....
```

  One advantage of declaring a constant instead of using a constant expression directly in a function or creating a subfunction (see below) that evaluates to a constant is that the constant is evaluated once at the beginning of a run, possibly updated occasionally when the working precision is changed. In contrast, a constant expression appearing anywhere else will be reevaluated each time Bertini evaluates the system during path tracking or other operations.

- **RANDOM and RANDOM_REAL**

  Many applications require the use of random numbers, either real or complex. The keywords `random` and `random_real` are used for this purpose. The following bit of code will produce a real number k and a complex number w, to be used in the subsequent polynomials.

```
...
random_real k;
random w;
...
```

  You can see these random numbers in the file `random_values` and in the copy of the polynomial system reported at the end of `main_data` after the run.

- **SUBFUNCTIONS**

  Redundant subexpressions are common in polynomial systems coming from applications. For example, the subexpression `x^2 + 1` may appear in each of 10 polynomials. To use a subfunction, simply choose a symbol, assign the expression to the symbol, and then use it in the functions. There is no need to declare subfunctions, as described in §A.4.

```
...
V = x^2 + 1;
...
f1 = 2*V^2 + 3.7;
....
```

B.1.2 ▪ Allowable Names

Names (of functions, variables, etc.) must fit the regular expression

```
[a-zA-Z][a-zA-Z0-9_\[\]]*
```

In other words, names must begin with a letter (lowercase or capital), followed by any combination of letters, digits, underscores, and square brackets.

Exceptions. Names cannot be any of the following:

- keywords, such as `variable` or `constant`;
- `I`, which is reserved for $\sqrt{-1}$;
- `Pi`, reserved for π; and
- special functions described in the next section, such as `sin`.

Names can be declared only once per file and can be assigned a value only once as well.

B.1.3 ▪ Constant and Parameter Expressions

Constant expressions can occur in the definitions of constants, functions, subfunctions, and, in the case of a user-defined homotopy, parameters. A constant expression is a number or an expression that evaluates to a number, that is, one that does not contain any variables. Parameter expressions, which may appear in parameter homotopies and user-defined homotopies, are like constant expressions, except they can additionally depend on the path variable.

Bertini will accept numbers in either standard notation (e.g., 3.14159 or 0.0023) or scientific notation (e.g., 0.314159e1 or 2.3e-3). No decimal point is needed in the case of an integer. To define complex numbers, use the reserved symbol `I` for $\sqrt{-1}$, e.g., `1.35 + 0.98*I`. Please note that the multiplication symbol `*` is always necessary.

As a technical aside, each number should fit the regular expression:

```
([0-9]+[\.]?([0-9]*)?([eE][-+]?[0-9]+)?)|([0-9]*\.[0-9]+([eE][-+]?[0-9]+)?)
```

When using Bertini, users should strive to include exact constants, lest the software interpret inaccurate constants as exact. For example,

```
1/3
```

is preferred to

```
0.333
```

which Bertini would interpret and use as $\frac{333}{1000}$, at any level of precision.

Since we want users to be as precise as possible, we have included in Bertini several functions that may be used when defining constants and parameters. These are

- `exp()`,
- `sin()`,
- `cos()`, and
- `tan()`.

In addition, for expressions a and b, one may use

- `a^b`, where `b` evaluates to a real number, and
- `pow(a,b)`, which is exactly the same as `a^b`.

These may be nested within each other, and the argument passed to the first four can be any constant or parameter expression. For the last two, additional restrictions are placed upon a depending on the integrality of b. If b is not an integer, then a must be a positive real constant. If b is a nonnegative integer, then a can be any valid expression, and $pow(a,0) = a^0 = 1$. Except for affine user-defined homotopies, a must be a nonzero real constant when b is a negative integer. For affine user-defined homotopies, a can be any valid nonzero expression.

Bertini will read in all provided digits and will make use of as many as possible in computations, depending on the working precision level. If the working precision level exceeds the number of digits provided for a floating-point number, all further digits are assumed to be 0 (i.e., the input is always assumed to be exact). This is the natural choice, as there is no way for Bertini to know the intended values of any digits not specified. However, the user should be aware that truncating coefficients can have an impact on the corresponding algebraic set—see §8.5 and §3.2.4. Rational numbers are stored exactly and approximated to the working precision when needed.

B.1.4 ▪ Common Syntax Errors in the `input` File

The following are common errors that occur in the `input` file:

- All statements must end with a semicolon. This also applies to `CONFIG ... END;` and `INPUT ... END;` block structures.
- Bertini is case-sensitive for keywords and names defined by the user.
- The symbol for $\sqrt{-1}$ is `I`, not `i` (but one may use `i` after including the statement "`i=I;`").
- In scientific notation, the base is represented by either `e` or `E`, for example, `1.23e-5` means 1.23×10^{-5}.
- For multiplication, `*` is necessary (concatenation is not good enough).
- Exponentiation is designated by `^` and not `**`.
- All symbols (except subfunctions) must be declared.
- No symbol may be declared twice.
- Both a path variable and at least one parameter are needed for user-defined homotopies.

B.2 ▪ The `start` File

In many situations, Bertini will automatically create a homotopy, including a start system, and solve the start system to find the points at the beginning of each path. The only exceptional case is when the user provides a homotopy. In that case, since the user is free to define the start system, even a nonpolynomial one, the user must provide a set of starting points for the homotopy. These points must be provided in a `start` file.

The structure of this file is very simple, though it is extremely specific. In the following, `[]` represents a blank line:

```
[integer number of points in the file]
[]
[pt 1, coord 1, real] [pt 1, coord 1, imag]
[pt 1, coord 2, real] [pt 1, coord 2, imag]
...
[pt 1, last coord, real], [pt 1, last coord, imag]
[]
[pt 2, coord 1, real] [pt 2, coord 1, imag]
[pt 2, coord 2, real] [pt 2, coord 2, imag]
...
[pt 2, last coord, real], [pt 2, last coord, imag]
[]
 ...
[]
[last pt, coord 1, real] [last pt, coord 1, imag]
[last pt, coord 2, real] [last pt, coord 2, imag]
...
[last pt, last coord, real], [last pt, last coord, imag]
```

Any additional lines beyond the last coordinate of the last point are ignored, so the user can freely add comments, etc., at the bottom of the file.

Each number in the file must be presented as a floating-point number (in the working precision). For example, the following is a `start` file for the two points $(1, 2-3i, 7+i)$, $(2.3, 1/4, 60000)$:

```
2

1  0
2 -3
7  1

2.3 0
0.25 0
6e4 0
```

B.3 ▪ Other Input Files

In the case of component membership testing (see §8.4), the `member_points` file is needed in addition to the `input` file. This file, which contains a list of all points to be tested, is identical in structure to the `start` file and must be named `member_points`.

In order to perform either component sampling (see §8.3) or component membership testing (see §8.4), a file must be provided containing the witness data of an algebraic set (i.e., the witness points and the linear slices, among other bits of information). This file, which must be named `witness_data`, is a compact, machine-readable file. Although in principle it could be produced manually, the best way to create it is by executing a positive-dimensional Bertini run. It is highly unlikely that a user would want to create this file by hand, so we do not record the rather complicated structure here.

B.4 ▪ Command-Line Input for Sampling and Sharpening

Bertini is interactive in three settings: component sampling (see §8.3), endpoint sharpening (see §7.2), and witness set printing (see §11.1). In each case, the user navigates through a short series of menus, with rather obvious choices to be made at each stage. Details are not provided here as the menus are straightforward. Automation of these interactive modes is described briefly in §C.2.

Appendix C
Calling Options

Calling Bertini from the command line is simple, as described in §C.1 below. Though most types of Bertini runs use file-based input and output and are not interactive, §C.2 describes three interactive modes. Advice on the automation of Bertini calls is given in §C.3.

C.1 ▪ Basic Bertini Calls

At the command line, the basic structure of a Bertini call is of the form

```
» bertini <input file>
```

or, if needed,

```
» bertini <input file> <start file>
```

where the default names are `input` and `start`, respectively. If the default names are used, they can be omitted from the command line. Bertini stores its output file in the current working directory, and any other files, e.g., `member_points`, must be in the current working directory.

Of course, one may call Bertini from another directory to store all of the output files in that directory. It is often useful to place a copy of Bertini in a fixed location, say, a `bin` directory, then produce a new directory for both input and output files for each run.

C.2 ▪ Interactive Modes

Bertini has three interactive modes: endpoint sharpening (`SharpenOnly:1`), component sampling (`TrackType:2`), and witness set printing (`TrackType:4`).

In each case, a call to Bertini, as above, will result in a sequence of queries asking exactly what the user wants done and/or where to find various data. This slightly complicates the automation of such calls, as described in the next section.

Below we describe the menus and queries that arise in each of the interactive modes.

Endpoint sharpening

Endpoint sharpening, invoked with the setting `SharpenOnly:1;`, makes use of one basic menu, with subsequent queries as needed:

```
Sharpening options:
  1. Sharpen all endpoints
  2. Sharpen endpoints listed in a file
  3. Manually input endpoints to sharpen
  4. Recreate output (i.e., run the postprocessor)
  5. Change the number of sharpening digits (currently 14 digits)
  6. Change the sharpening method (currently Newton's method)
  7. Quit
Please enter your option:
```

After option 5 or 6 is used, Bertini returns to this menu to allow the user to initiate a sharpening operation.

Here is a brief description of each option:

Option 1. This option uses all endpoints from output file `raw_data`.

Option 2. This option leads to the query

```
Please input the name of the file that lists the
endpoints to sharpen or type 'quit' or 'exit' (max of
255 characters):
```

after which Bertini runs all points listed in that file.

Option 3. Bertini reads in all files in `raw_data` and asks the user to specify a path number via a query of the form

```
The path numbers range between 0 and 239.  Please enter
a path number to sharpen (-1 to quit and -9 to display
the list of path numbers):
```

Option 4. This option reproduces the output files for the run.

Option 5. Bertini queries the user for the number of digits of accuracy desired.

Option 6. This options leads to the query

```
Sharpening method:
  1. Newton's method
  2. Endgame
Please enter your option:
```

after which Bertini returns to the main menu.

Option 7. This terminates the Bertini run.

One typical use of endpoint sharpening includes an increase in the number of sharpening digits (Option 5), followed by some list of endpoints to sharpen (Options 1–3). Another is to recreate the output files after changing some of the postprocessing configurations (Option 4).

Component sampling

When sampling is invoked with configuration setting `TrackType:2`, Bertini displays a sequence of queries, which can be summarized as follows:

```
Choose a dimension:
Choose a component:
How many points would you like on this component?
Enter 0 to write to a file, 1 for the screen:
Enter a file name (if file selected in previous query):
```

For the first two queries (dimension and component), Bertini provides a useful list of all dimensions and components (including degrees) that are available.

Witness set printing

The setting `TrackType:4`, which prints out a witness set, initiates an interactive mode that is very similar to component sampling. Bertini asks for the dimension and component, as above, then two file names. The first file name is for the witness points of the selected component. The second file name is for the linear slicing polynomials.

C.3 ▪ Automation of Bertini Calls

It is relatively simple to call Bertini from a shell script or from another program. All that is required is the creation of the necessary input file(s), a simple command line call, then the (somewhat more technical) parsing of some of the output files, as described in Appendix D.

To automate the interactive modes described above, simply create a text file with the response to each query or menu on its own line. Then redirect the input to this file when calling Bertini.

For example, suppose we want to print the witness data for the third 11-dimensional component of some algebraic set. If we want the points printed to a file named `pts` and the linear slicing polynomials printed to a file named `lins`, the text file (named `queryResponses`, for example) must contain exactly

```
11
2
pts
lins
```

where the 2 indicates the third component, as Bertini numbers components starting with 0. The call to Bertini would then be of the form

```
» bertini input < queryResponses
```

Appendix D
Output Files

Almost all Bertini output is written to files, and there are several files—some easily read by humans, others designed for straightforward mechanical parsing—produced by the most common types of runs. The goal of this appendix is to describe the human-readable and the most useful machine-readable output files. A few files, particularly those that would be useful only for debugging Bertini, are not described here.

The output for zero-dimensional runs is provided in §D.1, while the output for positive-dimensional runs is provided in §D.2.

D.1 ▪ Zero-Dimensional Runs

This section details the output from zero-dimensional runs, i.e., those runs used to find isolated solutions. Output to the screen is discussed briefly in the first subsection, followed by an overview of the various output files and specific output formats in the second and third subsections, respectively.

D.1.1 ▪ Output to the Screen

Aside from the interactive mode of endpoint sharpening (`SharpenOnly:1`), the output to the screen consists of a running list of which path is currently being tracked (see the related configuration `PrintPathProgress` in §E.11.3), followed by various statistics about the run and the endpoints of the paths. In the case of regeneration, the information during the run is organized a bit differently, but the user should be able to understand it after reading the section on regeneration (§5.4).

Perhaps the most important bit of information on the screen is at the end, namely, the number of "failed" paths, i.e., the number of paths for which the path variable failed to reach 0. This may happen for a number of reasons, and there is typically a bit of advice after the number of failed paths about how to get these paths to converge. If there is nothing listed after the number of paths tracked, then there were no path failures. The default option in Bertini (`SecurityLevel:0;`) allows paths that appear to be heading to infinity to be truncated early. Accordingly, such paths will be categorized as "failed" even though this is the action Bertini has been asked to perform. See §4.2 and §E.6.1 for more on this.

Similarly, the number of path crossings is reported to the screen at the end of the run. If it appears that paths crossed during the run, the user is advised to try the run again with tighter tracking tolerances. This anomaly is reported when two paths have nearly the same value at the start of the endgame (default of $t = 0.1$ using the configuration `EndgameBdry` described in §E.5.5). This could be a fluke in that two paths happen to be close at the start of the endgame, or it could mean that the tracker jumped from one path to another, resulting in two paths seemingly leading to the same endpoint while missing the endpoint of the jumped path. Tightening the tracking tolerances typically fixes this.

In the affine case, just after the running list of path numbers comes a table summarizing the number of finite solutions, classified as nonsingular/singular and real/nonreal. This is followed by a table listing the number of real and nonreal solutions by multiplicity. Here, *multiplicity* means the number of paths ending at the same endpoint. If infinite endpoints were computed, a similar summary is presented for those endpoints as well. The user-defined case is identical.

The projective case, in which all variables appear in a `hom_variable_group`, is similar to the affine case, but the tables list all solutions found, because there is no well-defined sense of finite versus infinite solutions.

D.1.2 ▪ Overview of Output Files

The following files are produced by each of the basic zero-dimensional runs (affine, projective, user-defined):

- `main_data`
- `raw_solutions`
- `raw_data`
- `nonsingular_solutions`
- `singular_solutions`

In addition, the files `real_finite_solutions` and `finite_solutions` are produced in the affine case. The file `real_solutions` is produced in the projective and user-defined cases.

Aside from the first three in the list above, the contents of each file are exactly what the name indicates. The file `main_data` is the main human-readable file for all zero-dimensional runs. The files `raw_solutions` and `raw_data` are machine-readable. The former simply lists all endpoints of all paths, while the latter lists all the information that can be found in `main_data`, but in a machine-readable format.

The contents of each file are specified in the next subsection.

D.1.3 ▪ Format of Specific Files

This technical section specifies the format of each zero-dimensional output file. We use the following notation:

`[type,description]`: data type and (optionally) brief description of data
`%`: Comment
. . .: Skipped text
+++: New line not actually in output file

main_data

```
Number of variables:  [int]
Variables:  [list of strings, variable names]
Rank:  [int, rank of the system]
- - - - - - - - - - - - - - - - - - - - - - - - - -
Solution [integer, index] (path number [int])
Estimated condition number:  [float]
Function residual:  [float]
Latest Newton residual:  [float]
T value at final sample point:  [float]
Maximum precision utilized:  [int]
T value of first precision increase:  [float]
Accuracy estimate, internal coordinates (difference of last
two
    +++ endpoint estimates):  [float]
Accuracy estimate, user's coordinates (after dehomogenization,
    +++ if applicable):  [float]
Cycle number:  [int]
[float, real pt of coord 1] [float, imag pt of coord 1]
[float, real pt of coord 2] [float, imag pt of coord 2]
...
[float, real pt of final coord] [float, imag pt of final coord]
Paths with the same endpoint, to the prescribed tolerance:  [list
   +++ of ints, all path numbers with same endpoint]
Multiplicity:  [int, number of ints in list on previous line]
% The previous block (beginning with the line of dashes)
% is repeated for each endpoint.
- - - - - - - - - - - - - - - - - - - - - - - - - -
At tol=[float], there appear to be [int] solutions.


********** input file needed to reproduce this run **********
% This is a copy of the input file that produced this output,
% including the random seed used for the run.

% The file ends with Bertini version information.
```

raw_solutions

```
[int, number of points]

[int, path number]
[float, real pt of coord 1] [float, imag pt of coord 1]
[float, real pt of coord 2] [float, imag pt of coord 2]
...
[float, real pt of final coord] [float, imag pt of final coord]
% The previous block (beginning with the path number)
% is repeated for each point.  One blank line separates
% consecutive points.
```

raw_data

```
[int, number of variables]
% The number of variables in this file is exactly the number
```

```
% specified in the projective case but is more than the number
% specified in the affine case, to account for the homogenizing
% variable(s).
0
% The following is a block, repeated for each point.
% There is no delimiter between consecutive blocks.
[int, path number]
[int, precision used for the endpoint]
[float, real pt of coord 1] [float, imag pt of coord 1]
[float, real pt of coord 2] [float, imag pt of coord 2]
...
[float, real pt of final coord] [float, imag pt of final coord]
[float, function residual]
[float, condition number of the Jacobian matrix]
[float, final Newton residual]
[float, final pathvariable value]
[float, difference between last two estimates at t=0]
[float, pathvariable value of first precision increase]
[int, cycle number from endgame]
[int, 1 if successful path else 0]
% End of block repeated for each point.
-1

[int, 0 for double; 1 for multiprecision; 2 for rational]
% Previous line describes format for remainder of data.
% Type "num" below is used to represent any of these options.
[int, number of variable groups] [int, number of variables]
% File ends with a list (whose length is the product of the
% integers on the previous line) of the coefficients of the
% patch equations used after homogenization
% Each line has the following format:
[num, real pt] [num, imag pt]
```

All other files

All other output files of interest to users have the same form as `raw_solutions` with one exception: the path number line is omitted from each point. These files intentionally have exactly the same structure as Bertini `start` files: such files often contain points to be used as starting points for subsequent runs.

D.2 ▪ Positive-Dimensional Runs

For positive-dimensional runs, particularly `TrackType:1` runs, the most interesting output is displayed on the screen. This information is described in the next subsection. The various output files for such runs are described in §D.2.2, while the output files for all other positive-dimensional runs are described in §D.2.3.

D.2.1 ▪ Output to the Screen

Output to the screen is, for most users, the most useful part of `TrackType:1` runs, i.e., numerical irreducible decompositions. Assuming the use of the default algorithm, i.e., the regenerative cascade, the screen output begins with a running list of which paths are being run in which codimension. This is followed by a table, Regenerative Cascade Summary providing information about the paths in each codimension. The columns of this table are mostly straightforward, and this output will not be of particular interest to most users.

After that table is printed, Bertini attempts to remove junk points in each codimension other than the top, after which it is known how many actual witness points are present in each codimension. Bertini presents these data in a straightforward table, Witness Set Summary.

After this, Bertini executes a pure-dimensional decomposition in each nonempty codimension, followed by the two tables of output that will interest most users. First comes the table Witness Set Decomposition, listing the number of components and points in each of the nonempty codimensions. The final bit of output is a table, Decomposition by Degree, that lists the dimension and degree of each component.

For one example consisting of a degree 2 surface, three lines, a cubic curve, and a point, the final table is as follows:

```
************** Decomposition by Degree **************

Dimension 2:  1 classified component
-----------------------------------
degree 2:  1 component

Dimension 1:  4 classified components
-----------------------------------
degree 1:  3 components
degree 3:  1 component

Dimension 0:  1 classified component
-----------------------------------
degree 1:  1 component

*****************************************************
```

D.2.2 ▪ Overview of Output Files

Aside from a wide array of intermediate files and logs filled with details of the tracking of each path, Bertini produces two output files that are of interest to most users, `main_data` and `witness_data`. Here are the specific formats for these two files, using the same notation as in the zero-dimensional case:

main_data

```
Number of variables:  [int]
Variables:  [list of strings, variable names]
Rank:  [int, rank of the system]

-------DIMENSION [int]-------

NONSINGULAR SOLUTIONS
----------
Path number:  [int]
Component number:  [int]
Estimated condition number:  [float]
[float, real pt of coord 1] [float, imag pt of coord 1]
[float, real pt of coord 2] [float, imag pt of coord 2]
...
[float, real pt of final coord] [float, imag pt of final coord]
```

```
Multiplicity:  [int]
Deflations needed:  [int]
% Previous block (beginning with line of dashes) repeated for
% each nonsingular endpoint in this dimension.
%
% List of singular endpoints is similar and begins with
% header SINGULAR SOLUTIONS.
%
% Dimension block (everything above here, up to the line
% containing the word DIMENSION) repeated for each
% nonempty dimension.
%
% Input file and version information conclude the file, as
% in the zero-dimensional case.
```

witness_data

```
[int, number of variables]
% Previous line includes additional homogenizing variable(s),
% as appropriate.
[int, number of nonempty codims]
% Following block represents a single codim; repeated for each.
[int, codim]
[int, total number of points in codim]
% Following block represents a single point; repeated for each.
[int, precision used for the endpoint]
[float, real pt of coord 1] [float, imag pt of coord 1]
[float, real pt of coord 2] [float, imag pt of coord 2]
...
[float, real pt of final coord] [float, imag pt of final coord]
[int, precision used for the last approximation]
% The next point is the last approximation of the point on the
path before convergence.
[float, real pt of coord 1] [float, imag pt of coord 1]
[float, real pt of coord 2] [float, imag pt of coord 2]
...
[float, real pt of final coord] [float, imag pt of final coord]
[float, condition number]
[int, corank of Jacobian at this point]
[float, smallest nonzero sing val of Jacobian]
[float, largest zero sing val of Jacobian]
[int, type]
[int, multiplicity]
[int, component number for this point]
[int, deflations needed for this point]
% End of block for point.
% End of block for codim.
-1 % designates end of witness points

% Remaining data is related to slices, randomization,
% homogenization, and patches.
[int, 0 for double; 1 for multiprecision; 2 for rational]
% Previous line describes format for remainder of data.
% Type "num" below is used to represent any of these options.
```

```
% The following block is repeated for each nonempty codim.
% First, matrix A used for randomization:
[int, number of rows] [int, number of columns]
% List of entries of A, each a complex number of type num.

% Second, matrix W used to maintain proper homogenization.
% List of entries of W, each an integer, of same length as A.

% Third, vector H used for homogenization, random if projective
input.
[int, size of H]
% List of entries of H, each a complex number of type num.

% Fourth, a number homVarConst
% (0 for affine, random for projective).
[num, real part] [num, imag part]

% Fifth, matrix B for linear slice coefficients.
[int, number of rows] [int, number of columns]
% List of entries of B, each a complex number of type num.

% Sixth and finally, vector p for patch coefficients.
[int, size of p]
% List of entries of p, each a complex number of type num.
% End of block repeated for each nonempty codim.
```

D.2.3 ▪ Output for other Types of Positive-Dimensional Runs

Component sampling

The output of component sampling (`TrackType:2`) is a list of points on the indicated component, either written to a file or displayed on the screen. If printed to a file, these points will be listed using the standard one-coordinate-per-line format used for all Bertini output.

Component membership

There is output both to the screen and to a file for component membership (`TrackType:3`). The screen output is easily human-readable and consists of statements about the component on which each point in `member_points` lies.

The file produced, `incidence_matrix`, is a machine-readable binary matrix indicating the component on which each point lies. Here is the exact format:

incidence_matrix

```
[int, number of nonempty codims]
% List of data in format on next line, one line per
% nonempty codim.
[int, codim] [int, number of components]

[int, number of points]

% Binary matrix with one row per point, columns corresponding
% to components.  0 if given point is not on given component;
% else 1.
```

Witness set printing

In this mode (`TrackType:4`), Bertini produces two files, one containing the witness points for the specified component, the other containing the linear slice information.

The file containing the witness points has the same format as a `start` file. The file containing the linear slice information is in the format of an `input` file with the linear polynomials listed as functions.

Appendix E
Configuration Settings

Configuration settings make Bertini highly configurable and permit a wide range of applications. This chapter provides a description of the configurations, acceptable values, and default values. Since the names of the configurations are case-insensitive, the names are presented in mixed case for improved readability. For easy reference, the configurations are presented in 11 categories and listed in Tables E.1 and E.2.

E.1 ▪ Track Type Settings

The `TrackType` configuration determines the computation that Bertini performs, which we summarize in ascending order.

E.1.1 ▪ TrackType

- Datatype: Integer
- Acceptable values: Integers between −4 and 6
- Default value: 0

−4 : Bertini evaluates the input system at the given points. If the input system is a user-defined or parameter homotopy, the value of the path variable for each point is read from the file `start_time`. If `start_time` does not exist, the path variable is set to 0.

−3 : Bertini evaluates the input system and its Jacobian with respect to the variables at the given points. If the input system is a user-defined or parameter homotopy, the Jacobian with respect to the parameters is also computed. The Jacobians are printed using a row major ordering. In the homotopy case, the value of the path variable for each point is read from the file `start_time`. If `start_time` does not exist, the path variable is set to 0.

−2 : Bertini performs a Newton iteration on the input system at the given points (see §7.3.2). If the input system is a user-defined or parameter homotopy, the value of the path variable for each point is read from the file `start_time`. If `start_time` does not exist, the path variable is set to 0.

Table E.1. *List of configurations.*

Category	Configuration name
§E.1: Track type	`TrackType`
§E.2: Multiple precision	`MPType`
	`Precision`
§E.3: Adaptive precision	`CoeffBound`
	`DegreeBound`
	`AMPMaxPrec`
	`AMPSafetyDigits1`
	`AMPSafetyDigits2`
§E.4: Tracker	`TrackTolBeforeEG`
	`TrackTolDuringEG`
	`ODEPredictor`
	`MaxNewtonIts`
	`MaxStepsBeforeNewton`
	`MaxStepSize`
	`MinStepSizeBeforeEG`
	`MinStepSizeDuringEG`
	`MaxNumberSteps`
	`StepsForIncrease`
	`StepFailFactor`
	`StepSuccessFactor`
	`PathTruncationThreshold`
§E.5: Endgame	`FinalTol`
	`EndgameNum`
	`SampleFactor`
	`NumSamplePoints`
	`EndgameBdry`
	`MinCycleTrackback`
	`NbhdRadius`
	`MaxCycleNum`
	`CycleTimeCutoff`
	`RatioTimeCutoff`
§E.6: Security	`SecurityLevel`
	`SecurityMaxNorm`
§E.7: Postprocessing	`TargetTolMultiplier`
	`FunctionTolerance`
	`RatioTolerance`
	`ImagThreshold`
	`CondNumThreshold`
	`EndpointFiniteThreshold`
	`SharpenDigits`
	`SharpenOnly`

−1 : Bertini performs a Newton iteration and estimates the condition number, using the spectral norm, of the Jacobian evaluated at each of the given points (see §7.3.2). If the input system is a user-defined or parameter homotopy, the value of the path variable for each point is read from the file `start_time`. If `start_time` does not exist, the path variable is set to `0`.

Table E.2. *List of configurations (continued).*

Category	Configuration name
§E.8: User-defined and parameter homotopies	`UserHomotopy`
	`ParameterHomotopy`
§E.9: Numerical irreducible decomposition	`WitnessGenType`
	`MaxCodimension`
	`SpecificCodimension`
	`JunkRemovalTest`
	`MaxLDTDepth`
	`WitnessSupersetOnly`
	`MultOneOnly`
	`MaxNumPtsForTrace`
	`MaxNumMonoLoops`
	`MaxNumBadMonoLoops`
	`ConstructWitnessSet`
§E.10: Regeneration	`UseRegeneration`
	`SliceTolBeforeEG`
	`SliceTolDuringEG`
	`SliceFinalTol`
	`RegenStartLevel`
	`RegenRemoveInf`
	`RegenHigherDimTest`
§E.11: Miscellaneous	`UseDiagonal`
	`OutputLevel`
	`PrintPathProgress`
	`ScreenOut`
	`RandomSeed`
	`IntrinsicMultiplier`
	`SingValZeroTol`
	`DeleteTempFiles`
	`NeedToDiff`

`0`: Bertini either computes a set of solutions of the input polynomial system or tracks solution paths defined by a user-defined or parameter homotopy (see Chapter 6). For a polynomial system, if either `UseRegeneration` (see §E.10.1) or `UseDiagonal` (see §E.11.1) is set to 1, Bertini computes the set of isolated nonsingular solutions. Otherwise, for polynomial systems, Bertini computes a superset of the isolated solutions (see Chapter 2).

`1`: Bertini computes a numerical irreducible decomposition of the input polynomial system (see Part II). See §E.9 for additional configuration settings.

`2`: Bertini samples a computed irreducible component of the solution set of the input polynomial system (see §8.3).

`3`: Bertini performs a membership test on the points in `member_points` with the computed irreducible components in the solution set of the input polynomial system (see §8.4).

4 : Bertini prints the linear system and witness points of a witness set for either a computed irreducible component or the union of the computed irreducible components of a given dimension (see §8.2).

5 : Bertini computes a witness set for the projection of a computed generically reduced irreducible component of the solution set of the input polynomial system (see Chapter 16). The projection is described in the file `projection`.

6 : Bertini performs an isosingular stabilization test (see §13.2) for the input polynomial system at the given point in `start`. The test checks whether the point is a smooth point on an irreducible component of multiplicity one. If it is and `ConstructWitnessSet` is set to 1, then Bertini performs monodromy loops (see §10.2.2) to generate additional points in a witness point set for the component and in this manner tries to find a complete witness point set. The trace test (see §10.2.1) is used to determine when the set is complete.

E.2 ▪ Multiple Precision Settings

Every algorithm implemented in Bertini can be executed using fixed double precision or fixed higher precision arithmetic, while path tracking algorithms are capable of these and adaptive precision (see §3.2). This is controlled by the `MPType` configuration. When using fixed higher precision, the `Precision` configuration determines the number of bits used. Bertini uses fixed precision for computations which do not actually track paths such as function and Jacobian evaluation and Newton iterations. In this case, Bertini defaults to double precision when adaptive precision path tracking is requested.

Additional configurations used for the adaptive precision path tracking algorithms are described in §E.3.

E.2.1 ▪ MPType

- Datatype: Integer
- Acceptable values: 0, 1, and 2
- Default value: 2

0 : Bertini uses fixed double precision arithmetic.

1 : Bertini uses fixed higher precision arithmetic with the precision determined by `Precision` (see §E.2.2).

2 : Bertini uses adaptive precision arithmetic for path tracking (see §3.2). See §E.3 for additional configurations.

E.2.2 ▪ Precision

- Datatype: Integer
- Acceptable values: Integers greater than or equal to 64
- Default value: 96

The value is the precision, in bits, when using fixed higher precision arithmetic. In general, N bits provide $\lfloor N \log_{10} 2 \rfloor$ decimal digits.

E.3 ▪ Adaptive Precision Settings

The adaptive precision path tracking algorithms discussed in §3.2 depend on information extracted from the input system and the following configurations.

E.3.1 ▪ CoeffBound

- Datatype: Double
- Acceptable values: Positive doubles
- Default value: 1000 for user-defined homotopies (estimated for all others)

The value is a bound on the coefficients of the input system which is used to adjust the precision in the adaptive precision path tracking algorithms (see §3.2). For all systems but user-defined homotopies, Bertini estimates a bound based on evaluating the system and its Jacobian at a random point. The user-provided value takes precedence over the internally approximated value.

E.3.2 ▪ DegreeBound

- Datatype: Integer
- Acceptable values: Positive integers
- Default value: 5 for user-defined homotopies (computed for all others)

The value is a bound on the degree of the input system which is used to adjust the precision in the adaptive precision path tracking algorithms (see §3.2). For all systems but user-defined homotopies, Bertini computes the degree bound during parsing. The user-provided value takes precedence over the internally computed value.

E.3.3 ▪ AMPMaxPrec

- Datatype: Integer
- Acceptable values: Integers greater than or equal to 64
- Default value: 1024

The value is the maximum precision used by the adaptive path tracking algorithms (see §3.2). Tracking of a path is stopped when the precision requested by the algorithm exceeds this maximum.

E.3.4 ▪ AMPSafetyDigits1

- Datatype: Integer
- Acceptable values: All integers
- Default value: 1

The value is the number of the first type of safety digits used in the adaptive precision path tracking algorithms (see §3.2 and [23]). Positive values increase precision to yield more security at some cost in computation time. Negative values speed computation by keeping the precision lower, with the possibility that the precision remains too low to accurately track the path.

E.3.5 ▪ AMPSafetyDigits2

- Datatype: Integer
- Acceptable values: All integers
- Default value: 1

The value is the number of the second type of safety digits used in the adaptive precision path tracking algorithms (see §3.2 and [23]). Positive values increase precision to yield more security at some cost in computation time. Negative values speed computation by keeping the precision lower, with the possibility that the precision remains too low to accurately track the path.

E.4 ▪ Tracker Settings

The tracking configuration settings are used to tune the path tracker (see §2.2) to the input system. Settings include changing the tracking tolerances, the predictor method, the number of Newton iterations per step, and minimum and maximum step size.

E.4.1 ▪ TrackTolBeforeEG

- Datatype: Double
- Acceptable values: Positive double
- Default value: 10^{-5}

The value is the tracking tolerance before the endgame (see §3.3) is started. In order to declare a step successful, the estimate for the error between the point and the path needs to be smaller than the tracking tolerance.

E.4.2 ▪ TrackTolDuringEG

- Datatype: Double
- Acceptable values: Positive double
- Default value: 10^{-6}

The value is the tracking tolerance during the endgame (see §3.3). In order to declare a step successful, the estimate for the error between the point and the path needs to be smaller than the tracking tolerance.

E.4.3 ▪ ODEPredictor

- Datatype: Integer
- Acceptable values: Integer between −1 and 8
- Default value: 5

−1 : Bertini uses the zeroth-order (constant) prediction method.

0 : Bertini uses the first-order Euler prediction method.

1 : Bertini uses the second-order Heun prediction method.

2 : Bertini uses the classical fourth-order Runge–Kutta prediction method.

3 : Bertini uses the second-order Heun prediction method with the first-order Euler method for error estimation.

4 : Bertini uses the fourth-order Runge–Kutta–Norsett prediction method with a third-order error estimation.

5 : Bertini uses the fifth-order Runge–Kutta–Fehlberg prediction method with a fourth-order error estimation.

6 : Bertini uses the fifth-order Runge–Kutta–Cash–Karp prediction method with a fourth-order error estimation.

7 : Bertini uses the sixth-order Runge–Kutta–Dormand–Prince prediction method with a fifth-order error estimation.

8 : Bertini uses the seventh-order Runge–Kutta–Verner prediction method with a sixth-order error estimation.

See §2.2 for more details regarding prediction methods.

E.4.4 ▪ MaxNewtonIts

- Datatype: Integer
- Acceptable values: Nonnegative integers
- Default value: 2

The value is the maximum number of Newton corrector iterations used per step. If set to 0, the steps use a prediction-only tracking approach. See §2.2 for more details on predictor-corrector path tracking.

E.4.5 ▪ MaxStepsBeforeNewton

- Datatype: Integer
- Acceptable values: Nonnegative integers
- Default value: 1

The value is the maximum number of consecutive successful steps before applying a Newton iteration when using an error-estimating prediction method. If set to 0, Newton iterations are performed after each prediction regardless of the estimated error.

E.4.6 ▪ MaxStepSize

- Datatype: Double
- Acceptable values: Positive doubles less than 1
- Default value: 0.1

The value is the maximum step size that Bertini will allow during the path tracking. This value is also the size of the first attempted step.

E.4.7 ▪ MinStepSizeBeforeEG

- Datatype: Double
- Acceptable values: Positive doubles no larger than the maximum step size
- Default value: 10^{-14} when using fixed precision and 10^{-250} when using adaptive precision

The value is the minimum step size that Bertini will use before the endgame (see §3.3) is started. This configuration should not be changed when using adaptive precision path tracking (see §3.2).

E.4.8 ▪ MinStepSizeDuringEG

- Datatype: Double
- Acceptable values: Positive doubles no larger than the maximum step size
- Default value: 10^{-15} when using fixed precision and 10^{-250} when using adaptive precision

The value is the minimum step size that Bertini will use during the endgame (see §3.3). This configuration should not be changed when using adaptive precision path tracking (see §3.2).

E.4.9 ▪ MaxNumberSteps

- Datatype: Integer
- Acceptable values: Positive integers
- Default value: 10000

The value is the maximum number of steps on each section of the path.

E.4.10 ▪ StepsForIncrease

- Datatype: Integer
- Acceptable values: Positive integers
- Default value: 5

The value is the number of consecutive successful steps in order for the step size to be increased.

E.4.11 ▪ StepFailFactor

- Datatype: Double
- Acceptable values: Positive doubles less than 1
- Default value: 0.5

The value is the scaling factor used to reduce the step size upon failure.

E.4.12 ▪ StepSuccessFactor

- Datatype: Double
- Acceptable values: Doubles at least 1
- Default value: 2.0

The value is the scaling factor used to increase the step size after having the consecutive number of successful steps.

E.4.13 ▪ PathTruncationThreshold

- Datatype: Double
- Acceptable values: Positive doubles
- Default value: 10^5

A path is truncated if the infinity norm of the path in the endgame (see §3.3) exceeds this bound.

E.5 ▪ Endgame Settings

The endgame configuration settings are used to tune the endgame (see §3.3) to the input system. Settings include changing the final tolerance, endgame algorithm, sample factor, and where the endgame algorithm starts.

E.5.1 ▪ FinalTol

- Datatype: Double
- Acceptable values: Positive double
- Default value: 10^{-11}

The value is the tolerance used by the endgame (see §3.3) to estimate the endpoint of the path.

E.5.2 ▪ EndgameNum

- Datatype: Integer
- Acceptable values: 1, 2, and 3
- Default value: 1

1 : Bertini uses the power-series endgame (see §3.3.2).

2 : Bertini uses the Cauchy endgame (see §3.3.3).

3 : Bertini uses the trackback Cauchy endgame (see §3.3.3).

E.5.3 ▪ SampleFactor

- Datatype: Double
- Acceptable values: Positive doubles less than 1
- Default value: 0.5

The value is the factor used to generate the geometrical sequence of endgame samples.

E.5.4 ▪ NumSamplePoints

- Datatype: Integer
- Acceptable values: Integers greater than or equal to 2
- Default value: 2

The value is the number of sample points used by the endgame to generate the endpoint approximation.

E.5.5 ▪ EndgameBdry

- Datatype: Double
- Acceptable values: Positive doubles less than 1
- Default value: 0.1

The value is the value of the path variable where the endgame algorithm starts.

E.5.6 ▪ MinCycleTrackback

- Datatype: Integer
- Acceptable values: Positive integers
- Default value: 4

When using the trackback Cauchy endgame (see §3.3.2), any path whose cycle number is at least this value will trigger a backtracking procedure.

E.5.7 ▪ NbhdRadius

- Datatype: Double
- Acceptable values: Positive doubles less than 1
- Default value: 10^{-100}

The value corresponds to the radius of the neighborhood surrounding the ending value of the path variable where the endgame is halted.

E.5.8 ▪ MaxCycleNum

- Datatype: Integer
- Acceptable values: Positive integers
- Default value: 6

The value is used by the endgames to provide a basic estimation of the maximum cycle number. For each path, the endgame algorithms also determine an upper bound on the cycle number with Bertini using the largest of the maximum cycle number configuration and the internally computed value for bounding the cycle number.

E.5.9 ▪ CycleTimeCutoff

- Datatype: Double
- Acceptable values: Nonnegative doubles less than 1
- Default value: 10^{-8}

The value is the minimum value where the Cauchy endgame stops checking the heuristic on the cycle (see §3.3.3), which is used to determine whether the path is currently in the endgame operating zone (see §3.3).

E.5.10 ▪ RatioTimeCutoff

- Datatype: Double
- Acceptable values: Nonnegative doubles less than 1
- Default value: 10^{-14}

The value is the minimum value where the Cauchy endgame stops checking the heuristic on the ratio of the minimum and maximum value around a Cauchy loop (see §3.3.3), which is used to determine whether the path is currently in the endgame operating zone (see §3.3).

E.6 ▪ Security Settings

The adaptive precision and tracking setting provide security in the path tracking. The security configuration settings below control the behavior of Bertini for paths which appear to be heading to infinity (see §4.2).

E.6.1 ▪ SecurityLevel

- Datatype: Integer
- Acceptable values: 0 and 1
- Default value: 0

0 : Bertini truncates paths for which two consecutive approximations of the endpoint computed by the endgame (see §3.3) are larger than `SecurityMaxNorm`.

1 : Bertini does not truncate paths based on approximations of the endpoint.

E.6.2 ▪ SecurityMaxNorm

- Datatype: Double
- Acceptable values: Positive doubles
- Default value: 10^4

If `SecurityLevel` is set to 0, a path is truncated if two consecutive approximations of the endpoint are larger than this value.

E.7 ▪ Postprocessing Settings

The classification of points computed by a homotopy designed for computing the isolated points (see Part I) is based on both data from the homotopy and configuration settings, including tolerances for real, singular, and infinite solutions. Additional postprocessing configurations are related to sharpening the endpoints (see §7.2).

E.7.1 ▪ TargetTolMultiplier

- Datatype: Double
- Acceptable values: Positive doubles at least 1
- Default value: 10

This value is multiplied by `FinalTol` to yield the tolerance used to numerically determine if two endpoints should be considered the same point.

E.7.2 ▪ FunctionTolerance

- Datatype: Double
- Acceptable values: Positive doubles less than 1
- Default value: 10^{-14} (double precision), $10^{2-p\log_{10}(2)}$ (p-bit fixed precision), 10^{-300} (adaptive precision)

A polynomial evaluated at a point is considered to be zero if the magnitude is smaller than this value. See also `RatioTolerance`.

E.7.3 ▪ RatioTolerance

- Datatype: Double
- Acceptable values: Positive doubles less than 1
- Default value: 0.99

  A computed value is considered to be zero if the ratio of two different approximations is smaller than this value. See also `FunctionTolerance`.

E.7.4 ▪ ImagThreshold

- Datatype: Double
- Acceptable values: Positive doubles
- Default value: 10^{-8}

  An endpoint is considered a real point if the infinity norm of the imaginary part is smaller than this value.

E.7.5 ▪ CondNumThreshold

- Datatype: Double
- Acceptable values: Positive doubles
- Default value: 10^{8}

  An endpoint is considered a singular point if it is either the endpoint of multiple paths or if an approximation of the condition number, in the spectral norm, is larger than this value.

E.7.6 ▪ EndpointFiniteThreshold

- Datatype: Double
- Acceptable values: Positive doubles
- Default value: 10^{5}

  An endpoint is considered at infinity if the infinity norm of the point is larger than this value.

E.7.7 ▪ SharpenDigits

- Datatype: Integer
- Acceptable values: Nonnegative integers
- Default value: 0

  If the value is positive, each endpoint is classified as singular or nonsingular based on the local behavior of Newton's method. If the point is classified as nonsingular, Newton iterations are applied to the point until the Newton residual indicates the number of correct digits in the approximation is larger than this value. See Chapter 7 for more information.

E.7.8 ▪ SharpenOnly

- Datatype: Integer
- Acceptable values: 0 and 1
- Default value: 0

If the value is 1, the sharpening module (see §7.2) is executed, which allows for sharpening the endpoints and rewriting the output files. See Chapter 7 for more information.

E.8 ▪ User-Defined and Parameter Homotopy Settings

The user-defined and parameter homotopy settings are used to allow Bertini to track specialized homotopies constructed by the user (see Chapter 6).

E.8.1 ▪ UserHomotopy

- Datatype: Integer
- Acceptable values: 0, 1, and 2
- Default value: 0

0: The input system is not a user-defined homotopy.

1: The input system is a user-defined homotopy with the paths tracked using affine coordinates, i.e., all variables are declared in a single `variable` statement.

2: The input system is a user-defined homotopy with the variables declared in any number of `variable_group` and `hom_variable_group` statements. Affine variables (those declared in a `variable_group`) will be homogenized, and the paths will be tracked in a product of projective spaces. Affine variables are then dehomogenized before reporting the results. (See Chapter 4 for information on projective spaces and homogenization/dehomogenization.)

E.8.2 ▪ ParameterHomotopy

- Datatype: Integer
- Acceptable values: 0, 1, and 2
- Default value: 0

0: The input system is not set up for a parameter homotopy.

1: The input system is a parameterized polynomial system and Bertini will solve the ab initio system using randomly generated parameter values (see Chapter 6).

2: The input system is a parameterized polynomial system and Bertini will track the starting points from the starting parameter values in `start_parameters` to the final parameter values in `final_parameters` (see Chapter 6).

E.9 ▪ Numerical Irreducible Decomposition Settings

The numerical irreducible decomposition configuration settings are used to determine and tune the algorithms used to compute the numerical irreducible decomposition (see Part II). Settings include the witness superset generation algorithm, the junk removal test, and the maximum number of monodromy loops.

E.9.1 ▪ WitnessGenType

- Datatype: Integer
- Acceptable values: 0, 1, and 2
- Default value: 2

0 : Bertini uses the cascade algorithm (see §9.4) to compute a witness superset.

1 : Bertini uses the dimension-by-dimension algorithm (see §9.3) to compute a witness superset.

2 : Bertini uses the regenerative cascade algorithm (see §9.4) to compute a witness superset.

E.9.2 ▪ MaxCodimension

- Datatype: Integer
- Acceptable values: Nonnegative integers
- Default value: 0

If the value is 0, all possible codimensions are explored. Otherwise, this value is the maximum codimension under consideration.

E.9.3 ▪ SpecificCodimension

- Datatype: Integer
- Acceptable values: Nonnegative integers
- Default value: 0

If the value is 0, all possible codimensions are explored. Otherwise, this value is the only codimension considered.

E.9.4 ▪ JunkRemovalTest

- Datatype: Integer
- Acceptable values: 0 and 1
- Default value: 1

0 : Bertini uses the homotopy membership test (see §8.4) to eliminate junk points from a witness superset.

1 : Bertini uses the truncated local dimension test (see §11.5) followed by the homotopy membership test to eliminate junk points from a witness superset.

E.9.5 ▪ MaxLDTDepth

- Datatype: Integer
- Acceptable values: Positive integers
- Default value: 3

This value is the maximum depth used in the local dimension test (see §11.5) to determine whether a point in a witness superset is a junk point.

E.9.6 ▪ WitnessSupersetOnly

- Datatype: Integer
- Acceptable values: 0 and 1
- Default value: 0

0: For `TrackType:1`, Bertini computes a witness point superset (see §8.2), then removes junk points (see §10.1) to obtain a witness point set, which it then decomposes (see §10.2) into irreducible witness point sets.

1: For `TrackType:1`, Bertini computes only a witness point superset (see §8.2).

E.9.7 ▪ MultOneOnly

- Datatype: Integer
- Acceptable values: 0 and 1
- Default value: 0

0: For `TrackType:1`, Bertini computes a numerical irreducible decomposition (see Chapter 8) for the input polynomial system.

1: For `TrackType:1`, Bertini computes a numerical irreducible decomposition for only the irreducible components of multiplicity one (see §8.1) of the input polynomial system.

E.9.8 ▪ MaxNumPtsForTrace

- Datatype: Integer
- Acceptable values: Positive integers
- Default value: 10

The value is the maximum number of remaining groups when the monodromy loops (see §10.2.2) are stopped and the combinatorial trace test (see §10.2.1) is used to complete the decomposition.

E.9.9 ▪ MaxNumMonoLoops

- Datatype: Integer
- Acceptable values: Nonnegative integers
- Default value: 1000000

The value is the maximum number of monodromy loops (see §10.2.2) used in the decomposition.

E.9.10 ▪ MaxNumBadMonoLoops

- Datatype: Integer
- Acceptable values: Nonnegative integers
- Default value: 10

The value is the maximum number of consecutive monodromy loops (see §10.2.2) that do not yield new information before switching to the combinatorial trace test (see §10.2.1).

E.9.11 ▪ ConstructWitnessSet

- Datatype: Integer
- Acceptable values: 0 or 1
- Default value: 0

0: For `TrackType:6`, Bertini performs only the isosingular stabilization test (see §13.2).

1: For `TrackType:6`, Bertini performs the isosingular stabilization test, and if successful, Bertini performs monodromy loops (see §10.2.2) with the trace test (see §10.2.1) with the goal of computing a witness set for the associated irreducible component.

E.10 ▪ Regeneration Settings

The regeneration configuration settings are used to tune the regeneration (see §5.4) and regenerative cascade (see §9.4) algorithms. Settings include the tracking tolerances for slice moving and the restarting of a regeneration-based algorithm.

E.10.1 ▪ UseRegeneration

- Datatype: Integer
- Acceptable values: 0 and 1
- Default value: 0

0: Bertini does not use the regeneration or the regenerative cascade algorithm (unless the witness superset generation configuration is set for the regenerative cascade algorithm).

1 : Depending on `TrackType`, Bertini uses the regeneration algorithm for computing isolated nonsingular solutions or the regenerative cascade algorithm for computing a witness superset.

E.10.2 ▪ SliceTolBeforeEG

- Datatype: Double
- Acceptable values: Positive double
- Default value: 10^{-7}

The value is the tracking tolerance before the endgame is started for the slice moving used by the regeneration-based algorithms. In order to declare a step successful, the estimate of the error between the point and the path needs to be smaller than the tracking tolerance.

E.10.3 ▪ SliceTolDuringEG

- Datatype: Double
- Acceptable values: Positive double
- Default value: 10^{-8}

The value is the tracking tolerance during the endgame for the slice moving used by the regeneration-based algorithms. In order to declare a step successful, the estimate of the error between the point and the path needs to be smaller than the tracking tolerance.

E.10.4 ▪ SliceFinalTol

- Datatype: Double
- Acceptable values: Positive double
- Default value: 10^{-11}

The value is the tolerance used by the endgame to estimate the endpoint of the path for the slice moving used by the regeneration-based algorithms.

E.10.5 ▪ RegenStartLevel

- Datatype: Integer
- Acceptable values: Nonnegative integers
- Default value: 0

If the value is positive, the value is where the regeneration-based algorithm will be restarted.

E.10.6 ▪ RegenRemoveInf

- Datatype: Integer
- Acceptable values: 0 and 1
- Default value: 1

0 : Bertini does not remove nonsingular infinite endpoints from the zero-dimensional regeneration.

1 : Bertini removes all infinite endpoints from the zero-dimensional regeneration.

E.10.7 ▪ RegenHigherDimTest

- Datatype: Integer
- Acceptable values: 0 and 1
- Default value: 1

0 : Bertini does not check to see whether any of the other polynomials vanish at each point computed during the zero-dimensional regeneration or the equation-by-equation approach with diagonal homotopies.

1 : Bertini does check to see whether any of the other polynomials vanish at each point computed and removes those which are zero, based on `FunctionTolerance` and `RatioTolerance`, during the zero-dimensional regeneration or the equation-by-equation approach with diagonal homotopies.

E.11 ▪ Miscellaneous Settings

The following configurations are used to adjust miscellaneous settings including using diagonal homotopies with an equation-by-equation approach (see §E.11.1) to compute isolated nonsingular solutions, adjusting the amount of output, and seeding the random number generator.

E.11.1 ▪ UseDiagonal

- Datatype: Integer
- Acceptable values: 0 and 1
- Default value: 0

0 : Bertini does not use diagonal homotopies for solving.

1 : If `TrackType` is set to 0, Bertini uses an equation-by-equation approach with diagonal homotopies (see §E.11.1) to compute the isolated nonsingular solutions of the input polynomial system. If the user has selected to use regeneration and the diagonal approach, Bertini will use regeneration.

E.11.2 ▪ OutputLevel

- Datatype: Integer
- Acceptable values: −1, 0, 1, 2, and 3
- Default value: 0

−1 : Bertini produces only the `raw_data` file.

0 : Bertini produces the standard output files and a summary of the paths.

1 : Bertini produces everything with a setting of 0 and additionally prints data regarding the value of the path variable at each step in `output`.

2 : Bertini produces everything with a setting of 1 and additionally prints data regarding the residuals at each step in `output`.

3 : Bertini produces everything with a setting of 2 and additionally prints the points along the path in `output`.

E.11.3 ▪ PrintPathProgress

- Datatype: Integer
- Acceptable values: Nonnegative integers
- Default value: 20

If positive, this is the number of paths tracked by Bertini between messages printed on the screen informing the user about Bertini's progress. This message is not printed when the value is 0.

E.11.4 ▪ ScreenOut

- Datatype: Integer
- Acceptable values: 0 and 1
- Default value: 0

0 : Bertini does not print data regarding each path to the screen.

1 : Bertini prints data regarding each path to the screen.

E.11.5 ▪ RandomSeed

- Datatype: Integer
- Acceptable values: Nonnegative integer
- Default value: 0

If positive, the value is used to seed the random number generator. If 0, Bertini produces a random seed based on the current time.

E.11.6 ▪ IntrinsicMultiplier

- Datatype: Double
- Acceptable values: Nonnegative double no larger than 1
- Default value: 0.75

The value is used to determine whether to use intrinsic or extrinsic slicing based on the ratio of the number of variables when using intrinsic and extrinsic slicing. If the value is 0, extrinsic slicing is always used, and intrinsic slicing is always used when the value is 1.

E.11.7 ▪ SingValZeroTol

- Datatype: Double
- Acceptable values: Positive doubles
- Default value: 10^{-12}

This value is used to determine when singular values should be considered 0 when performing a computation using fixed precision. See §7.6 for additional details regarding rank determination.

E.11.8 ▪ DeleteTempFiles

- Datatype: Integer
- Acceptable values: 0 and 1
- Default value: 1

0 : Bertini does not delete temporary files created during parsing.

1 : Bertini deletes the temporary files created during parsing.

E.11.9 ▪ NeedToDiff

- Datatype: Integer
- Acceptable values: 0, 1, and 2
- Default value: 1

0 : Bertini reuses the temporary files from a previous run to avoid differentiating the input system again.

1 : Bertini differentiates the input system.

2 : Bertini reuses the temporary files from a previous run but parses the input system to update the parameters (see §6.2).

Appendix F
Tips and Tricks

One key aspect of numerical computations is tuning the algorithm and software to meet the demands of the problem. This chapter provides some background on several different wrenches the user may employ to perform this tuning with some fundamental concepts that were already provided in Chapter 7.

The three sections of this appendix follow a standard work flow for solving a problem. In particular, the first section (§F.1) focuses on various formulations of the problem that the analyst wants to solve. This can be considered a "preconditioning" step since the problem is transformed into a theoretically equivalent system that can be solved more efficiently. In §F.2, we consider ways to adjust the configurations in Bertini to meet the demands of solving the given polynomial system. The final section presents tips for modifying the tolerances used in the postprocessor.

F.1 ▪ Formulating Polynomial Systems

The presentation of a polynomial system in Bertini can have far-reaching consequences in terms of efficiency. Some examples of ways to increase efficiency are through the use of subfunctions to reduce the number of operations needed to evaluate the system and its Jacobian, working intrinsically on certain linear spaces, rounding coefficients and rescaling the polynomial system, and exploiting structure. The following subsections focus on each of these topics in turn.

F.1.1 ▪ Reducing Operations

One way to reduce the number of operations needed to evaluate a polynomial system is to take advantage of the straight-line program format utilized in Bertini (see §1.1.5). In many applications, the problem to be solved is built by combining various subexpressions, e.g., the nine-point problem in §5.5.2, where the same expression can be used several times in the polynomial system. By incorporating such subexpressions into subfunctions (see §A.4), the polynomial system and its Jacobian can be evaluated using fewer operations. This can have a profound impact on the time it takes to solve the system, since function and Jacobian evaluation typically account for roughly half the computational cost of a Bertini run.

The nine-point problem described in §5.5.2 also has another feature, namely, that the eight polynomials all have the same form presented in (5.35). Such a structure is also common in polynomial systems resulting from discretizing differential equations discussed in Chapter 17. In particular, the big polynomial systems are constructed from only a few basic polynomials using different arguments. Although we will not fully document how to use the feature here since it is still in a development and testing phase, we mention that these basic polynomials can be precompiled into a straight-line program and then are incorporated into the system in `input` using the declaration `definedSubfunction`. These are different from incorporating subexpressions into subfunctions described above since they have arguments, each of which can be any valid expression. Apart from reducing typing, this can drastically reduce the cost of constructing a straight-line program for the input polynomial system and evaluating the resulting system.

F.1.2 ▪ Intrinsic Slicing

When a polynomial system contains one or more linear polynomials, the user should consider whether to use extrinsic or intrinsic slicing. Extrinsic slicing uses the full set of variables and the linear polynomials, while intrinsic slicing uses a set of basis vectors for the solution set of the linear polynomials with a smaller set of variables. For numerical stability, an orthonormal basis is better. Moreover, by using a straight-line program format in Bertini, solving intrinsically often requires only minor modifications to the input system. That is, one should not substitute the expressions into the system and expand, as this increases the number of operations needed to evaluate and could introduce additional numerical error by rounding coefficients. Due to the cubic complexity of (dense) linear solving via Gaussian elimination, intrinsic slicing can have a drastic impact on the time it takes to solve the system, especially when the system needs to be randomized by Bertini.

Intrinsic slicing is used in regeneration (see §5.4), especially in the early stages when there are many linear polynomials. In fact, as discussed in that section, the ordering of the polynomials can have an impact on solving the system. If constructing an intrinsic slicing scheme is infeasible, it is often advantageous to put the linear polynomials first, especially when they are nongeneric linear polynomials, so that all of the further computations are performed on that linear space.

F.1.3 ▪ Rounding and Rescaling

The set of actual coefficients of the polynomial system that the user wants to solve is a point in a parameter space. As discussed in §6.5, this point may be on a subset of the full parameter space where the number of isolated solutions is fewer or the irreducible decomposition is different than for a general point in the full parameter space. Therefore, it is imperative that the coefficients are known to accuracy high enough to reveal this special structure. Exact representations of the coefficients, e.g., as demonstrated in §7.8, is best, so that Bertini can compute them to any desired accuracy.

Rounding errors in the coefficients can also be prevented by not expanding polynomial systems, i.e., using a straight-line format, as discussed in §F.1.1.

Since Bertini uses numerical approaches, coefficient scaling can have a drastic impact on solving. If all the coefficients are of roughly the same order, then simply rescaling all of them to be of unit magnitude is often the best practice. If the coefficients are widely scaled, one should employ variable and function scaling as discussed in §7.7.

F.1.4 ▪ Formulations with Structure

Especially when formulating large-scale polynomial systems, the user would be wise to think about the structure present in the system. Some examples include multihomogeneous structure (see §4.5 and §5.1), linear- and polynomial-product structures (see §5.2), and monomial structure (see §5.3). An example of this is the use of isotropic coordinates in §5.5.2, which, for example, replaces the quadratic polynomial $x^2 + y^2$ with the bilinear polynomial $z\bar{z}$ by taking $z = x + iy$ and $\bar{z} = x - iy$. The development and exploitation of structure in the polynomial system can strikingly improve the efficiency of solving.

In §H.3, we list some other software packages that can identify and exploit structure not currently developed in Bertini. In short, solving various systems may require a variety of techniques, some of which Bertini does efficiently and some of which are performed well in these other software packages.

Special structure is fragile: It is often easy to destroy structure. One simple example of this is that the use of intrinsic slicing often drastically increases the number of monomials, which can have a drastic impact on the mixed volume (see §5.3). That is, when using intrinsic slicing, the algebraic relationships between the coefficients remain, but the mixed volume is sharp only when the system has general coefficients. Multihomogeneous structure can also be destroyed by using intrinsic slicing when the linear polynomials are actually multilinear when considered with respect to the variable groups.

F.2 ▪ Solving Tips

Once the polynomial system is formulated, the next task is to consider tuning the configurations in Bertini. Until a fully functional artificial intelligence module is completed for Bertini, this requires human interaction guided by an understanding of the polynomial system being solved. The six topics of this section are related to changing the tracking tolerances, configuring the endgame, understanding truncation, tuning the adaptive precision rules, using sharpening to your advantage, and adjusting tests used by regeneration.

F.2.1 ▪ Tracking Tolerances

The tracking tolerances, along with the tolerance for computing the endpoint of the path, have a profound impact on the performance of Bertini, in terms of both reliability and efficiency. For even random systems, as the number of variables or the degrees of the polynomials (or both) increase, tighter tracking tolerances are needed. Especially for systems with large or widely scaled coefficients, this means decreasing the configurations `TrackTolBeforeEG` and `TrackTolDuringEG` a few orders of magnitude, as in §5.5.2. When using regeneration (see §5.4) or the regenerative cascade (see §9.4), the tolerances used for moving linear slices, namely, `SliceTolBeforeEG` and `SliceTolDuringEG`, should also be changed accordingly.

F.2.2 ▪ Endgame Configurations

Apart from choosing which endgame to use, which is discussed in §3.3, and the final tolerance for computing the endpoint, two other key factors play a role in configuring the endgame. The first is the order of the prediction used by the endgame, which is controlled based on the number of sample points used, namely, `NumSamplePoints`. Generally, higher-order predictions will result in faster convergence of the endgame approximations at the cost of generating the additional sample points. However, such predictions can lead one into incorrectly determining that one is in the endgame operating zone (see §3.3.1) so that the approximation of the endpoint may not be within the desired tolerance of the actual endpoint. However, the increased convergence rate for the higher-order prediction methods can be of tremendous benefit, especially for endpoints with high cycle numbers.

The second key factor is the scaling factor for the geometric sequence where the predictions are made, which is defined by the configuration `SampleFactor`. For the power-series endgame (see §3.3.2), decreasing the sampling factor means that the points used in the extrapolation are further apart, which increases numerical stability. In general, this means that fewer useless predictions will be made outside of the endgame operating zone. However, one should not decrease this configuration too much since this will mean sampling more points and making more predictions in ill-conditioned areas around singular solutions. A sampling factor somewhere between $1/2$ (default) and $1/100$ is generally a good rule of thumb.

In summary, if one expects many singular solutions, increasing the prediction order of the endgame by increasing the number of sample points used and decreasing the sampling factor could improve the efficiency of the endgame.

F.2.3 ▪ Issues with Path Truncation

Bertini uses path truncation to stop tracking paths which appear to be diverging to a point at infinity. When employing a user-defined homotopy in affine coordinates (see §7.5.1), a homotopy path may have infinite length so truncation is essential for Bertini to finish in a finite amount of time. In all other cases, Bertini tracks paths internally on some product of projective spaces (see Chapter 4) to ensure that each path has finite length. However, endpoints of paths that diverge to infinity are often highly singular, requiring a substantial amount of computational time to achieve accurate approximations. Thus, truncation can be used to minimize the wasteful computational costs associated with such endpoints. Nonetheless, truncation should be used with care and an understanding of its ramifications, especially when solutions with a large norm are expected. We note that techniques for deciding whether a path is truly diverging to infinity are developed in [102] using Puiseux series.

There are a few configurations related to path truncation that we shall highlight here. The configuration `PathTruncationThreshold` is used to truncate paths by looking at the infinity norm of the points on the path. This will mainly be used by Bertini to truncate paths for a user-defined homotopy in affine coordinates. The configuration `SecurityMaxNorm` is used to truncate paths by looking at the approximations of the endpoint obtained by running the endgame. Each approximation will be dehomogenized to test the point based on the user-specified coordinates. This truncation test can be turned on and off by using the configuration `SecurityLevel`.

It is a good idea to increase `SecurityMaxNorm` to be a few orders of magnitude larger than the norms of the solutions of interest.

F.2.4 ▪ Configuring the Adaptive Precision Rules

As discussed in Chapter 3, Bertini relies upon rules for adapting the precision based on the local conditioning of the path. Since the default configuration settings on the number of safety digits, namely, `AMPSafetyDigits1` and `AMPSafetyDigits2`, are positive, Bertini by default takes a guarded and pessimistic approach to improve reliability. That is, Bertini might increase the precision higher than is actually needed based on the numerical error of the local computations. This comes at the expense of speed.

A cavalier approach would be to set these configurations to negative values, resulting in the longer use of lower precision. This could speed up the computations but could also lead to trouble if the precision used is not high enough to overcome numerical errors.

Most users would be wise to leave these configurations at their default levels, for the sake of security.

F.2.5 ▪ Advantage of Sharpening

In Chapter 7, several sharpening techniques were presented. Here, we focus on using Newton's method during the course of running a homotopy. In short, after the endgame has successfully approximated to the endpoint based on `FinalTol`, sharpening is used to increase the accuracy of the approximation. Sharpening during a homotopy run, which is accomplished by using the configuration `SharpenDigits`, will perform Newton iterations immediately after the endgame has successfully approximated the endpoint. When the endpoint is nonsingular and the approximation is sufficiently accurate, Newton iterations starting at the approximation will converge quadratically to the endpoint. For a singular solution or a poor approximation of the endpoint, the sharpening module will return the approximation generated by the endgame.

One advantage of using such sharpening, especially when nonsingular points and multiplicity one components are the main interest, is that all points that are successfully refined by the sharpening module will be returned with a similar accuracy and in the same precision based on `SharpenDigits`. This increased accuracy, i.e., accuracy beyond `FinalTol`, comes at the expense of running the sharpening module. However, this refinement and standardization of the solutions can make future computations and decisions easier, e.g., sorting the solutions, which is discussed in §F.3.

F.2.6 ▪ Adjusting Tests Used by Regeneration

Regeneration (see §5.4), as implemented in Bertini, uses three tests to remove points from being *regenerated* to the next level, namely, finite versus infinite, higher-dimensional test, and singular versus nonsingular. By default, points for which the infinity norm in the original coordinates is larger than `EndpointFiniteThreshold` do not proceed to the next level. When one expects intermediate points computed during regeneration to be quite large, consider turning this test off by using the setting `RegenRemoveInf:0`. The user should also consider employing homogeneous coordinates (see Chapter 4) and explore related issues dealing with truncation in §F.2.3.

For a square system, if a computed point satisfies a polynomial which will be used later in the regeneration, then this point cannot lead to isolated solutions. Even with several evaluations in possibly different precision with various approximations of the point, numerical issues such as poor scaling (see §7.7), rounding errors (see §7.8), and high degrees can still lead to incorrectly determining whether a polynomial evaluated at a point is zero. Due to this, Bertini permits this test to be turned off by using the setting `RegenHigherDimTest:0`.

Each singular endpoint is removed since Bertini currently does not employ deflation (see §11.6) with regeneration. The approach for deciding which endpoints are singular is based on §7.6. The potential for similar numerical issues as with the previous test also arise with this approach. In addition to reformulating the polynomial system, the user should consider decreasing `FinalTol` since a more accurate approximation of each endpoint will produce a clearer picture of the behavior at the actual solution.

F.3 ▪ Additional Postprocessing

After approximating the endpoint of a homotopy path, Bertini needs to make decisions about the true endpoint. For a zero-dimensional run, the onscreen display which classifies points based on finite versus infinite, real versus nonreal, and singular versus nonsingular depend on the configurations `EndpointFiniteThreshold`, `ImagThreshold`, and `CondNumThreshold`, respectively.

One efficient way for the user to intervene to more carefully examine the points is to set `EndpointFiniteThreshold` very large so that all computed points are considered to be finite. Then, one uses the sharpening module to obtain more accurate approximations of the endpoints. As in §9.1, by comparing the norms of the approximations, norms of the imaginary part of the approximations, and minimum singular values of the Jacobian matrices for the approximations in a logarithmic scale, one is more likely able to decide which points are truly finite, real, and nonsingular, respectively.

As discussed in §14.3, nonsingular solutions of square systems can be rigorously certified as real or nonreal using `alphaCertified` [95, 96]. (See also §H.3.)

Appendix G
Parallel Computing

Many algorithms in numerical algebraic geometry, especially the fundamental algorithm of path tracking, can be efficiently parallelized. The continued growth in the development of multicore processors, graphical processing units, and larger networks of computers ensures that the future of parallel computing in Bertini, and numerical algebraic geometry as a whole, will be based on exploiting hybrid parallelism on heterogeneous parallel computers using both distributed memory and shared memory models. Even though Bertini is far from complete on this front, we will highlight some progress in this direction.

The main parallelism employed by Bertini is a distributed memory parallelism model based on the message passing interface (MPI). Bertini has been used with various implementations of MPI, including MPICH, MPICH2, and Open MPI. However, due to some minor differences in the implementations, one should consult the user's manual for their specific implementation to ensure the proper functionality.

In Bertini, all the MPI specific code is contained in conditional preprocessor directives (`#ifdef... #endif`) that are controlled by a compile-time constant named `_HAVE_MPI`. That is, if `_HAVE_MPI` is not defined while Bertini is being compiled, the compiler will skip over the parallel code and run sequentially. An MPI-based compiler must be used when `_HAVE_MPI` is defined at compile time. By simply modifying the `Makefile`, the user can switch to an MPI-based compiler and indicate to the compiler that the constant `_HAVE_MPI` is to be defined.

Once compiled, parallel instances of Bertini using MPI are started using an MPI parallel job launcher, typically called `mpirun` or `mpiexec`. See the user's manual of the implementation for the proper syntax of the command line options for the MPI parallel job launcher, which includes changing the total number of processes, the name of the hosts to utilize, and the number of processes to run on each host.

Some functions in Bertini also support a shared memory parallelism model based on OpenMP. The OpenMP specific code is similarly controlled by a compile-time constant named `_OPENMP`. When using an OpenMP-based compiler, the `GMP` and `MPFR` [57] libraries used by Bertini must also be compiled to be thread-safe. We leave specific details to the respective user's manuals.

Appendix H
Related Software

Given that polynomial systems show up in many diverse areas of mathematics, science, and engineering, we envision Bertini as an engine that could fit inside a number of vehicles. Moreover, not all new algorithms in the field are implemented inside Bertini. Thus, there are a number of software packages and projects related to Bertini.

In this brief appendix, we list those packages and provide some basic details.

H.1 ▪ Interfaces between Bertini and Other Software

The following independent software packages have built-in interfaces to Bertini. For each, we list the name of the package, its developers, the website for the package, any references specific to the Bertini interface, and a very brief synopsis of the connection to Bertini.

ApCoCoA
Developers: M. Kreuzer and the rest of the ApCoCoA Team (interface: E. Ullah)
Website: `www.apcocoa.org`
Synopsis: ApCoCoA is an acronym for Applied Computations in Computer Algebra and is a part of the larger CoCoA (Computations in Commutative Algebra) software package. Functionality includes calls to Bertini to find isolated solutions of a polynomial system. Includes nonhomogeneous, homogeneous, and multihomogeneous options, and users may specify some configuration settings within ApCoCoA.

Bermeja
Developers: P. Rostalski
Website: `math.berkeley.edu/~philipp/Software/Software`
Synopsis: Bermeja is MATLAB software for computations in convex algebraic geometry. It ties together several packages, including Yalmip, SeDuMi, MPT, Bertini, SparsePOP, and alphaCertified. Bermeja includes MATLAB M-files that interface with Bertini for solving, sampling, and membership testing.

DGP-Solver
Developers: G. Abbasnejad and M. Carricato
Website: `grab.diem.unibo.it/DGP-Solver/`
Synopsis: DGP-Solver is narrowly focused on solving for the equilibrium positions of a platform suspended by $n \leq 6$ cables [1, 2, 41]. It calls Bertini to solve the various cases by parameter homotopy.

Macaulay2
Developers: D. R. Grayson and M. E. Stillman (interface: D. J. Bates, E. Gross, A. Leykin, and J. Rodriguez)
Website: See Macaulay2 help files.
Reference: [16]
Synopsis: Macaulay2 is computational algebraic geometry software. Bertini.m2 is the Macaulay2 interface to Bertini and one part of the Leykin-led Numerical Algebraic Geometry software project within Macaulay2. Bertini functionality available within Macaulay2 includes zero-dimensional and positive-dimensional solving, parameter homotopies, user-defined homotopies, sampling, and membership testing.

Maple, MATLAB, *Mathematica*
The Bertini team does not maintain official Maple, MATLAB, or *Mathematica* interfaces. However, members of the team have interfaced to Bertini from these programs many times. If you are in need of code for writing input files or, more importantly, reading in output files, please contact us.

Singular
In the preparation of the article [14], we became familiar with the strength of Singular [48] and its usefulness to users of Bertini. Singular is general-purpose symbolic software well adapted to the full range of algebraic and algebraic-geometric computations. No interface between Singular and Bertini exists at this time.

H.2 ▪ Modules Using Bertini as a Library

The following software packages use Bertini as a library.

LocalDimFinder
Developers: J. D. Hauenstein, A. J. Sommese, and C. W. Wampler
Website: `bertini.nd.edu/LocalDimFinder`
Reference: [92]
Synopsis: LocalDimFinder is an implementation of the local dimension test discussed in §11.5. In the article above, this method was applied to the problem of mechanism mobility.

Paramotopy
Developers: D. J. Bates, D. Brake, and M. Niemerg
Website: `www.paramotopy.com`
Reference: [12]
Synopsis: Paramotopy is a C++ software package that automates the use of Bertini's parameter homotopy run type. The website also has MATLAB scripts for collecting and displaying this data.

H.3 ▪ Software Packages Similar to Bertini

The following software packages are similar to Bertini in various ways. If Bertini does not suit your needs, perhaps one of these will work.

alphaCertified
Developers: J. D. Hauenstein and F. Sottile
Website: `www.math.tamu.edu/~sottile/research/stories/alphaCertified`
Reference: [96]

Synopsis: alphaCertified is software for certifying numerical solutions to polynomial equations. It is an implementation of Smale's α-theory and can be used to certify some of the output of Bertini runs.

HOM4PS-2.0
Developers: T. L. Lee, T.-Y. Li, and C. H. Tsai
Website: `www.math.nsysu.edu.tw/~leetsung/works/HOM4PS_soft.htm`
Reference: [114]
Synopsis: HOM4PS-2.0 is a fast implementation of the polyhedral homotopy method. A parallel version, HOM4PS-2.0para, is also evolving.

KhRo
Developers: D. J. Bates and F. Sottile
Website: `www.math.tamu.edu/~sottile/research/stories/Rolle`
Reference: [28]
Synopsis: KhRo is a set of Maple scripts implementing Gale duality and Khovanskii–Rolle continuation for finding real solutions of polynomial systems via entirely real computations.

Multiplicity
Developers: W. Hao, A. J. Sommese, and Z. Zeng
Website: `www.nd.edu/~whao/MulLab.html`
Reference: [85]
Synopsis: Multiplicity is MATLAB code for finding the multiplicity structure of nonlinear systems, including polynomial systems. It uses the equation-by-equation closedness subspace methods, which makes it very efficient.

PHCpack
Developers: J. Verschelde
Website: `homepages.math.uic.edu/~jan/PHCpack/phcpack.html`
Reference: [172]
Synopsis: PHCpack is an implementation of many of the numerical algebraic geometry methods that have been developed, including a number that are not available in Bertini. It includes many options and many homotopy types, including polyhedral homotopies.

POLSYS_GLP
Developers: H.-J. Su, J. M. McCarthy, M. Sosonkina, and L. T. Watson
Website:
`www.umbc.edu/engineering/me/vrml/research/software/polsysglp/polsysglp`
Reference: [169]
Synopsis: POLSYS_GLP is aimed at finding isolated solutions of square systems. It makes use of the sophisticated *general linear product* homotopy type, which is not implemented anywhere else. The codes were adapted from POLSYS_PLP by S. M. Wise, A. J. Sommese, and L. T. Watson [185].

Bibliography

[1] G. ABBASNEJAD AND M. CARRICATO, *Real solutions of the direct geometrico-static problem of under-constrained cable-driven parallel robots with 3 cables: A numerical investigation*, Meccanica, 47 (2012), pp. 1761–1773. (Cited on p. 331)

[2] ———, *Direct geometrico-static problem of underconstrained cable-driven parallel robots with 5 cables*, in Computational Kinematics: Proceedings of the 6th International Workshop on Computational Kinematics (CK2013), F. Thomas and A. Pérez Gracia, eds., Springer, 2013. (Cited on p. 331)

[3] A. N. AL-KHATEEB, J. M. POWERS, S. PAOLUCCI, A. J. SOMMESE, J. A. DILLER, J. D. HAUENSTEIN, AND J. MENGERS, *One-dimensional slow invariant manifolds for spatially homogeneous reactive systems*, J. Chem. Phys., 131 (2009), 024118. (Cited on pp. xvi, 120)

[4] E. L. ALLGOWER, D. J. BATES, A. J. SOMMESE, AND C. W. WAMPLER, *Solution of polynomial systems derived from differential equations*, Computing, 76 (2010), pp. 1–10. (Cited on pp. xvi, 261, 262, 263, 268)

[5] E. L. ALLGOWER, S. G. CRUCEANU, AND S. TAVENER, *Application of numerical continuation to compute all solutions of semilinear elliptic equations*, Adv. Geom., 9 (2009), pp. 371–400. (Cited on pp. xvi, 262)

[6] ———, *Turning points and bifurcations for homotopies of analytic maps*, in Interactions of Classical and Numerical Algebraic Geometry, Contemp. Math. 496, AMS, Providence, RI, 2009, pp. 1–10. (Cited on pp. xvi, 262)

[7] E. L. ALLGOWER AND K. GEORG, *Introduction to Numerical Continuation Methods*, Classics in Appl. Math. 45, SIAM, Philadelphia, PA, 2003. (Cited on p. 34)

[8] E. ALLMAN AND J. RHODES, *Phylogenetic ideals and varieties for the general Markov model*, Adv. in Appl. Math., 40 (2008), pp. 127–148. (Cited on p. 246)

[9] H. ALT, *Über die Erzeugung gegebener ebener Kurven mit Hilfe des Gelenkviereckes*, Z. Angew. Math. Mech., 3 (1923), pp. 13–19. (Cited on pp. xvi, 95)

[10] V. I. ARNOL'D, *Singularities of smooth mappings.*, Russian Math. Surveys, 23 (1968), pp. 1–43. (Cited on p. 216)

[11] P. AUBRY, F. ROUILLIER, AND M. SAFEY EL DIN, *Real solving for positive dimensional systems*, J. Symbolic Comput., 34 (2002), pp. 543–560. (Cited on p. 237)

[12] D. J. BATES, D. A. BRAKE, AND M. E. NIEMERG, *Paramotopy: Parameter homotopies in parallel.* Available at `www.math.colostate.edu/~bates/preprints/BBN_paramotopy.pdf`. (Cited on p. 332)

[13] D. J. BATES, B. DAVIS, D. EKLUND, E. HANSON, AND C. PETERSON, *Perturbed regeneration for singular solutions of polynomial systems.* In preparation. (Cited on p. 90)

[14] D. J. BATES, W. DECKER, J. D. HAUENSTEIN, C. PETERSON, G. PFISTER, F.-O. SCHERYER, A. J. SOMMESE, AND C. W. WAMPLER, *Comparison of probabilistic algorithms for analyzing the components of an affine algebraic variety*. Available at `www.nd.edu/~sommese/preprints/bdhppsswComparison.pdf`. (Cited on pp. 248, 332)

[15] D. J. BATES, D. EKLUND, AND C. PETERSON, *Computing intersection numbers of Chern classes*, J. Symbolic Comput., 50 (2013), pp. 493–507. (Cited on p. 245)

[16] D. J. BATES, E. GROSS, A. LEYKIN, AND J. RODRIGUEZ, *Bertini.m2: Macaulay2 interface for Bertini*. In preparation. (Cited on p. 332)

[17] D. J. BATES, J. D. HAUENSTEIN, T. M. MCCOY, C. PETERSON, AND A. J. SOMMESE, *Recovering exact results from inexact numerical data in algebraic geometry*, Experiment. Math., 22 (2013), pp. 38–50. (Cited on pp. 177, 248)

[18] D. J. BATES, J. D. HAUENSTEIN, C. PETERSON, AND A. J. SOMMESE, *Numerical decomposition of the rank-deficiency set of a matrix of multivariate polynomials*, in Approximate Commutative Algebra, Texts Monogr. Symbol. Comput., Springer, New York, 2009, pp. 55–77. (Cited on p. 210)

[19] ——, *A numerical local dimension test for points on the solution set of a system of polynomial equations*, SIAM J. Numer. Anal., 47 (2009), pp. 3608–3623. (Cited on pp. 173, 187)

[20] D. J. BATES, J. D. HAUENSTEIN, AND A. J. SOMMESE, *Efficient path tracking methods*, Numer. Algorithms, 58 (2011), pp. 451–459. (Cited on p. 35)

[21] ——, *A parallel endgame*, in Randomization, Relaxation, and Complexity in Polynomial Equation Solving, Contemp. Math. 556, AMS, Providence, RI, 2011, pp. 25–35. (Cited on pp. 47, 53)

[22] D. J. BATES, J. D. HAUENSTEIN, A. J. SOMMESE, AND C. W. WAMPLER, *Bertini: Software for numerical algebraic geometry*. Available at `bertini.nd.edu`. (Cited on pp. xv, 3)

[23] ——, *Adaptive multiprecision path tracking*, SIAM J. Numer. Anal., 46 (2008), pp. 722–746. (Cited on pp. 45, 46, 306)

[24] ——, *Stepsize control for path tracking*, in Interactions of Classical and Numerical Algebraic Geometry, Contemp. Math. 496, AMS, Providence, RI, 2009, pp. 21–31. (Cited on p. 46)

[25] D. J. BATES AND L. OEDING, *Toward a salmon conjecture*, Experiment. Math., 20 (2011), pp. 358–370. (Cited on p. 246)

[26] D. J. BATES, C. PETERSON, AND A. J. SOMMESE, *Applications of a numerical version of Terracini's lemma for secants and joins*, in Algorithms in Algebraic Geometry, A. Dickenstein, F.-O. Schreyer, and A. J. Sommese, eds., Springer, New York, 2008, pp. 1–14. (Cited on p. 245)

[27] D. J. BATES, C. PETERSON, A. J. SOMMESE, AND C. W. WAMPLER, *Numerical computation of the genus of an irreducible curve within an algebraic set*, J. Pure Appl. Algebra, 215 (2011), pp. 1844–1851. (Cited on pp. 243, 244)

[28] D. J. BATES AND F. SOTTILE, *Khovanskii–Rolle continuation for real solutions*, Found. Comput. Math., 11 (2011), pp. 563–587. (Cited on p. 333)

[29] C. BELTRÁN AND A. LEYKIN, *Certified numerical homotopy tracking*, Experiment. Math., 21 (2012), pp. 69–83. (Cited on p. 248)

[30] ——, *Robust certified numerical homotopy tracking*, Found. Comput. Math., 13 (2013), pp. 253–295. (Cited on p. 248)

[31] C. BELTRÁN AND L. M. PARDO, *On Smale's 17th problem: A probabilistic positive solution*, Found. Comput. Math., 8 (2008), pp. 1–43. (Cited on p. 248)

[32] C. BELTRÁN AND M. SHUB, *The Complexity and Geometry of Numerically Solving Polynomial Systems*, arXiv:1211.1528, 2012. (Cited on p. 248)

[33] D. N. BERNSTEIN, *The number of roots of a system of equations*, Funct. Anal. Appl., 9 (1975), pp. 183–185. (Cited on p. 82)

[34] G. M. BESANA, S. DI ROCCO, J. D. HAUENSTEIN, A. J. SOMMESE, AND C. W. WAMPLER, *Cell decomposition of almost smooth real algebraic surfaces*, Numer. Algorithms, (2012), pp. 1–34. (Cited on pp. 165, 232, 233, 237)

[35] G. BLEKHERMAN, J. HAUENSTEIN, J. C. OTTEM, K. RANESTAD, AND B. STURMFELS, *Algebraic boundaries of Hilbert's SOS cones*, Compos. Math., 148 (2012), pp. 1717–1735. (Cited on p. 256)

[36] L. BLUM, F. CUCKER, M. SHUB, AND S. SMALE, *Complexity and Real Computation*, Springer, New York, 1998. (Cited on pp. 123, 226, 248)

[37] O. BOTTEMA AND B. ROTH, *Theoretical Kinematics*, Dover Publications, New York, 1990. (Cited on pp. xvi, 99)

[38] D. A. BRANNAN, M. F. ESPLEN, AND J. J. GRAY, *Geometry*, Cambridge University Press, Cambridge, UK, 1999. (Cited on p. 66)

[39] R. L. BURDEN AND J. D. FAIRES, *Numerical Analysis*, Brooks/Cole, Pacific Grove, CA, 2010. (Cited on p. 34)

[40] P. BÜRGISSER AND F. CUCKER, *On a problem posed by Steve Smale*, Ann. of Math. (2), 174 (2011), pp. 1785–1836. (Cited on p. 248)

[41] M. CARRICATO AND G. ABBASNEJAD, *Direct geometrico-static analysis of under-constrained cable-driven parallel robots with 4 cables*, in Cable-Driven Parallel Robots, T. Bruckmann and A. Pott, eds., Springer-Verlag, Berlin, 2013, pp. 269–285. (Cited on p. 331)

[42] A. K. CLINE, C. B. MOLER, G. W. STEWART, AND J. H. WILKINSON, *An estimate for the condition number of a matrix*, SIAM J. Numer. Anal., 16 (1979), pp. 368–375. (Cited on p. 43)

[43] N. J. CORNISH, *WMAP Observatory: Lagrange points* (2012). Available at map.gsfc.nasa.gov/mission/observatory_l2.html. (Cited on pp. 94, 95)

[44] D. COX, J. LITTLE, AND D. O'SHEA, *Ideals, Varieties, and Algorithms: An Introduction to Computational Algebraic Geometry and Commutative Algebra*, 2nd ed., Undergraduate Texts in Mathematics, Springer, New York, 1997. (Cited on pp. 18, 146, 147, 150)

[45] D. F. DAVIDENKO, *On a new method of numerical solution of systems of nonlinear equations*, Dokl. Akad. Nauk SSSR (N.S.), 88 (1953), pp. 601–602. (Cited on pp. 32, 33)

[46] ———, *On approximate solution of systems of nonlinear equations*, Ukraïn. Mat. Zh., 5 (1953), pp. 196–206. (Cited on pp. 32, 33)

[47] B. H. DAYTON AND Z. ZENG, *Computing the multiplicity structure in solving polynomial systems*, in Proceedings of ISSAC 2005, New York, ACM, 2005, pp. 166–123. (Cited on pp. 173, 187, 190)

[48] W. DECKER, G.-M. GREUEL, G. PFISTER, AND H. SCHÖNEMANN, SINGULAR *3-1-6: A computer algebra system for polynomial computations* (2012). Available at www.singular.uni-kl.de. (Cited on p. 332)

[49] S. Di Rocco, D. Eklund, C. Peterson, and A. J. Sommese, *Chern numbers of smooth varieties via homotopy continuation and intersection theory*, J. Symbolic Comput., 46 (2011), pp. 23–33. (Cited on p. 245)

[50] P. Dietmaier, *The Stewart-Gough platform of general geometry can have* 40 *real postures*, in Advances in Robot Kinematics: Analysis and Control, J. Lenarcic and M. L. Husty, eds., Kluwer Academic, Dordrecht, Netherlands, 1998, pp. 7–16. (Cited on p. 248)

[51] J. Draisma and J. Rodriguez, *Maximum Likelihood Duality for Determinantal Varieties*, `arXiv:1211.3196`, 2012. (Cited on p. 247)

[52] M. Drton, B. Sturmfels, and S. Sullivant, *Lectures on Algebraic Statistics*, Springer, New York, 2009. (Cited on p. 246)

[53] D. Eisenbud, *Commutative Algebra with a View Toward Algebraic Geometry*, Graduate Texts in Mathematics 150, Springer, New York, 1995. (Cited on p. 19)

[54] D. Eklund, C. Jost, and C. Peterson, *A method to compute Segre classes of subschemes of projective space*, J. Algebra Appl., 12 (2013). (Cited on p. 245)

[55] R. T. Farouki and H. Pottmann, *Exact Minkowski products of N complex disks*, Reliab. Comput., 8 (2002), pp. 43–66. (Cited on p. 66)

[56] J. C. Faugére and D. Lazard, *The combinatorial classes of parallel manipulators*, Mech. Mach. Theory, 30 (1995), pp. 765–776. (Cited on p. 112)

[57] L. Fousse, G. Hanrot, V. Lefèvre, P. Pélissier, and P. Zimmermann, *MPFR: A multiple-precision binary floating-point library with correct rounding*, ACM Trans. Math. Software, 33 (2007). (Cited on pp. 45, 273, 329)

[58] S. Friedland and E. Gross, *A proof of the set-theoretic version of the salmon conjecture*, J. Algebra, 356 (2012), pp. 374–379. (Cited on p. 246)

[59] A. Friedman and B. Hu, *Bifurcation from stability to instability for a free boundary problem arising in a tumor model*, Arch. Ration. Mech. Anal., 180 (2006), pp. 293–330. (Cited on p. 264)

[60] ——, *Stability and instability of Liapounov-Schmidt and Hopf bifurcation for a free boundary problem arising in a tumor model*, Trans. Amer. Math. Soc., 360 (2008), pp. 5291–5342. (Cited on p. 264)

[61] L. D. García-Puente, N. Hein, C. Hillar, A. Martín del Campo, J. Ruffo, F. Sottile, and Z. Teitler, *The secant conjecture in the real Schubert calculus*, Experiment. Math., 21 (2012), pp. 252–265. (Cited on p. 247)

[62] I. Gargantini and P. Henrici, *Circular arithmetic and the determination of polynomial zeros*, Numer. Math., 18 (1971/72), pp. 305–320. (Cited on p. 44)

[63] K. Georg, *Improving the efficiency of exclusion algorithms*, Adv. Geom., 1 (2001), pp. 193–210. (Cited on p. 224)

[64] ——, *A new exclusion test*, J. Comput. Appl. Math., 152 (2003), pp. 147–160. (Cited on p. 224)

[65] L. Göttsche, *Hilbert Schemes of Zero-Dimensional Subschemes of Smooth Varieties*, Lecture Notes in Math. 1572, Springer-Verlag, Berlin, 1994. (Cited on p. 42)

[66] D. R. Grayson and M. E. Stillman, *Macaulay2: A software system for research in algebraic geometry*. Available at `www.math.uiuc.edu/Macaulay2/`. (Cited on p. 243)

[67] G.-M. GREUEL AND G. PFISTER, *A* **Singular** *Introduction to Commutative Algebra*, Springer-Verlag, Berlin, 2002. (Cited on pp. 243, 248)

[68] A. GRIEWANK AND M. R. OSBORNE, *Analysis of Newton's method at irregular singularities*, SIAM J. Numer. Anal., 20 (1983), pp. 747–773. (Cited on pp. 10, 11, 276)

[69] Z. A. GRIFFIN AND J. D. HAUENSTEIN, *Real solutions to systems of polynomial equations and parameter continuation*, Adv. Geometry, to appear. Available at `www.math.ncsu.edu/~jdhauens/preprints/ghRealSolving.pdf`. (Cited on pp. 227, 248)

[70] M. GRIFFIS AND J. DUFFY, *Method and Apparatus for Controlling Geometrically Simple Parallel Mechanisms with Distinctive Connections*, US Patent 5,179,525, 1993. (Cited on p. 155)

[71] J. GUCKENHEIMER AND Y. XIANG, *Defining equations for bifurcations and singularities*, Mosc. Math. J., 3 (2003), pp. 935–946, 1199. (Cited on p. 216)

[72] R. C. GUNNING AND H. ROSSI, *Analytic Functions of Several Complex Variables*, Prentice-Hall, Englewood Cliffs, NJ, 1965. (Cited on pp. 42, 252)

[73] C. HAN, J. KIM, J. KIM, AND F. C. PARK, *Kinematic sensitivity analysis of the 3-UPU parallel mechanism*, Mech. Mach. Theory, 37 (2002), pp. 787–798. (Cited on p. 188)

[74] W. HAO, J. D. HAUENSTEIN, B. HU, Y. LIU, A. J. SOMMESE, AND Y.-T. ZHANG, *Multiple stable steady states of a reaction-diffusion model on zebrafish dorsal-ventral patterning*, Discrete Contin. Dyn. Syst. Ser. S, 4 (2011), pp. 1413–1428. (Cited on pp. xvi, 262, 264, 265, 266)

[75] ———, *Bifurcation for a free boundary problem modeling the growth of a tumor with a necrotic core*, Nonlinear Anal. Real World Appl., 13 (2012), pp. 694–709. (Cited on pp. xvi, 262, 264, 266)

[76] ———, *Continuation along bifurcation branches for a tumor model with a necrotic core*, J. Sci. Comput., 53 (2012), pp. 395–413. (Cited on pp. xvi, 262, 264, 265, 266)

[77] W. HAO, J. D. HAUENSTEIN, B. HU, T. MCCOY, AND A. J. SOMMESE, *Computing steady-state solutions for a free boundary problem modeling tumor growth by Stokes equation*, J. Comput. Appl. Math., 237 (2013), pp. 326–334. (Cited on pp. xvi, 262, 264, 266)

[78] W. HAO, J. D. HAUENSTEIN, B. HU, AND A. J. SOMMESE, *A domain decomposition algorithm for computing multiple steady states of differential equations*. Available at `www.nd.edu/~sommese/preprints/bootstrapping.pdf`. (Cited on pp. xvi, 261, 262, 264, 268)

[79] ———, *A three-dimensional steady-state tumor system*, Appl. Math. Comput., 218 (2011), pp. 2661–2669. (Cited on pp. xvi, 262, 264, 266)

[80] W. HAO, J. D. HAUENSTEIN, C.-W. SHU, A. J. SOMMESE, Z. XU, AND Y.-T. ZHANG, *A homotopy method based on WENO schemes for solving steady state problems of hyperbolic conservation laws*, J. Comput. Phys., 250 (2013), pp. 332–346. (Cited on pp. xvi, 262, 266, 269)

[81] W. HAO, B. HU, AND A. J. SOMMESE, *Numerical algebraic geometry and differential equations*, to appear in Future Vision and Trends on Shapes, Geometry and Algebra, R. De Amicis and G. Conti, eds., Springer, Berlin. Available at `www.nd.edu/~sommese/preprints/SAGA_HHS.pdf`. (Cited on pp. xvi, 261, 262)

[82] ———, *Cell cycle control and bifurcation for a free boundary problem modeling tissue growth*, J. Sci. Comput., 56 (2013), pp. 350–365. (Cited on pp. xvi, 262, 264, 266, 267)

[83] W. HAO, O. T. KERNELL, AND A. J. SOMMESE, *A Chebyshev method for a free boundary problem modeling tumor growth*. Available at www.nd.edu/~sommese/preprints/HKS.pdf. (Cited on p. 265)

[84] W. HAO, G. LIN, Z. XU, E. ROSEN, A. J. SOMMESE, AND M. ALBER, *Effect of fitted reaction rates on predicting thrombin production*. Available at www.nd.edu/~sommese/preprints/coagulation.pdf. (Cited on pp. 262, 264)

[85] W. HAO, A. J. SOMMESE, AND Z. ZENG, *An algorithm and software for computing multiplicity structures at zeros of nonlinear systems*, ACM Trans. Math. Software. To appear. (Cited on pp. 173, 187, 333)

[86] J. HARRIS, *Galois groups of enumerative problems*, Duke Math. J., 46 (1979), pp. 685–724. (Cited on p. 247)

[87] J. D. HAUENSTEIN, *Numerically computing real points on algebraic sets*, Acta Appl. Math., (2012), pp. 1–15. (Cited on p. 237)

[88] J. D. HAUENSTEIN, Y.-H. HE, AND D. MEHTA, *Numerical elimination and moduli space of vacua*, J. High Energy Phys., to appear. Available at www.math.ncsu.edu/~jdhauens/preprints/hhmElim.pdf. (Cited on p. 251)

[89] J. D. HAUENSTEIN, J. RODRIGUEZ, AND B. STURMFELS, *Maximum likelihood for matrices with rank constraints*. Available at www.math.ncsu.edu/~jdhauens/preprints/hrsMLE.pdf. (Cited on pp. 246, 247)

[90] J. D. HAUENSTEIN AND A. J. SOMMESE, *Witness sets of projections*, Appl. Math. Comput., 217 (2010), pp. 3349–3354. (Cited on pp. 251, 253)

[91] ———, *Membership tests for images of algebraic sets by linear projections*, Appl. Math. Comput., 219 (2013), pp. 6809–6818. (Cited on pp. 251, 256, 260)

[92] J. D. HAUENSTEIN, A. J. SOMMESE, AND C. W. WAMPLER, *Mechanism mobility and a local dimension test*, Mech. Mach. Theory, 46 (2011), pp. 1193–1206. (Cited on p. 332)

[93] ———, *Regeneration homotopies for solving systems of polynomials*, Math. Comput., 80 (2011), pp. 345–377. (Cited on pp. xvi, 89, 190, 263, 264)

[94] ———, *Regenerative cascade homotopies for solving polynomial systems*, Appl. Math. Comput., 218 (2011), pp. 1240–1246. (Cited on p. 167)

[95] J. D. HAUENSTEIN AND F. SOTTILE, *alphaCertified: Software for certifying numerical solutions to polynomial equations*. Available at www.math.tamu.edu/~sottile/research/stories/alphaCertified. (Cited on pp. 226, 241, 248, 328)

[96] ———, *Algorithm 921: alphaCertified: Certifying solutions to polynomial systems*, ACM Trans. Math. Software, 38 (2012), p. 28. (Cited on pp. 123, 226, 227, 241, 248, 328, 332)

[97] J. D. HAUENSTEIN AND C. W. WAMPLER, *Numerically intersecting algebraic varieties via witness sets*, Appl. Math. Comput., 217 (2013), pp. 5730–5742. (Cited on pp. 200, 221)

[98] ———, *Isosingular sets and deflation*, Found. Comput. Math., 13 (2013), pp. 371–403. (Cited on pp. 200, 210, 216)

[99] ———, *Numerical algebraic intersection using regeneration*. Available at www.math.ncsu.edu/~jdhauens/preprints/hwGeneralIntersection.pdf. (Cited on pp. 195, 203)

[100] B. HUBER AND B. STURMFELS, *A polyhedral method for solving sparse polynomial systems*, Math. Comp., 64 (1995), pp. 1541–1555. (Cited on p. 83)

[101] ——, *Bernstein's theorem in affine space*, Discrete Comput. Geom., 17 (1997), pp. 137–141. (Cited on p. 83)

[102] B. HUBER AND J. VERSCHELDE, *Polyhedral end games for polynomial continuation*, Numer. Algorithms, 18 (1998), pp. 91–108. (Cited on pp. 47, 326)

[103] M. L. HUSTY, *An algorithm for solving the direct kinematics of general Stewart-Gough platforms*, Mech. Mach. Theory, 31 (1996), pp. 365–380. (Cited on p. 109)

[104] M. L. HUSTY AND A. KARGER, *Self-motions of Griffis-Duffy type parallel manipulators*, in Proceedings of the 2000 IEEE International Conference on Robotics and Automation, San Francisco, CA, 2000, pp. 7–12. (Cited on p. 155)

[105] M. L. HUSTY, A. KARGER, H. SACHS, AND W. STEINHILPER, *Kinematik und Robotik*, Springer-Verlag, Berlin, 1997. (Cited on p. xvi)

[106] M. L. HUSTY, M. PFURNER, H.-P. SCHRÖCKER, AND K. BRUNNTHALER, *Algebraic methods in mechanism analysis and synthesis*, Robotica, 25 (2007), pp. 661–675. (Cited on p. 59)

[107] F. JOHN, *Extremum problems with inequalities as subsidiary conditions*, in Studies and Essays Presented to R. Courant on his 60th Birthday, January 8, 1948, Interscience Publishers, New York, 1948, pp. 187–204. (Cited on p. 238)

[108] D. E. KNUTH, *The Art of Computer Programming: Seminumerical Algorithms*, vol. 2, 2nd ed., Addison-Wesley Series in Computer Science and Information Processing, Addison-Wesley, Reading, MA, 1981. (Cited on p. 45)

[109] Y. C. KUO AND T. Y. LI, *Determining dimension of the solution component that contains a computed zero of a polynomial system*, J. Math. Anal. Appl., 338 (2008), pp. 840–851. (Cited on p. 173)

[110] J.-L. LAGRANGE, *Essai sur le problème des trois corps*, in Œuvres de Lagrange, tome 6, Gauthier-Villars, Paris, 1772, pp. 229–332. Available at `archive.org/details/oeuvresdelagrang06lagr`. (Cited on p. 90)

[111] J. B. LASSERRE, M. LAURENT, AND P. ROSTALSKI, *Semidefinite characterization and computation of zero-dimensional real radical ideals*, Found. Comput. Math., 8 (2008), pp. 607–647. (Cited on p. 240)

[112] ——, *A prolongation-projection algorithm for computing the finite real variety of an ideal*, Theoret. Comput. Sci., 410 (2009), pp. 2685–2700. (Cited on p. 240)

[113] G. LECERF, *Quadratic Newton iteration for systems with multiplicity*, Found. Comput. Math., 2 (2002), pp. 247–293. (Cited on p. 190)

[114] T.-L. LEE, T. Y. LI, AND C. H. TSAI, *HOM4PS-2.0: A software package for solving polynomial systems by the polyhedral homotopy continuation method*, Computing, 83 (2008), pp. 109–133. (Cited on pp. 83, 333)

[115] A. LEYKIN, *Numerical primary decomposition*, in ISSAC 2008, ACM, New York, 2008, pp. 165–172. (Cited on p. 209)

[116] A. LEYKIN AND F. SOTTILE, *Galois groups of Schubert problems via homotopy continuation*, Math. Comput., 78 (2009), pp. 1749–1765. (Cited on p. 247)

[117] A. LEYKIN, J. VERSCHELDE, AND A. ZHAO, *Newton's method with deflation for isolated singularities of polynomial systems*, Theoret. Comput. Sci, 359 (2006), pp. 111–122. (Cited on p. 190)

[118] ——, *Higher-order deflation for polynomial systems with isolated singular solutions*, in Algorithms in Algebraic Geometry, A. Dickenstein, F.-O. Schreyer, and A. J. Sommese, eds., Springer, New York, 2008, pp. 79–97. (Cited on p. 190)

[119] T. Y. LI, *Numerical Solution of Multivariate Polynomial Systems by Homotopy Continuation Methods*, Acta Numer. 6, Cambridge University Press, Cambridge, UK, 1997, pp. 399–436. (Cited on p. 83)

[120] ——, *Numerical solution of polynomial systems by homotopy continuation methods*, in Finite Element Methods, Handb. Numer. Anal. 11, North-Holland, Amsterdam, 2003, pp. 209–304. (Cited on p. 83)

[121] T. Y. LI, T. SAUER, AND J. A. YORKE, *The cheater's homotopy: An efficient procedure for solving systems of polynomial equations*, SIAM J. Numer. Anal., 26 (1989), pp. 1241–1251. (Cited on p. 90)

[122] X.-D. LIU, S. OSHER, AND T. CHAN, *Weighted essentially non-oscillatory schemes*, J. Comput. Phys., 115 (1994), pp. 200–212. (Cited on p. xvi)

[123] A. J. LOTKA, *Undamped oscillations derived from the laws of mass action*, J. Amer. Chem. Soc., 42 (1920), pp. 1595–1599. (Cited on p. 263)

[124] Y. LU, D. J. BATES, A. J. SOMMESE, AND C. W. WAMPLER, *Finding all real points of a complex curve*, in Proceedings of the Midwest Algebra, Geometry and Its Interactions Conference, Contemp. Math. 448, AMS, Providence, RI, 2007, pp. 183–205. (Cited on pp. 165, 233)

[125] F. S. MACAULAY, *The Algebraic Theory of Modular Systems*, Cambridge University Press, Cambridge, MA, 1916. (Cited on p. 186)

[126] J. M. MCCARTHY, *Geometric Design of Linkages*, Springer, New York, 2000. (Cited on p. xvi)

[127] J.-P. MERLET, *A parser for the interval evaluation of analytical functions and its applications to engineering problems*, J. Symbolic Comput., 31 (2001), pp. 475–486. (Cited on p. 224)

[128] ——, *Interval analysis for certified numerical solution of problems in robotics*, Int. J. Appl. Math. Comput. Sci., 19 (2009), pp. 399–412. (Cited on p. 224)

[129] R. MIRANDA, *Algebraic Curves and Riemann Surfaces*, AMS, Providence, RI, 1995. (Cited on p. 243)

[130] A. P. MORGAN, *A transformation to avoid solutions at infinity for polynomial systems*, Appl. Math. Comput., 18 (1986), pp. 77–86. (Cited on p. 63)

[131] ——, *Solving Polynomial Systems Using Continuation for Engineering and Scientific Problems*, Prentice-Hall, Englewood Cliffs, NJ, 1987. (Cited on pp. xvi, 16, 120, 137)

[132] A. P. MORGAN AND A. J. SOMMESE, *A homotopy for solving general polynomial systems that respects m-homogeneous structures*, Appl. Math. Comput., 24 (1987), pp. 101–113. (Cited on pp. 32, 79)

[133] ——, *Coefficient-parameter polynomial continuation*, Appl. Math. Comput., 29 (1989), pp. 123–160. *Errata*, Appl. Math. Comput., 51 (1992), p. 207. (Cited on pp. xviii, 103)

[134] A. P. MORGAN, A. J. SOMMESE, AND C. W. WAMPLER, *Polynomial continuation for mechanism design problems*, in Computational Solution of Nonlinear Systems of Equations (Fort Collins, CO, 1988), Lecture Notes in Appl. Math. 26, AMS, Providence, RI, 1990, pp. 495–517. (Cited on p. xviii)

[135] ——, *Computing singular solutions to nonlinear analytic systems*, Numer. Math., 58 (1991), pp. 669–684. (Cited on p. 47)

[136] ——, *Computing singular solutions to polynomial systems*, Adv. in Appl. Math., 13 (1992), pp. 305–327. (Cited on p. 47)

[137] ——, *A power series method for computing singular solutions to nonlinear analytic systems*, Numer. Math., 63 (1992), pp. 391–409. (Cited on p. 47)

[138] ——, *A product-decomposition bound for Bezout numbers*, SIAM J. Numer. Anal., 32 (1995), pp. 1308–1325. (Cited on p. 81)

[139] B. MOURRAIN AND J. PAVONE, *Subdivision methods for solving polynomial equations*, J. Symbolic Comput., 44 (2009), pp. 292–306. (Cited on p. 224)

[140] D. MUMFORD, *Algebraic Geometry: Complex Projective Varieties*, vol. 1, Classics in Mathematics, Springer-Verlag, Berlin, 1995. (Cited on p. 18)

[141] T. OJIKA, *Modified deflation algorithm for the solution of singular problems. I. A system of nonlinear algebraic equations*, J. Math. Anal. Appl., 123 (1987), pp. 199–221. (Cited on p. 190)

[142] T. OJIKA, S. WATANABE, AND T. MITSUI, *Deflation algorithm for the multiple roots of a system of nonlinear equations*, J. Math. Anal. Appl., 96 (1983), pp. 463–479. (Cited on p. 190)

[143] L. PACHTER AND B. STURMFELS, *Algebraic Statistics for Computational Biology*, Cambridge University Press, Cambridge, UK, 2005. (Cited on p. 245)

[144] D.L. PIEPER, *The Kinematics of Manipulators Under Computer Control*, Ph.D. thesis, Computer Science Department, Stanford University, Stanford, CA, 1968. (Cited on p. 111)

[145] J. M. PORTA, L. ROS, AND F. THOMAS, *A linear relaxation technique for the position analysis of multiloop linkages*, IEEE Trans. Robotics, 25 (2009), pp. 225–239. (Cited on p. 224)

[146] M. RAGHAVAN AND B. ROTH, *Solving polynomial systems for the kinematic analysis and synthesis of mechanisms and robot manipulators*, ASME J. Mech. Design, 117 (1995), pp. 71–79. (Cited on p. xvi)

[147] S. ROBERTS, *On three-bar motion in plane space*, Proc. London Math. Soc., 7 (1875), pp. 14–23. (Cited on pp. 99, 114)

[148] B. ROTH, *A Generalization of Burmester Theory: Nine-Point Path Generation of Geared Five-Bar Mechanisms with Gear Ratio Plus and Minus One*, Ph.D. thesis, Columbia University, New York, 1962. (Cited on p. xvi)

[149] B. ROTH AND F. FREUDENSTEIN, *Synthesis of path-generating mechanisms by numerical means*, ASME J. Eng. Industry, 85 (1963), pp. 298–306. (Cited on pp. xvi, 95)

[150] F. ROUILLIER, M.-F. ROY, AND M. SAFEY EL DIN, *Finding at least one point in each connected component of a real algebraic set defined by a single equation*, J. Complexity, 16 (2000), pp. 716–750. (Cited on p. 237)

[151] D. RUPPRECHT, *Semi-numerical absolute factorization of polynomials with integer coefficients*, J. Symbolic Comput., 37 (2004), pp. 557–574. (Cited on p. 173)

[152] A. SEIDENBERG, *A new decision method for elementary algebra*, Ann. of Math. (2), 60 (1954), pp. 365–374. (Cited on p. 237)

[153] B. SHIFFMAN AND A. J. SOMMESE, *Vanishing Theorems on Complex Manifolds*, Progress in Mathematics 56, Birkhäuser Boston, Boston, MA, 1985. (Cited on p. 252)

[154] A. J. SOMMESE AND J. VERSCHELDE, *Numerical homotopies to compute generic points on positive dimensional algebraic sets*, J. Complexity, 16 (2000), pp. 572–602. (Cited on pp. 19, 42, 167)

[155] A. J. SOMMESE, J. VERSCHELDE, AND C. W. WAMPLER, *Numerical decomposition of the solution sets of polynomial systems into irreducible components*, SIAM J. Numer. Anal., 38 (2001), pp. 2022–2046. (Cited on pp. 19, 144, 172, 276)

[156] ——, *Using monodromy to decompose solution sets of polynomial systems into irreducible components*, in Applications of Algebraic Geometry to Coding Theory, Physics and Computation (Eilat, 2001), NATO Sci. Ser. II Math. Phys. Chem. 36, Kluwer Academic, Dordrecht, Netherlands, 2001, pp. 297–315. (Cited on p. 19)

[157] ——, *A method for tracking singular paths with application to the numerical irreducible decomposition*, in Algebraic Geometry, de Gruyter, Berlin, 2002, pp. 329–345. (Cited on p. 19)

[158] ——, *Symmetric functions applied to decomposing solution sets of polynomial systems*, SIAM J. Numer. Anal., 40 (2002), pp. 2026–2046. (Cited on pp. 19, 173)

[159] ——, *Homotopies for intersecting solution components of polynomial systems*, SIAM J. Numer. Anal., 42 (2004), pp. 1552–1571. (Cited on pp. 195, 197)

[160] A. J. SOMMESE AND C. W. WAMPLER, *Numerical algebraic geometry*, in The Mathematics of Numerical Analysis (Park City, UT, 1995), Lecture Appl. Math. 32, AMS, Providence, RI, 1996, pp. 749–763. (Cited on pp. 19, 173)

[161] ——, *The Numerical Solution of Systems of Polynomials Arising in Engineering and Science*, World Scientific, Singapore, 2005. (Cited on pp. xvii, 3, 14, 17, 32, 47, 59, 75, 90, 103, 104, 112, 148, 162, 163, 167, 173, 191, 252)

[162] ——, *Exceptional sets and fiber products*, Found. Comput. Math., 28 (2008), pp. 171–196. (Cited on p. 205)

[163] M. SOSONKINA, L. T. WATSON, AND D. E. STEWART, *Note on the end game in homotopy zero curve tracking*, ACM Trans. Math. Software, 22 (1996), pp. 281–287. (Cited on p. 47)

[164] F. SOTTILE, *Real Solutions to Equations from Geometry*, University Lecture Series 57, AMS, Providence, RI, 2011. (Cited on p. 247)

[165] J. STOER AND R. BULIRSCH, *Introduction to Numerical Analysis*, 3rd ed., Texts in Applied Mathematics 12, Springer, New York, 2002. (Cited on p. 34)

[166] E. STUDY, *Von den Bewegungen und Umlegungen*, Math. Ann., 39 (1891), pp. 441–556. (Cited on p. 59)

[167] ——, *Geometrie der Dynamen*, B.G. Teubner, Leipzig, Germany, 1903. (Cited on p. 59)

[168] B. STURMFELS, *Solving Systems of Polynomial Equations*, CBMS Regional Conference Series in Mathematics 97, AMS, Providence, RI, 2002. (Cited on p. 248)

[169] H.-J. SU, J. M. MCCARTHY, M. SOSONKINA, AND L. T. WATSON, *Algorithm 857: POLSYS_GLP—a parallel general linear product homotopy code for solving polynomial systems of equations*, ACM Trans. Math. Software, 32 (2006), pp. 561–579. (Cited on pp. 82, 333)

[170] Y. TAO, H. LIANG, AND F. BAI, *A hybrid algorithm for multi-homogeneous Bézout number*, Appl. Math. Comput., 189 (2007), pp. 1755–1764. (Cited on p. 72)

[171] L.-W. TSAI AND A. P. MORGAN, *Solving the kinematics of the most general six- and five-degree-of-freedom manipulators by continuation methods*, ASME J. Mech., Trans., Auto. Design, 107 (1985), pp. 48–57. (Cited on p. xv)

[172] J. VERSCHELDE, *Algorithm 795: PHCpack: A general-purpose solver for polynomial systems by homotopy continuation*, ACM Trans. Math. Software, 25 (1999), pp. 251–276. (Cited on pp. 83, 247, 333)

[173] J. VERSCHELDE AND R. COOLS, *Symbolic homotopy construction*, Appl. Algebra Engrg. Comm. Comput., 4 (1993), pp. 169–183. (Cited on p. 81)

[174] J. VERSCHELDE, P. VERLINDEN, AND R. COOLS, *Homotopies exploiting Newton polytopes for solving sparse polynomial systems*, SIAM J. Numer. Anal., 31 (1994), pp. 915–930. (Cited on p. 83)

[175] V. VOLTERRA, *Variazionie fluttuazioni del numero d'individui in specie animali convivent*, Mem. Acad. Lincei., 2 (1926), pp. 31–113. (Cited on p. 263)

[176] R. J. WALKER, *Algebraic Curves*, Dover, New York, 1962. (Cited on p. 18)

[177] D. R. WALTER, M. L. HUSTY, AND M. PFURNER, *A complete kinematic analysis of the SNU 3-UPU parallel robot*, in Interactions of Classical and Numerical Algebraic Geometry, D. J. Bates, G. M. Besana, S. Di Rocco, and C. W. Wampler, eds., Contemp. Math. 496, AMS, Providence, RI, 2009, pp. 331–346. (Cited on p. 188)

[178] C. W. WAMPLER, *Bezout number calculations for multi-homogeneous polynomial systems*, Appl. Math. Comput., 51 (1992), pp. 143–157. (Cited on p. 72)

[179] ——, *Forward displacement analysis of general six-in-parallel SPS (Stewart) platform manipulators using soma coordinates*, Mech. Mach. Theory, 31 (1996), pp. 331–337. (Cited on pp. 59, 109)

[180] C. W. WAMPLER, J. D. HAUENSTEIN, AND A. J. SOMMESE, *Mechanism mobility and a local dimension test*, Mech. Mach. Theory, 46 (2011), pp. 1193–1206. (Cited on pp. 187, 188)

[181] C. W. WAMPLER AND A. P. MORGAN, *Solving the kinematics of general 6R manipulators using polynomial continuation*, in Robotics: Applied Mathematics and Computational Aspects (Loughborough, 1989), Inst. Math. Appl. Conf. Ser. New Ser. 41, Oxford University Press, New York, 1993, pp. 57–69. (Cited on pp. xv, 75)

[182] C. W. WAMPLER, A. P. MORGAN, AND A. J. SOMMESE, *Complete solution of the nine-point path synthesis problem for four-bar linkages*, ASME J. Mech. Design, 114 (1992), pp. 153–159. (Cited on pp. xvi, xviii, 95, 98, 99)

[183] ——, *Complete solution of the nine-point path synthesis problem for four-bar linkages—Closure*, ASME J. Mech. Design, 119 (1997), pp. 150–152. (Cited on p. xvi)

[184] C. W. WAMPLER AND A. J. SOMMESE, *Numerical algebraic geometry and algebraic kinematics*, Acta Numer., 20 (2011), pp. 469–567. (Cited on pp. xvi, xvii, 59)

[185] S. M. WISE, A. J. SOMMESE, AND L. T. WATSON, *Algorithm 801: POLSYS_PLP: A partitioned linear product homotopy code for solving polynomial systems of equations*, ACM Trans. Math. Software, 26 (2000), pp. 176–200. (Cited on pp. 82, 333)

[186] Z. ZENG, *Computing multiple roots of inexact polynomials*, Math. Comput., 74 (2005), pp. 869–903. (Cited on p. 46)

[187] ——, *The closedness subspace method for computing the multiplicity structure of a polynomial system*, in Interactions of Classical and Numerical Algebraic Geometry, Contemp. Math. 496, AMS, Providence, RI, 2009, pp. 347–362. (Cited on pp. 173, 187)

Software Index

Page numbers in boldface indicate where configuration settings can be found in the users manual portion of this book.

alphaCertified, 226, 227, 248, 328, 332
AMPMaxPrec, 180, **305**
AMPSafetyDigits1, **305**, 326
AMPSafetyDigits2, **306**, 326
ApCoCoA, 331

Bermeja, 331

CharacteristicClasses, 245
CoeffBound, 132, **305**
CondNumThreshold, 122, 123, 226, **313**, 328
ConstructWitnessSet, 219, **317**
CycleTimeCutoff, 53, **311**

DegreeBound, 132, **305**
DeleteTempFiles, 107, **321**
DGP-Solver, 331

EndgameBdry, 48, **310**
EndgameNum, 47, **310**
EndpointFiniteThreshold, 61, 99, 122, **313**, 327, 328

FinalTol, 13, 46, 51, 53, 67, 68, 119, 120, 122, 130, 131, 137, 174, 177, **309**, 327
FunctionTolerance, **312**

_HAVE_MPI, 329
HOM4PS-2.0, 333

I, 37, 285
ImagThreshold, 58, 122, 225, **313**, 328
IntrinsicMultiplier, **321**

JunkRemovalTest, **315**

KhRo, 333

LDTDepth, 173
LocalDimFinder, 187, 332

Macaulay2, 332
Makefile, 273, 329
MaxCodimension, 182, 183, **315**
MaxCycleNum, 50, **311**
MaxLDTDepth, 183, **316**
MaxNewtonIts, **307**
MaxNumBadMonoLoops, 177, **317**
MaxNumberSteps, **308**
MaxNumMonoLoops, 177, **317**
MaxNumPtsForTrace, 176, 177, **316**
MaxStepsBeforeNewton, **307**
MaxStepSize, **308**
MinCycleTrackback, 54, **310**
MinStepSizeBeforeEG, **308**
MinStepSizeDuringEG, **308**
MPType, 13, 14, 126, 128, 132, 136, 138, 180, **304**
Multiplicity, 333
MultOneOnly, **316**

NbhdRadius, 51, **311**
NeedToDiff, 107, **321**
NumSamplePoints, 51–53, **310**, 325

ODEPredictor, **307**
_OPENMP, 329
OutputLevel, **320**

ParameterHomotopy, 105, 106, 117, 130, 131, **314**

Paramotopy, 102, 332
PathTruncationThreshold, 99, **309**, 326
PHCpack, 333
Pi, 37, 285
POLSYS_GLP, 333
POLSYS_PLP, 333
Precision, 13, 14, 126, 132, 138, **304**
PrintPathProgress, **320**

RandomSeed, 6, 136, **320**
RatioTimeCutoff, 53, **311**
RatioTolerance, 135, **313**
RegenHigherDimTest, **319**, 327
RegenRemoveInf, **319**, 327
RegenStartLevel, **318**

SampleFactor, 50, 53, 99, **310**, 326
ScreenOut, **320**
SecurityLevel, 60, 61, 64, 134, 293, **312**, 326
SecurityMaxNorm, 61, 62, 64, 99, **312**, 326
SharpenDigits, 120, 121, 124, 138, 177, **313**, 327
SharpenOnly, 122–124, 289, **314**
Singular, 332
SingValZeroTol, 135, **321**
SliceFinalTol, **318**
SliceTolBeforeEG, 99, **318**, 325
SliceTolDuringEG, 99, **318**, 325
SpecificCodimension, 182–184, **315**
StepFailFactor, **309**
StepsForIncrease, **308**
StepSuccessFactor, **309**

TargetTolMultiplier, 122, 174, 177, **312**
TrackTolBeforeEG, 99, **306**, 325
TrackTolDuringEG, 99, **306**, 325
TrackType, 13, 14, 39, 89, 126–128, 132, 133, 136, 138, 157, 180, 290, 291, **301**

UseDiagonal, **319**
UseRegeneration, 84, 87, 89, 100, **317**
UserHomotopy, 37, 39, 40, 132–134, **314**

WitnessGenType, 166, 168, 183, 184, **315**
WitnessSupersetOnly, 89, 166, **316**

Subject Index

#, xiv
\, xiv
□, xiv, 4

$\overline{A}$, xiv
ab initio phase, 101, 103, 104
accuracy, 42
adaptive multiprecision, 41, 45
adjacent minors, 166, 168
affine complex algebraic set, 17
affine real algebraic set, 17
algebraic set, 17, 144; *see also* affine real algebraic set; constructible set; quasi-projective algebraic set
algebraic statistics, 246
alpha theory, 226
Alt's problem, *see* nine-point four-bar problem
approximate solution, 226
associated solution, 226

Bézout number, xviii, 73–76, 78, 98, 184
Bézout's theorem, 7
Bell number, 72
Bertini, 22
Bertini `doi`, 273
Bertini modules, *see* `LocalDimFinder`; `Paramotopy`
Bertini's theorem, 22, 163
bifurcation, 269
bootstrap method, 261, 267
branch point, 28, 48, 49, 244
Bricard six-revolute (6R) mechanism, 154

$\mathbb{C}, \mathbb{C}^*$, xiii
$\mathbb{C}[\,]$, xiv
Cauchy endgame, 47
cell decomposition, 227
certification, 226, 248
cheater's homotopy, 90
chemistry, xvi
Chern class, 245
classical cascade, 162, 167, 183
classical topology, 19
closure, 20
codimension, 148, 166
comment character, 5, 274, 281
complex dimension, *see* dimension
complex plane, 146
complex projective space, *see* projective space
complex topology, 19
complex versus real dimension, 224
condition number, 12
constructible set, 251, 252
continuation, 24
corank of a system, 162
critical points, 209, 231, 237
curve, 147
cycle number, 47, 48, 52, 244, 326

Darcy's law, 261
Davidenko differential equation, 32, 33
deflation, 149, 172, 189
  isosingular, 216
deflation sequence, 217
degree, xiv, 144
degrees of freedom, 155, 187
  infinitesimal, 187
dehomogenization, 61
Denavit–Hartenburg parameters, 75
dimension, 17, 19, 144, 147, 224
dimension-by-dimension solving, 183
discretization, 261

discriminant variety, 251
divergent path, 24
domain decomposition method, 261, 268

eigenvalue problem, 75
elimination theory, 251
endgame, 24, 34, 41, 47; *see also* Cauchy endgame; parallel Cauchy endgame; power-series endgame
  Cauchy integral, *see* Cauchy endgame
endgame operating zone, 48, 325
endgame setup, 48
endpoints, 24
Enneper surface, 255
enumerative geometry, 247
equation-by-equation, 71
equidimensional, 147
equilibrium states, xvi
Euler's method, 34
Euler–Newton step, 35
extrinsic slicing, 85

fiber product, 196
filtering method, 261, 267
first-order tracking, 34
fixed-point homotopy, 129
free boundary problems, 264
function evaluation, 39

gamma trick, 26, 32, 44, 45, 104, 176, 192
general linear product, 80
generalized eigenvalue problem, 74
generically reduced, 149
generically true, 21, 22
geometric genus, 259
good homotopy, 31, 129
gradient descent homotopy, 248
Grassmann variety, 210, 213
Grauert's direct image theorem, 42
Griewank–Osborne example, 10
Griffis–Duffy platform, *see* Stewart–Gough platform

higher-order methods, 35
homogeneous coordinates, xiv, 56
homogeneous polynomial, 56
homogenization, 61
homotopy, 24, 30; *see also* cheater's homotopy; linear product homotopy; multihomogeneous homotopy; parameter homotopy; polyhedral homotopy; polynomial-product homotopy; regeneration homotopy; total-degree homotopy; user-defined homotopy
homotopy continuation, 23, 24
Hurwitz formula, 243
hyperbolic conservation law, 266
hypersurface, 144, 162

implicitization problem, 251
intrinsic slicing, 85, 198, 324
inverse kinematics, 75
irreducible algebraic set, 17, 18, 145
irreducible component, 17, 145
irreducible decomposition, 19
irreducible polynomial, 17
isolated solution, 20
isosingular, 200
isosingular deflation, 216
isosingular local dimension, 219
isosingular set, 210, 214, 218
isosingular stabilization, 218, 219, 221, 304, 317

Jacobian, 20
join variety, 245
junk point, 171, 172
junk removal, 171

k-fold, 147
kinematics, 143

Lagrange points, 90, 94, 101, 109
likelihood function, 246
limaçon, 232
linear product homotopy, 71, 76, 80
linear products, 76
local dimension, 172, 187
  depth-bounded, 187

Macaulay dual space, 173, 186
  depth, 188
Macaulay matrix, 186

manifold point, 17, 145
marching method, 34
maximum likelihood degree, 246
maximum likelihood estimation, 246
mechanism analysis, xvi
mechanism mobility, 143
mechanism synthesis, xvi
membership testing, 160
Minkowski product, 65
Minkowski quotient, 66
Minkowski sum, 82
mixed volume, 82, 325
monodromy group, 176
monodromy loop, 173, 175, 176
MPFR, 45
multidegree, 73
multidegree notation, xiv
multidegree table, 73, 74, 77
multihomogeneous Bézout number, 73, 101
multihomogeneous homotopy, 71, 72, 78, 79, 100, 159
multihomogeneous structures, 64
multiplicity, 9, 12, 20, 32, 41, 42, 148, 184–186
    convergent power series, 184

nested parameter spaces, 111
Newton homotopy, 129, 130, 133
Newton polytope, 81
Newton's method, 9, 35, 43, 119, 189, 266
nine-point four-bar problem, 95, 109, 113, 323
nondimensionalized variables, 91
nonsingular, 9
nonsingular solution, 20
nonsolutions, 109, 168
nonsquare system, xviii
Nullstellensatz, 17
numerical algebraic geometry, xv, 3
numerical algebraic kinematics, xvii
numerical elimination theory, xx, 251
numerical irreducible decomposition, 19, 144, 150, 159
    specific dimension, 181

O( ), xiv
ODE predictors, 35
off-line phase, 103
on-line phase, 103
open ball, 20
overdetermined system, 14, 185

$\mathbb{P}$, xiv
parallel Cauchy endgame, 47
parameter homotopy, 101, 104, 105, 331
partitioned linear product, 80
path tracking, 31, 32, 120
path variable, 37
paths, 24
pejorative manifolds of Kahan, 46
perturbation approach, 89
phylogenetics, 245
point, 147
polar class, 245
polyhedral homotopy, 71, 82
polyhedral homotopy example, 83
polyhedral mixed volume, 101
polynomial, xiv
polynomial-product homotopy, 81
positive-dimensional, 143
positive-dimensional run, 276
power-series endgame, 47, 50, 326
precision, 42
predictor-corrector, 35
probability one, 15, 21, 22, 26, 43, 103
probability-zero event, 22
projective patch, 62, 68
projective plane, 56
projective space, 51, 55
pseudowitness point set, 253
pseudowitness set, 251, 253, 259
Puiseux expansion, 47, 326
pure-dimensional algebraic set, 19, 144, 147

quasi-projective algebraic set, 144

$\mathbb{R}$, xiii
radially symmetric solution, 264
ramification point, 28, 49, 244
random linear combination, 14
random patch, 51, 68
random seed, 6
randomized system, 42
randomizing a polynomial system, 22, 163
rank of a system, 162

real dimension, 17, 224
real numerical irreducible
decomposition, 223, 240
real radical ideal, 240
reduced algebraic set, 145, 149
reducible algebraic set, 145
reduction to the diagonal, 196
regeneration, 71, 83, 324, 327
regeneration homotopy, 80, 84, 100
regenerative cascade, 89, 162, 168, 183
regular point, 9, 145
rigid body, 58
robot kinematics, 75, 107, 109
robotics, xv
root count, 73, 102

salmon conjecture, 246
sampling, 159, 219, 290
Schubert problem, 247
secant variety, 245
serial-link robotic arm, xv
sextic example, 105
sharpening module, 121, 289, 327
singular point, 9, 145, 209, 218
singular set, 209
singular solution, 9, 20
singularity, 20
six-revolute (6R) serial-link robot, 75, 154
slice, 165
smooth point, 21, 145, 147, 149, 201, 209, 218, 221, 225, 304
Solar and Heliospheric Observatory satellite, 95
solution set, 144
square system, 14, 20, 184, 185
start system, 24
startpoint, 24, 105
step sizes, 35
Stewart–Gough platform
general, 107
Griffis–Duffy platform, 189
foldable, 156, 189
Type I, 155
Type II, 155
straight-line program, 15, 323, 324
strong entropy condition, 266
Study coordinates, 59, 155, 189
subfunctions, 15, 275, 276, 323
Sudoku puzzle, 248
surface, 145, 147

target system, 24
Terracini's lemma, 245
topology, *see* classical topology; complex topology; Zariski topology
total-degree homotopy, xviii, 24, 31, 32, 39, 72
trace, 173
trace test, 173
combinatorial, 175, 176
tracker settings, 35
tracking, 31
tumor-aggressiveness factor, 264, 269
tumor-growth model, 264, 269
twisted cubic curve, 145, 202, 277

unclassified point, 177
underdetermined system, 14
upper semicontinuity, 112, 115, 162, 217
user-defined homotopy, xvii, 39, 40, 131, 134

variable groups, 72

WENO scheme, xvi, 266
Wilkinson, 43
winding number, *see* cycle number
witness point, 150
witness point set, 150
witness set, 144, 150
print, 160, 179, 291

$\mathbb{Z}$, xiii
Zariski topology, 19